Selected papers from the Third Conference on Experimental Medicine and Surgery in Primates, Lyon, June 1972

Medical Primatology 1972

Part II

Experimental Surgery, Transplantation Immunology, Oral Medicine, and Neurophysiology and Experimental Psychology

Edited by

EDWARD I. GOLDSMITH
Cornell University Medical College, New York, N.Y.

J. MOOR-JANKOWSKI
New York University School of Medicine, New York, N.Y.

In collaboration with

C. R. Austin
H. Balner
N. Barnicot
R. Cortesini
F. Deinhardt
N. Gengozian
G. A. Gresham
J. E. Hamner III
S. S. Kalter
R. Naquet
M. Panigel
F. T. Perkins
J. Ruffié
A. Schrier
M. Vastesaeger
G. Zbinden
E. Zuckerkandl

With 190 figures and 50 tables

S. Karger · Basel · München · Paris · London · New York · Sydney 1972

Medical Primatology 1970
XXII+997 p., 284 fig., 174 tab., 1971
ISBN 3-8055-1163-9

S. Karger · Basel · München · Paris · London · New York · Sydney
Arnold-Böcklin-Strasse 25, CH–4011 Basel (Switzerland)

Printed in Switzerland by National-Zeitung AG, Basel
ISBN 3-8055-1487-5

To Frank Glenn, MD

Professor of Surgery Emeritus, Cornell University Medical College, whose pioneering efforts on behalf of Medical Primatology have guided and encouraged the editors of these volumes

Contents

Part II

Experimental Surgery

Transplantation Immunology

Oral Medicine

Neurophysiology and Experimental Psychology

Part I

General Primatology. Husbandry, Management, Breeding Choice of Species Conservation. Communication Media in Medical Primatology

Reproduction and Perinatal Studies

Genetics, Phylogenetics and Evolution

Part III

Infectious Diseases

Viral Oncology

Pharmacology and Toxicology

Foreword

The three volumes of 'Medical Primatology 1972' consist of papers selected from communications presented at the Third Conference on Experimental Medicine and Surgery in Primates, held in Lyon, France, 21st–23rd June 1972. The program of the Conference was designed to gather new information to complement the former volume *Medical Primatology 1970*, which resulted from the Second Conference.

Papers chosen for the present volumes passed two selection procedures: (1) 246 abstracts were submitted of which 176 were accepted to be presented as papers at the Third Conference; (2) these complete papers were then evaluated for validity and relevance of the results presented. With the exception of review articles, only papers containing new information were accepted. The selection process led to some difficult decisions, as no time was available for resubmission of rejected papers. The chosen 120 papers were edited for style, correct simian taxonomy, bibliography, and appropriate cross references.

Despite the rigorous standards applied, we expect that shortcomings will still be found since little time was available for editing in order to obtain publication within seven months after the Conference. We hope, however, that *Medical Primatology 1972* will satisfy the objectives set at the First Conference in 1967, namely, to provide periodically an up-to-date and complete handbook and reference source covering the broad field of Medical Primatology.

Mr. THOMAS KARGER has continued to back up the new field of medical primatology despite the commercial risk involved. Miss DENISE GREDER and Mr. ROLF STEINEBRUNNER of the Karger firm, have assisted us at the Conference and during the following months of hard editorial labor. Miss

GETTA SKOTCHDOPOLE, Assistant Editor, attended most ably to the coordination of editorial efforts and to proof-reading. The Editors alone, being nonsimian primates, must accept the blame for inadequacies.

EDWARD I. GOLDSMITH, MD
Professor of Surgery
Cornell University Medical College, New York

J. MOOR-JANKOWSKI, MD
Research Professor
New York University School of Medicine, New York

Laboratory for Experimental Medicine and Surgery in Primates (LEMSIP) New York, November 1972

Organization Committee of the Third Conference on Experimental Medicine and Surgery in Primates 1972

WHO Delegates	W. I. B. Beveridge, Geneva Z. Fejfar, Geneva
National Delegates	
Belgium	Dr. Robert Gortz University of Bruxelles Dr. J. Mortelmans Institut de Médecine tropicale, Antwerp
Canada	Dr. Samuel Solomon Royal Victoria Hospital, Montreal
France	Dr. Charles Merieux Institut Merieux, Lyon Dr. Robert Naquet Institut de Neuro-physiologie du CNRS, Marseille Dr. René Triau Fondation Merieux, Lyon
Germany	Dr. Hans-Jürg Kuhn Universität Frankfurt
Great Britain	Mr. K. R. Hobbs Medical Research Council, Carshalton Dr. F. T. Perkins National Institute for Medical Research, London
Holland	Dr. Hans Balner Radiobiological Institute TNO, Rijswijk
Israel	Dr. Uri Marinov National Council for Research and Development, Jerusalem

Italy	Dr. RAFFAELLO CORTESINI Università di Roma
Norway	Professor KARE BERG University of Oslo
Sweden	Professor ERNST BARANY Uppsala Universitet
Switzerland	Dr. HEINZ WEBER Sandoz AG, Basel Professor G. ZBINDEN Universität Zürich
United States	Dr. JULES S. CASS Veterans Administration, Washington, D.C. Dr. S. S. KALTER Southwest Foundation for Research and Education San Antonio, Texas Dr. C. H. KRATOCHVIL The Upjohn Company Kalamazoo
Scientific Secretary	Dr. J. RUFFIÉ Collège de France, Paris
Co-Chairmen	Dr. E. I. GOLDSMITH Cornell University Medical College, New York Dr. J. MOOR-JANKOWSKI New York University School of Medicine, New York

Sponsors and Benefactors of the Third Conference on Experimental Medicine and Surgery in Primates 1972

The Conference participants are greatly indebted to Dr. PIERRE DOROLLE, Deputy Director, World Health Organization, for the fruitful collaboration of WHO in the Conference.

The Conference was sponsored by the following international and national agencies:

International	European Society for the Study of Drug Toxicity International Association of Biologic Standardization
Belgium	Ministère de la Politique et Programmation Scientifique
Canada	McGill University, Montreal
France	Ministère de la Santé Publique et de la Sécurité Sociale Centre National de la Recherche Scientifique (CNRS) Institut National de la Santé et de la Recherche Médicale (INSERM) Université Paul Sabatier de Toulouse Université Claude Bernard de Lyon Fondation Merieux
Germany	German Pharmacological Society
Great Britain	The Royal College of Surgeons of England The Zoological Society of London
Holland	Netherlands Organization for Health Research
Israel	The Weizmann Institute

Italy	Italian Ministry of Health
Sweden	The Royal Swedish Academy of Sciences
Switzerland	Union of Swiss Societies of Experimental Biology
USA	New York University School of Medicine American Association for Laboratory Animal Science

The Conference Organization Committee is grateful to the governments and agencies which provided travel grants for their nationals to attend the Conference.

Our sincere thanks are due to Dr. CHARLES MERIEUX, the Merieux Foundation and the Merieux Institute. Their staff provided valuable scientific contributions to the Conference, assumed responsibility for the logistics, and offered warm hospitality to the participants, far beyond the usual role of hosts. The Merieux Foundation provided generously for the considerable expenditures of the local organization and helped many speakers to attend the Conference by defraying their expenses.

We wish to thank the following organizations, listed in alphabetical order, which contributed substantially to cover the expenses of the American participation in the Conference.

Major Contributors	*Baylor University* *Ciba-Geigy* *Generics Corporation of America* *Hoffmann-La Roche* *Schering Corporation* *The Upjohn Company*
Contributors	*Ortho Pharmaceutical Corporation* *Riker Laboratories, Inc.* *Sandoz Ltd. Basel* *Sherwood Medical Industries Inc.*

Experimental Surgery

Medical Primatology 1972. Proc. 3rd Conf. exp. Med. Surg. Primates, Lyon 1972, part II, pp. 2–11 (Karger, Basel 1972)

The Use of the Pregnant Baboon in Automotive Crash Injury Research

W. M. CROSBY

University of Oklahoma Health Sciences Center, Department of Gynecology and Obstetrics, Oklahoma City, Okla.

There are more than ten million births per year in the non-Communist developed countries of the world. Each pregnant woman in these countries travels in one way or another during her pregnancy, and presumably is subject to the same risk of accidental collision as anyone else. Automotive passenger restraint systems have been demonstrated to reduce the number of fatal injuries in automobile collisions, but none of these restraint systems has been designed for use during pregnancy. In 1964, RUBOVITZ [7] reported a pregnant woman, involved in an automobile collision, whose lap belt apparently was responsible for rupture of the pregnant uterus. The mother survived the injuries, but her fetus did not. Obstetricians have not had any scientific data upon which to make recommendations as to what type of restraint, if any, should be recommended for their patients. The present communication describes research using the pregnant baboon as the experimental animal to determine the most effective and least harmful restraint system for use by pregnant travellers.

Experimental Studies

The initial study was begun in 1965 and was supported by the Federal Aviation Administration, United States Department of Transportation [4]. The experimental animal was the Kenya baboon *(Papio cynocephalus)*. The impact apparatus was the Daisy decelerator located at the Holloman Air Force Base, Alamagordo, N.M. This consisted of a sled driven by compressed air which coasts along a 50-ft track and impacts into a brake. By altering the entrance velocity into the brake, the amount of decelerative force and the length of time over which it is exerted can be varied. The baboons were in the second half of pregnancy and were shipped by air from the breeding colony to the track. On the morning of the experiment, the animals were anesthetized. Intrauterine and intraperitoneal pressure transducers were inserted surgically, and the electrocardiograph leads were sutured into the fetal shoulder and elbow. Amniotic loss during surgery was replaced with physiologic saline. Following closure, the lead wires were brought out through the vagina or subcutaneously to the thigh. Anesthesia consisted of pentobarbital (one animal) fentanyl and droperodol (two animals) or phencyclidine (eight animals). With the exception of pentobarbital, the drugs did not seem to significantly alter physiologic response to impact.

Ten pregnant baboons were used in 11 deceleration experiments. Seven underwent impact facing forward at 20 G with lap belt restraint, including one that was impacted later in the day with the lap belt-shoulder harness combination (three-point restraint). An additional animal had a single impact at 20 G with three-point restraint. Two animals were impacted facing rearward, one at 20 G and the other at 40 G. In addition, three pregnant control animals had similar surgical procedures without deceleration.

Following forward-facing impact in animals with lap belt restraints, the chest thrusts out and the head lags behind, producing extension of the cervical spine. By the time the torso is horizontal, the abdomen impinges upon the seat pan, and the body of the mother folds over the pregnant uterus, compressing it between the lap belt and the maternal spine. The upper torso then rebounds upward and backward until it strikes the back of the seat. The entire sequence takes 0.5–0.75 sec.

When restrained by the three-point system, the chest impinges on the diagonal strap and rebounds back to the seat. Forward flexion over the pregnant uterus is prevented, as it is with rear-facing impact with either type of restraint. In one animal under three-point restraint, the diagonal strap was improperly placed, and the animal underwent violent rotation about the strap before rebounding back to the normal position.

In animals impacted with lap restraint alone, uterine pressure begins to rise as the uterus flatterns against the anterior abdominal wall before flexion of the mother's body occurs (fig. 1). A second pressure rise occurs as the mother flexes over the lap belt and compresses the uterus even further. This pressure is partially compensated by an increase in intraperitoneal pressure, but in every case the intrauterine pressure exceeded the intraperitoneal pressure by more than 50%. In animals subjected to violent motion the maternal heart rate invariably slowed during the first minute following impact, but returned to baseline level by the end of the second minute. The maternal blood pressure dropped in seven of the eight animals, but fell to shock levels in only two. Fetal bradycardia, with a 50-percent reduction in rate, occurred in every animal subjected to violent motion. This bradycardia lasted for five to ten minutes following impact and was a constant feature. In three animals, violent motion of the body was prevented by three-point

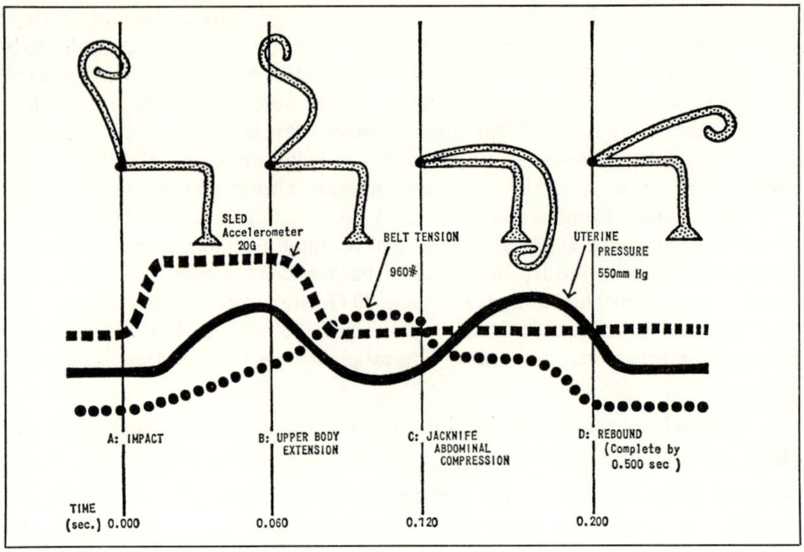

Fig. 1.

restraint or rear-facing impact. In these, maternal bradycardia did not occur and fetal bradycardia was less severe and of shorter duration. There was no change in maternal blood pressure following impact in the absence of violent motion.

None of the adult animals died as a direct result of impact, but three had significant traumatic injuries (tables I–III). One had an extensive broad ligament hematoma, one had a placental separation, and one had both a subdural hemorrhage and placental separation. In an additional animal, placental separation resulted in fetal death prior to impact. This was probably related to surgical implantation of a flow transducer on the hypogastric artery. None of the fetuses survived the experiment, and three of ten sustained injuries. Two died from cerebral hemorrhage, (No. 3313 and 3364) and one had a depressed skull fracture with laceration of the underlying brain (No. 3317). Two uninjured fetuses died in association with maternal shock (No. 3062 and 3310) and two as a result of placental separation (No. 2888 and 3592). One fetus died with the mother (No. 3596) and in two, the cause of fetal death was unknown (No. 3013 and 3365). One fetus was liveborn but died soon after birth of prematurity and atelectasis. The high rate of fetal and placental injury in these animals showed that flexion of the mother's body

Table I. Lap belt restraint. Foreward facing, 20 G

No.	Days of gestation	Presentation	Seat angle	Surgical procedure	Maternal autopsy	Fetal autopsy	Comments
2888	160	cephalic	45	hysterotomy, FECG	1/3 placental separation	no trauma	autopsy 20 min postimpact
3013	140	cephalic	45	hysterotomy, FECG	living	no trauma	
3062	110	breech	45	hysterotomy, FECG	minor trauma	no trauma	mother died in shock 24 h post impact
3311	154	cephalic	20	laparotomy, arteriotomy	hematoma, broad ligament	no trauma	second impact of animal No. 3310
3313	166	cephalic	20	arteriotomy	minor trauma	cerebral hemorrhage	fetus died 11 min postimpact
3592	103	cephalic	20	thoracotomy, laparotomy	1/4 placental separation	died prior to impact	placental separation due to surgery, not impact
3596	103	breech	20	thoracotomy, laparotomy, arteriotomy	minor trauma (aspiration)	no trauma	mother died 18 h postimpact of aspiration

Table II. Three-point restraint. Forward facing, 20 G

No.	Days gestation	Presentation	Seat angle[1]	Surgical procedure	Maternal autopsy	Fetal autopsy	Comments
3310[1]	154	cephalic	20	laparotomy, arteriotomy	hematoma, broad ligament	no trauma	animal rotated around shoulder strap
3314	153	cephalic	20	arteriotomy	minor trauma	no trauma	fetus born alive, died of prematurity

1 Animal impacted two hours later with lap restraint (No. 3311).

Table III. Rear facing impact

No.	Days gestation	Presentation	Seat angle, °	Surgical procedure	Maternal autopsy	Fetal autopsy	Comments
3364	145	breech	20	laparotomy, arteriotomy	subdural hematoma, placental separation	cerebral hemorrhage	40-G impact
3365	110	breech	20	arteriotomy	minor trauma	no trauma	20-G impact

over the lap belt was associated with fetal death as a result of maternal shock, placental separation, or direct fetal skull injury. The apparent diminution of physiologic response to impact in animals in which violent motion was prevented by properly-placed three-point restraint or rear-facing impact implied that prevention of forward flexion might reduce the harmful effect of sudden deceleration for both mother and fetus. However, surgical procedures required to place the electrocardiographic leads in the fetus and the pressure transducers in the uterus also procedured fetal death in the absence of impact. In subsequent tests, these surgical procedures were eliminated.

The next series of experiments were designed to test the difference between the lap belt and the three-point restraint system, and were carried out as a co-operative venture by the Department of Transportation, Wayne State University, and the University of Oklahoma [3]. The animals were shipped by air from the breeding colony in Oklahoma City to the test facility in Detroit, Michigan, where they were housed in individual cages. The morning of impact, the animals were anesthetized with phencyclidine (1 mg/kg) and placed in restraint chairs and carried to the track. Blood pressure was measured with a blood pressure manometer and stethoscope, and the fetal and maternal heart rates were obtained by the use of an ultrasound fetal heart detector. After the animal had recovered from anesthesia to the point that it was alert and responding to its surroundings, a muzzle was made of plastic tape and the arms and legs were taped together to prevent interference with body motion during impact. Analgesia during this brief period was provided by inhalation of 70% nitrous oxide and 30% oxygen. The anesthetic mask was removed immediately prior to sled acceleration. Follwing impact, the animal was removed from the sled while still in a state of confusion and replaced in the restraint chair. At this time the animal could be handled without additional anesthesia. The fetal heart rate, maternal pulse and blood pressure were taken at five- to 30-min intervals during the rest of the day. At the end of the working day, the animals were re-anesthetized with phencyclidine and examined. The fetal heart rate, maternal heart rate and maternal blood pressure were

Table IV. Lap belt restraint

No.	Peak G	Time, msec (over 15 G)	Total belt load, Lbs	Presen-tation	Days of gesta-tion	Outcome mother	Outcome fetus	Remarks
229	26.0	55	1,297	vertex	147	belt brn	died 7 h	animal hit nose on floor of sled; autopsy of mother was negative
224	26.5	60	932	vertex	156	fx. nose belt brn	died 1 h	animal hit nose on floor of sled; placenta 85% separated; autopsy of mother was negative
293	26.0	60	812	vertex	153	belt brn	died 5 h	
241	25.0	60	876	breech	119	belt brn	lived	
210_a	25.0	–	870	vertex	168	normal	died 4 days	
226	26.0	58	1,130	breech	164	belt brn	lived	
228	28.0	55	1,070	vertex	142	belt brn	lived	
239	28.5	48	960	vertex	151	fx. ulna	lived	
230	28.5	55	930	vertex	140	belt brn	lived	
209	28.5	58	930	vertex	142	velt brn	died 2 days	
212	26.3	55	1,120	vertex	140	quadri-plegia	died 8 days	not included in survival figures

recorded. The first two animals in which the fetus died were sacrificed and autopsies performed. No maternal injuries were found. Following this, whenever the fetal heart disappeared following impact, a cesarean section was done but the mother was not sacrificed. In three instances, the fetus died after the final examination and was discovered to have delivered at a later date. In these instances, the fetus and placenta had been partially eaten by the mother and were not available for study.

The results of these studies are indicated in tables IV and V. Table IV tabulates the data on 11 animals subjected to impact under lap restraint alone. The animals ranged in length of gestation from 119 to 158 days. Two fetuses were in breech presentation; nine presented cephalically. The latter is the normal configuration for the baboon fetus but the abnormal presentation did not apparently influence the outcome as both fetuses survived the impact. In no case was the fetal head fixed in the maternal pelvis

Table V. Three -point restraint

No.	Peak G	Time, (msec over 15 G)	Total belt load, number	Shld. load	Presentation	Days of gestation	Outcome mother	Outcome fetus	Remarks
223	26.0	60	704	500	breech	141	chest and belt bruise	died	delivered stillborn after last examination; half eaten
209	recorded failure				vertex	154	belt bruise	lived[1]	recorder failed to write data
55	26.0	60	940	560	vertex	141	normal	lived	
225	25.0	60	808	525	vertex	147	normal	lived	
227	25.0	50	780	680	vertex	157	normal	lived	
208	26.0	50	760	600	vertex	155	normal	lived	
219	26.0	50	820	560	vertex	157	normal	lived	
211	26.5	60	670	520	vertex	155	normal	lived[1]	
124	72+	50	1,300	850	vertex	160	normal	lived[2]	
221	27.0	58	650	460	breech	138	normal	lived	
294	27.5	60	840	500	vertex	136	normal	lived	
210_b	27.5	64	1,000	660	vertex	147	normal	lived	

1 Fetus died during labor four to five weeks later.
2 Peak G over 72 for an instant, due to erroneous setting of decelerating cylinder.

before or after impact. The peak deceleration ranged from 24.5 to 29 G with the exception of one animal in which an erroneous setting of the brake resulted in a peak of over 72 G for an instant. The mother and fetus survived without injury. In each animal, there were abrasions about the area of the lower abdomen where the belt rubbed against the skin. There were also abrasions of the elbows and wrists where the arms of the animals came in contact with the floor of the sled. One animal had a fracture of the ulna and another fractured a nasal bone. Both of these healed without incident and were felt to be unrelated to the outcome of the fetus. In one animal there was a spinal cord injury. The impact was quite similar to that for the rest of the animals. Moving pictures and data recordings indicated no particular deviation from the norms as described above. The fetus died and was stillborn approximately a week after impact. This fetal loss was felt to be due more to the mother's condition than the impact itself. The mother was sacrificed three weeks following impact after she had developed quadraplegia. An autopsy demonstrated compression of the cervical spinal cord without fracture of the vertebrae. The latter animal has been excluded from the fetal death rate calculations. Among lap-belted mothers, five of ten fetuses died following impact. The length of survival ranged from one hour to four days. The fetus that died within one hour of impact had complete separation of the placenta which clearly accounted for its death and the vaginal bleeding that was observed in the mother. In the remainder, there is no clear explanation of the cause of the fetal death. Each recovered from the initial depression of fetal heart rate noted immediately after impact. Later the fetal heart rate either declined slowly or was found to have unexpectedly disappeared. Only three placentas were available for study; two were normal and one had been separated.

Table V indicates the data recorded from 12 animals impacted with three-point restraints. These animals ranged in length of gestation from 136 to 157 days. There were two fetuses in breech presentation, the rest were normal. In each case, review of the movies revealed that the animal had impacted against the shoulder harness in proper position. Few of the animals had abrasions corresponding to belt burns across the chest or the abdomen. Examination following impact indicated depression of the fetal heart rate during the first five minutes with recovery to normal within 15 min. One fetus died of unknown cause from one and a half to two days following impact, and delivered during the night. The fetus was partially eaten as was the placenta, and there is no explanation for the death except for the temporal relationship between fetal death and impact. Two animals delivered stillborn fetuses four weeks following impact. Examination of the fetuses immediately following delivery did not reveal any evidence of degeneration, as would have been seen had the fetuses died soon after impact and then delivered three to four weeks later. It was concluded that these two fetuses died as a result of the labor process and not as a result of the impact.

Discussion

Lap belt restraint has been demonstrated to reduce the number and severity of injuries resulting from automobile collisions [6]. The advisability of lap belt restraint for pregnant passengers has been questioned from concern that uterine compression produced by body flexion over the lap belt might intensify rather than reduce fetal mortality. In one study, human pregnant victims of automobile collisions were followed prospectively. There was no statistically significant change in either fetal or maternal death rate when the victim wore a lap belt, but ejected victims of severe collisions had a higher maternal and fetal death rate than victims that remained within the vehicle [2]. Since the lap belt's primary mode of action is to prevent ejection, it was concluded that the lap belt restraint should be used by pregnant travellers. Nonprospective reports in humans have shown that lap belt restraint can be associated with uterine, placental and fetal injury [1, 8]. The initial animal studies in the pregnant baboon showed that the fetus frequently dies following maternal deceleration in the absence of demonstrable maternal trauma. In these studies, however, surgical instrumentation of the fetus *in utero* was felt to be at least partly responsible for the death of the fetus, because control animals, instrumented but not subjected to impact, also died. There remained the question whether deceleration alone was responsible for fetal death in the absence of maternal trauma or whether the major factor is forward flexion of the mother's body over the lap belt.

The second series of impacts was designed to answer that question. Tables IV and V show that each of the latter series of impacts were in the same general range of physical severity, both with regards to peak G and duration of G. Five of ten lap-belted mothers lost their fetuses as a direct effect of impact. When forward flexion was prevented by the addition of a shoulder harness to the lap belt, only one of 12 fetuses died as a result of the impact. The difference in fetal death rate between these two series is statistically significant ($p<0.05$). It is clear that prevention of forward flexion by maternal three-point restraint offers greater protection for the fetus during a collision than lap belt restraint alone.

Automobile crash injury research is hampered by lack of standardization between automobile accidents. Each collision is virtually unique. Conclusions drawn from study of automobile accidents as they occur are thus subject to considerable error. Experimental impacts suffer from lack of a subject whose anatomical kinematics are suitably close to that of the human accident victims to serve as a surrogate. Anthropomorphic dummies have provided much data in experimental impacts, but they can only provide an approximation of biomechanical trends occurring in living subjects. Research using human volunteers is necessarily confined to low level impact force which may not compare with higher forces developed in serious automobile collisions. The use of animals in automotive research is similarly hampered by subjective interpretation of minor injuries produced by impact. Forces required to produce death in animals far exceeds those encountered in survivable automotive collisions, yet the meaning of hairline fractures, bruises and petechiae produced by impacts of less decelerative force is difficult to ascertain.

From these experiments it is shown that the baboon fetus dies as a result of levels of decelerative force comparable to that encountered in the 'typical' automotive collision [5]. The fetal death rate in pregnant baboons subjected to controlled deceleration may be a better indicator of restraint system efficacy than experimental impacts using anthropomorphic dummies, human volunteers or nonpregnant animals.

References

1 Crosby, W. M.: Pathology of obstetrical injuries in Brinkhous 'Accident Pathology', pp. 204–207 (Government Printing Office, Washington, D. C. 1970).

2 Crosby, W. M. and Costiloe, J. P.: Safety of lap belt restraint for pregnant victims of automobile collisions. New Engl. J. Med. *284:* 632–636 (1971).

3 Crosby, W. M.; King, A. I., and Stout, L. C.: Fetal survival following impact. Improvement with shoulder harness restraint. Amer. J. Obstet. Gynec. *112:* 1101–1106 (1972).

4 Crosby, W. M.; Snyder, R. G.; Snow, C. C., and Hansen, P. G.: Impact injuries in pregnancy. I. Experimental studies. Amer. J. Obstet. Gynec. *101:* 100–110 (1968).

5 Egle, D. M.: Unpubl. data.

6 Huelke, D. F. and Gikas, P. W.: Causes of deaths in a automobile accidents. J. Amer. Med. Ass. *203:* 1100–1107 (1968).

7 Rubovitz, F. E.: Traumatic rupture of the pregnant uterus from 'seat belt' injury. Amer. J. Obstet. Gynec. *90:* 828–829 (1964).

8 Snyder, R. G.; Snow, C. C.; Young, J. C.; Crosby, W. M., and Price, G. T.: Pathology of trauma attributed to restraint systems in crash impacts. Aerospace Med. *39:* 812–829 (1968).

Author's address: Warren M. Crosby, M. D., Professor, Department of Gynecology and Obstetrics, University of Oklahoma Health Sciences Center, 800 N.E. 13th St., *Oklahoma City, OK 73190* (USA)

Medical Primatology 1972. Proc. 3rd Conf. exp. Med. Surg. Primates, Lyon 1972, part II, pp. 12–15 (Karger, Basel 1972)

Microsurgery of Anterior Segment of Baboon Eye

P. H. DE LAAGE

Laboratory for Experimental Medicine and Surgery in Primates of New York University Medical Center, New York, N.Y.

The characteristics of the anterior segment of the baboon eye recommend it to experimental microsurgery. The yellow baboon *(Papio cynocephalus)* was choosen because the curvature of the cornea is very comparable to that of man.

Our measurements were made on a series of 30 young adult baboons (weighing 12–18 kg); the following figures are averages of the data collected.

The head of the cynocephalus is characterised by a long nose of 88 mm, an orbitonasal angle of 120–130° and a high protuberant brow ridge which contributes to a deep orbital socket. The inside diameter of the orbit is about 25 mm, and the distance between orbits is 6.5 mm. The distance between the medial and lateral canthi of the eye is very small, 19 mm; so, a canthotomy is often necessary to facilitate surgery.

The eyeball of the cynocephalus is slightly smaller than a human eyeball: its vertical diameter at the equator is 22 mm; the horizontal diameter is 23 mm; the anteroposterior is only 20 mm.

The vertical diameter of the *cornea* is 11.5 mm, the horizontal diameter 12.5 mm. The center is thinner (0.45 mm) than the periphery (0.60 mm). The anterior chamber depth is about 2.8 mm.

The limbus is marked by a smooth indentation of the sclera and by a pigmented conjunctival ring of 2.5–3 mm width.

The sclera is thickest between the cornea and the rectus muscle insertions, behind which it becomes very thin. The muscles are inserted very close to the limbus: 3.5, 4, 4.5 and 5 mm are the respective measurements for the medial, inferior, lateral and superior recti.

The cornea is steeper than a human cornea. We found a radius of 6.9 mm at the vertical meridian and 7 mm at the horizontal – that is to say, a keratometer reading of 48 × 48.75 d, showing a with-the-rule-astigmatism of 0.75 d.

Histologic examination disclosed six layers in the cornea.

The epithelium is nonkeratinized and stratified, consisting of five or six layers of cells. It becomes suddenly thicker at the limbus, marking the transition with the conjunctiva, and a strong pigmentation appears as pigment granules in the cells.

The epithelium lies on a smooth *basal membrane*, which is intensively periodine-oxide reactive.

In spite of some opinions which state that the Bowman membrane is lacking in the *Catarrhinae* [1], we found just under the epithelium a cell-free homogeneous layer, which we consider to be a Bowman membrane. This layer is about 10 μm thick, distinct from the basal membrane and from the stromal lamellae. It takes a light coloration with Vernoeff's stain, demonstrating its difference from the stromal lamellae, and, peripherally, it ends abruptly, as a Bowman membrane does.

The stroma, very comparable to that of man, consists of about 50 layers of lamellae, between which many cells can be seen.

The Descemet membrane of cynocephalus has an intense periodine reaction, as in all primates. About 8 μm thick, it detaches readily from the stroma but adheres well to the endothelium. It ends at the limbus, thinning out and merging into the trabecular meshwork, to the formation of which it appears to contribute.

The endothelium completely covers the Descemet membrane. It consists of one layer of flat, polygonal cells. At the periphery, it covers each lamella of the trabeculum.

The iris shows the same elements as in a human eye: an irregular and very thin endothelium, the nature of which has been often disputed [2], a stroma with a muscle layer, and a posterior pigmented epithelium.

The stroma is very thick and composed of two layers, separated by the so-called cleft of Fuchs. In the anterior layer the chromatophores are very numerous and made up of either cells with slender processes or round and isolated cells. The posterior layer has a much poorer pigmentation. The muscle layer is very well developed, with a large sphincter which occupies almost one third of the iris.

The root of the iris is very thin, attached to the ciliary body by the continuation of the epithelium, the dilatator and only a few stromal cells.

The chamber angle shows the same highly-differentiated structure as that found in the human eye: A Schlemm's canal is present with its special inner wall, as well as a trabecular meshwork with a complicated system of lamellae organized in 10–15 superimposed trabeculae. The pectinate ligament was completely absent on all the slides we observed.

The ciliary muscle is well developed and attached directly to the trabecular meshwork, with which the fibers seem to be in continuity.

In *the posterior segment*, three facts caught our attention:

(1) the thinness of the sclera at the equator;

(2) the heavy pigmentation of the choroid;

(3) the remanent of hyaloid artery often attached to the optic disc.

Three kinds of microsurgery have been performed on baboon eyes: penetrating keratoplasty, crescentic keratectomy, and tenotomy, with the aim of adjusting operative techniques, adapting the instrumentation, and studying the postoperative astigmatism.

Before each operation an objective eye examination was made, including an examination under the slit lamp, a keratometer reading, a tonometry with the Schiotz tonometer, an examination of the fundus, and a mold of the cornea using the same material as that used for contact lens fitting.

Because of the shape of the baboon head, we had to design a special head rest in order to hold the nose at a suitable angle. When the surgeon is seated on the same side as the eye to be operated on, the microscope is well positioned, and it is possible to take pictures from the other side.

General anesthesia was administered, using an endotracheal tube.

The preparation of the eyelids of both eyes and the nose was done with antiseptic solution.

The conjunctiva was washed with saline, and antibiotics were instilled.

It is a simple matter to fix the eye with a 4.0 suture under the superior rectus and to hold the lids with a Barraquer-Colibri speculum.

When the anterior chamber has to be entered, some hemorrhages appear, coming from the diametrically opposite point of the chamber angle, but they stop quickly. The secondary aqueous humor is thicker than the human and richer in fibrin.

The iris is thick and hard to cut, and liberates many pigment granules. Therefore, the iridectomy has to be specially checked to be sure that it is completly through the iris. There is a tendency towards iris bombé, which is difficult to control, in spite of heparin, which maintains the aqueous humor fluidity very well. Two drops of pilocarpine 2%, instilled directly on the iris, bring it back down rapidly, flat and in myosis.

The cornea is thinner than a human cornea, so precautions have to be taken in suturing. We used 22 μm monofilament nylon suture, with an interrupted stitch and, when it was a question of a penetrating keratoplasty, an eight-passage continuous suture was also used. All the knots are pushed into the corneal stroma just under the Bowman's membrane.

At the end of the procedure, saline is injected in order to reform the anterior chamber, and 40 mg of Depomedrol is injected under the conjunctiva.

To prevent the baboon from rubbing its eye after surgery, it is fitted with a mask made of plaster and individually designed to fit the baboon head. This mask is completed just after the surgery, while the baboon is still under general anesthesia.

A piece of metallic grill in front of the eye permits instillation of drops during the postoperative course.

References

1 ROHEN, J. W.: Sehorgan. (The organ of vision); in HOFER, SCHULTZ and STARCK Primatologia, vol. 2. (Karger, Basel 1962).

2 VRABEC, F.: Sur la question de l'épithélium de la face antérieure de l'iris. (On the question of the endothelium of the anterior face of the iris.) Ophthalmologica *123:* 20–30 (1952).

Author's address: P. H. DE LAAGE, 6, bd Wilson, *F–06600 Antibes* (France)

Medical Primatology 1972. Proc. 3rd Conf. exp. Med. Surg. Primates, Lyon 1972, part II, pp. 16–28 (Karger, Basel 1972)

Gallstone Research in Baboons

C. K. McSherry, K. P. Morrissey, N. B. Javitt, Patricia May and F. Glenn

New York Hospital-Cornell Medical Center, New York, N.Y.

Although gallstones are one of man's most common afflictions, progress in understanding the pathogenesis of biliary calculi has been slow. The absence of suitable experimental models among animals has been a major factor in limiting the advancement of knowledge in this field. In the past, most investigations concerning gallstone formation were performed on dogs. Since gallstones are extremely rare in dogs and when present are of the calcific type, the application of such data to humans is subject to many limitations. The biliary system varies greatly in different species. Some animals, such as the rat, deer and horse, do not have a gallbladder; while in others, for example the dog, multiple ducts are present between the liver and the common bile duct. These anatomic variations may in part account for some of the discrepancies reported concerning etiologic factors of importance in gallstone formation.

With respect to nonhuman primates, cholesterol cholelithiasis has been described in both marmosets and baboons. Hamerton [5] described a marmoset that died of acute cholecystitis and calculous obstruction of the common bile duct. In 1963, Lapin and Yakovleva [6] reported the results of *post mortem* examination in 417 animals at the Sukhumi Animal House, USSR. Biliary calculi, composed largely of cholesterol, were found in two marmosets and three baboons. In 1970, Glenn and McSherry [4] reported two baboons *(Papio anubis* and *Papio cynocephalus*) with cholesterol cholelithiasis at the Laboratory for Experimental Medicine and Surgery in Primates (LEMSIP), New York (fig. 1). The cholesterol content of the calculi in these animals was 85 and 96% with the balance composed of mixed bile salts. Information concerning the occurrence of biliary calculi in baboons in their natural habitat is scarce. However, Osman Hill has received a number of

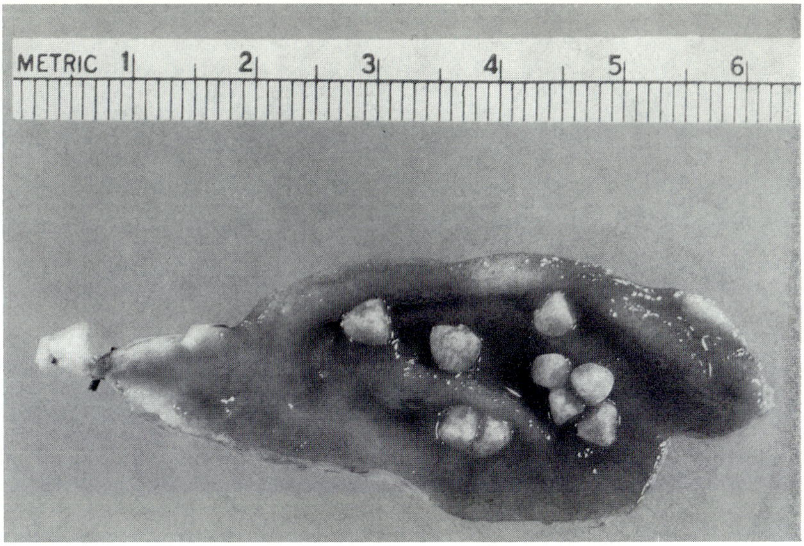

Fig. 1. Cholesterol stones in a gallbladder removed from an *anubis* baboon.

gallstones from East African natives which were said to have been removed from *anubis* baboons [8]. It was his opinion that gallstones in certain parts of Africa were common in wild baboons [10].

Materials and Methods

Since 1969, studies of hepatobiliary function have been performed in 54 female baboons at the Laboratory for Experimental Medicine and Surgery in Primates. Female baboons were selected for investigation on the basis of previous clinical studies by the authors [3] which documented a close relationship between pregnancy and the onset of symptoms believed to be due to biliary calculi. Of the 54 animals, 40 were normal, nine were pregnant and five had cholesterol cholelithiasis. Laparotomy was performed in these animals and samples of gallbladder and hepatic bile were obtained. Bile was analyzed for cholesterol, phospholipid and bile salt content by methods previously described [7]. In most of the animals the liver was also biopsied. Plasma, obtained immediately prior to operation, was also analyzed for cholesterol and phospholipid concentrations.

More than 90% of the dry weight of bile is contributed by cholesterol, phospholipids (principally lecithin) and bile salts. Bile salts are water soluble detergent-like molecules that form small aggregates in aqueous solutions called micelles. Lecithin, although insoluble in water, swells to form liquid crystals. Cholesterol is also insoluble in aqueous systems but is incorporated into the lecithin liquid crystals. These insoluble liquid crystals are then bro-

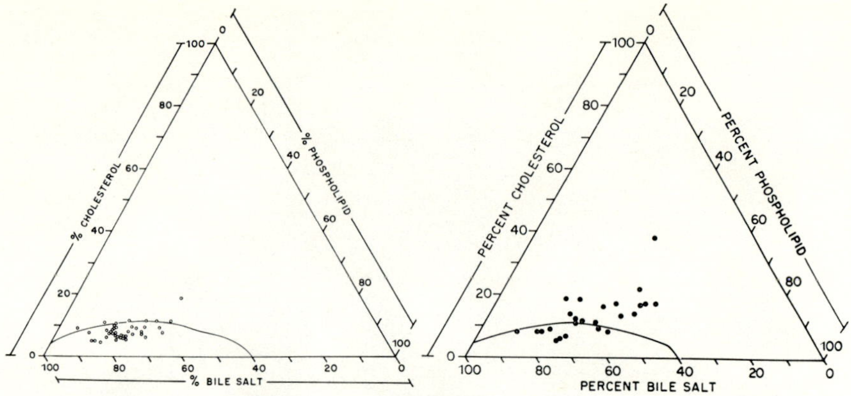

Fig. 2. The cholesterol, phospholipid and bile salt content of bile samples from normal baboons is plotted as a percentage of the total quantity of all three components as a single point on triangular coordinates. The area of the triangular graph below the solid line represents the zone of micellar solubility for cholesterol in bile. *a* Gallbladder (n = 41) and *b* hepatic (n = 25) bile.

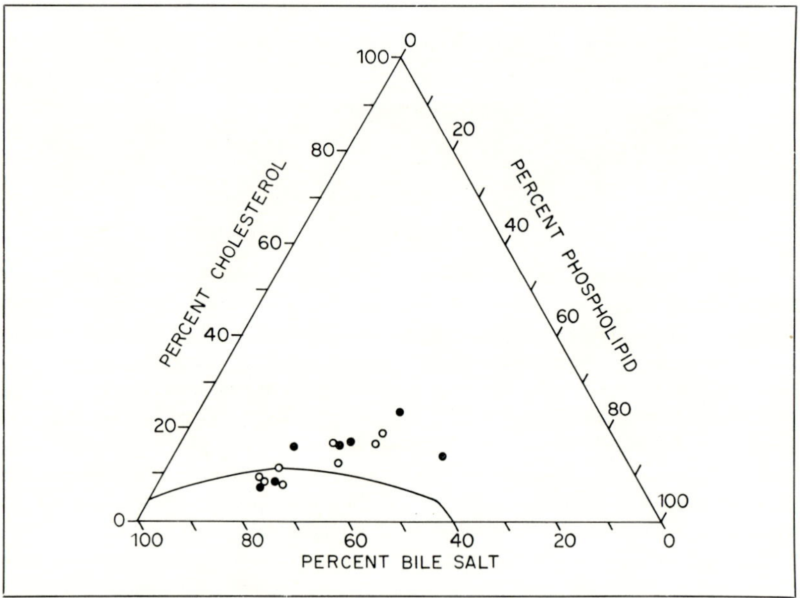

Fig. 3. Triangular phase diagram depicting the lipid composition of gallbladder (○) and hepatic (●) bile samples from pregnant baboons.

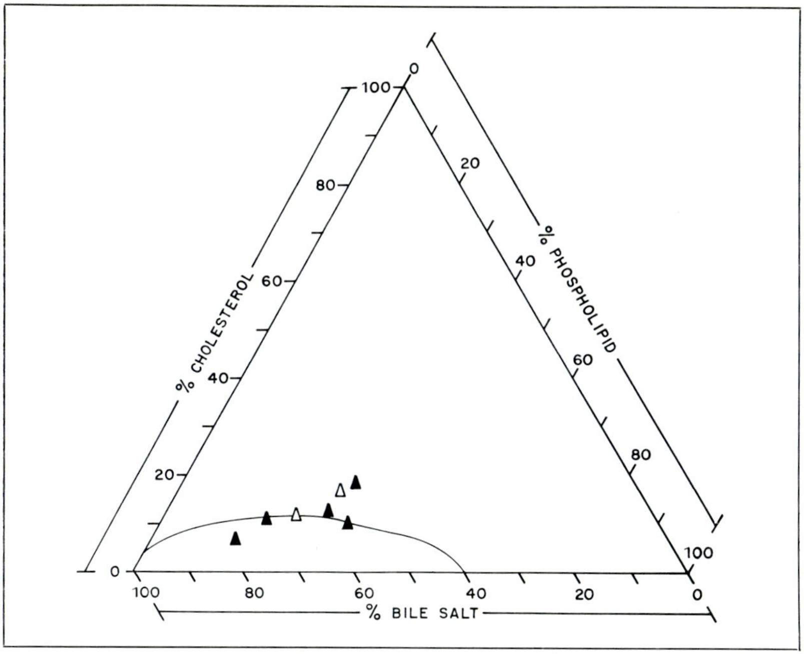

Fig. 4. The lipid content of hepatic (▲, n = 5) and gallbladder (□, n = 2) bile in baboons with gallstones.

ken up by bile salts into small soluble aggregates called mixed micelles. The capacity of the mixed micelle to solubilize cholesterol is related to the relative amounts of both lecithin and bile salts. In 1968, ADMIRAND and SMALL [1] proposed the use of a triangular phase diagram for depicting the relative concentrations of cholesterol, phospholipid and bile salts in individual bile samples. This system allowed the prediction of the maximum amount of cholesterol that could be solubilized in solutions containing varying proportions of bile salt and phospholipid.

In figure 2 the relative quantity of cholesterol, bile salts and phospholipids in each bile sample as a percentage of the total quantity of all three components is plotted as a single point on triangular coordinates. These bile samples, both hepatic and gallbladder, were obtained from control nonpregnant baboons that did not have calculi. It is evident from this figure that most normal baboons, 16 of 25 so studied, produced an hepatic bile in which cholesterol was out of the micellar zone and, therefore, present in a supersaturated state. In contrast, the cholesterol content of gallbladder bile, 34 of 41 samples analyzed, was usually within the micellar zone and, therefore, less than saturated.

Figure 3 depicts the triangular phase diagram for the hepatic and gallbladder bile samples from pregnant baboons. In five of seven baboons in which hepatic bile samples were available, cholesterol was present in a supersaturated state. With respect to gallblad-

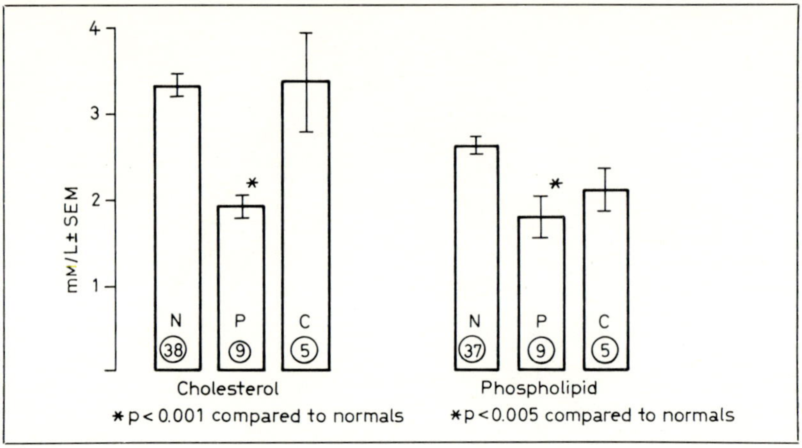

Fig. 5. The cholesterol and phospholipid content of plasma in baboons. Specimens obtained from normals (N), pregnant baboons (P) and baboons with cholelithiasis (C). ○ Number of animals.

der bile samples, only three of eight were within the micellar zone while the remaining five were supersaturated with cholesterol. It appears, therefore, that pregnancy may influence the solubility of cholesterol both in the gallbladder and in hepatic bile. A larger number of pregnant animals must be studied before this conclusion can be established.

Of the five baboons with cholesterol gallstones (fig. 4), hepatic bile was saturated with cholesterol in two and supersaturated in two other animals. In the remaining baboon the cholesterol content of hepatic bile was within the micellar zone. Only two baboons with gallstones had a sufficient quantity of bile present in the gallbladder to permit chemical analysis of its lipid content. In these animals the bile was saturated with cholesterol in one and supersaturated in the other.

The cholesterol and phospholipid content of the plasma in the three groups of baboons are depicted in figure 5. In the pregnant animals, the plasma cholesterol and phospholipid concentrations are significantly less than in the normals. This observation of the effect of pregnancy on serum lipid concentrations in baboons has previously been reported by van Zyl [11]. Similar data obtained from rhesus monkeys have been documented by Wolf *et al.* [12]. In contrast, the serum cholesterol increases during pregnancy in humans [9]. To date there is no satisfactory explanation for the serum cholesterol response to pregnancy in humans or other primate species. The serum lipids in the baboons with gallstones do not differ significantly from the normal animals without stones.

These observations on the lipid composition of bile and plasma suggest that we are unable to distinguish by chemical analysis that population of baboons which has cholesterol gallstones from the normal. Most normal baboons, at least under the conditions of this study, appear to produce an hepatic bile which is saturated or supersaturated with cholesterol. The same is true for the animals with gallstones.

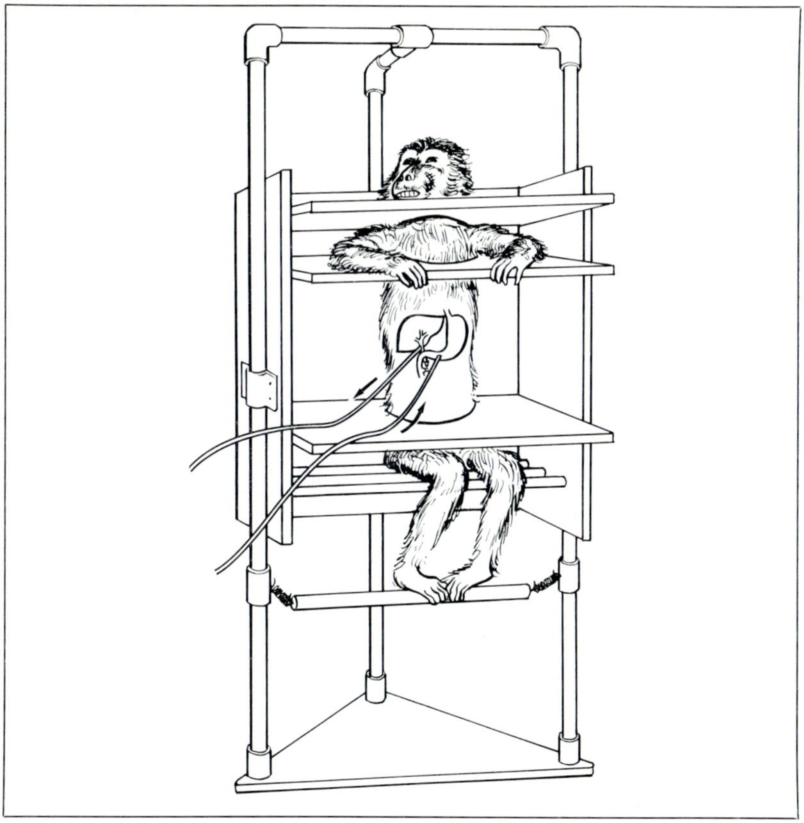

Fig. 6. Artist's sketch of baboon with permanent biliary fistula trained to remain in restraining chair for prolonged periods of time.

In order to assess the role of the liver and the importance of cholesterol solubility in hepatic bile in relation to the pathogenesis of gallstones, additional investigations were performed. An experimental model was utilized to study bile flow and composition over 24-hour periods and to evaluate the response of hepatic bile composition to various stimuli.

Adult female baboons have been trained to live for prolonged periods in restraining chairs of the type depicted diagramatically in figure 6. Approximately three to four weeks of progressive conditioning are required before the animals can be maintained without interruption in the apparatus. When accustomed to the chair, the baboons are operated upon and a permanent biliary fistula is created. The gallbladder is removed and a small T-tube (No. 8F or 10F) is sutured into the common bile duct. The common duct distal to the T-tube is divided and the ends sutured closed. The long limb of the T-tube exits through the abdominal wall via a stab wound in the right upper quadrant. A malecot catheter is used to

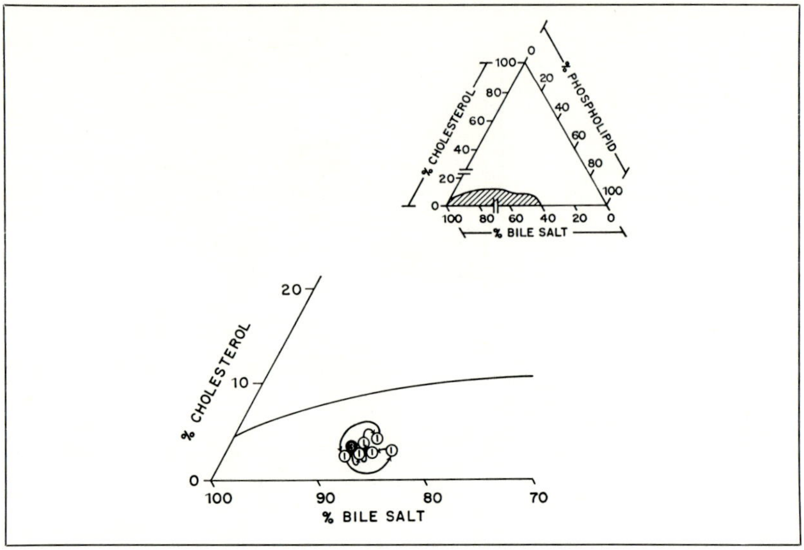

Fig. 7. Baboon No. 242. The effect of 5% interruption of the enterohepatic circulation on the lipid composition of hepatic bile. The triangle at the upper right represents the three-component system and the shaded area, the micellar zone. ● Control, hepatic bile; ○ number of samples of hepatic bile (each sample obtained at successive 30- and 60-min intervals).

establish a gastrostomy in the left upper quadrant of the abdomen. Following recovery from anesthesia, the baboon is returned to the restraining chair. Total bile flow may then be measured and aliquots of bile obtained for lipid analysis. To accomplish this, bile flow is directed through an electronic fraction collector [2]. This apparatus permits one to sample a predetermined percentage of drops of bile while returning the remainder of the unsampled bile to the animal via the gastrostomy tube. The latter is effected by a radiofrequency level detector and an infusion pump. To date seven baboons have been successfully trained and adapted to the restraining chair. Following creation of the biliary fistula, they have been studied for periods ranging from 75 to 90 days. The most common complications encountered have been accidental dislodgement of the T-tube and the formation of choledochoduodenal fistulas. Attempts to re-establish the bile fistulas in these animals have usually but not always been unsuccessful.

Figure 7 depicts the changes in the relative proportions of cholesterol, phospholipid and bile salt in the hepatic bile of one baboon observed over a 24-hour period. Utilizing the fraction collector, 5% of the bile flow was removed from the enterohepatic circulation for analysis. Samples of bile were collected every hour for the first 6 h and then every 2 h for the remaining 18 h. During this 24-hour period, the baboon was allowed both fluids and solid food *ad libitum*. It is of interest that during the 24-hour period the lipid composition of hepatic bile remained relatively constant and cholesterol was always within the micellar

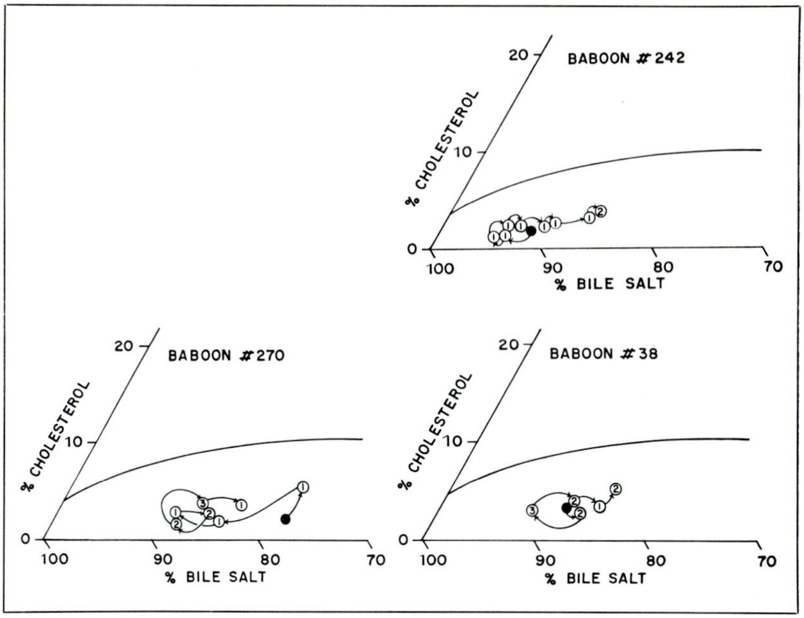

Fig. 8. The effect of bile salt (sodium taurocholate) on the lipid composition and cholesterol solubility of hepatic bile in three baboons. ● = Control – hepatic bile; ○ number of samples of hepatic bile (each sample obtained at successive 30- or 60-min intervals).

zone. The total volume of bile flow during the 24-hour period was 254 ml. Comparable experiments have been performed with 10, 20 and 25% interruption of the enterohepatic circulation and similar results in terms of lipid composition and bile volume have been observed. These studies indicate that the liver in the baboon is capable of increasing bile salt synthesis to maintain bile salt secretion and pool size at normal levels with up to 25% interruption of the enterohepatic circulation.

In order to investigate the possible mechanisms responsible for the regulation of cholesterol solubility in hepatic bile, additional experiments were performed. In three animals, human gallbladder bile (10 ml), obtained by needle aspiration in the course of cholecystectomy, was fed via the gastrostomy tube and changes in the lipid composition of hepatic bile monitored. Samples of hepatic bile were collected at 30-min intervals for 4 h and then 60-min intervals for an additional 4 h. In all animals there was a prompt increase in bile flow and an increased bile salt excretion in hepatic bile. As a consequence of the increased amount of bile salt relative to cholesterol and phospholipid in the hepatic bile, cholesterol solubility increased.

In subsequent experiments in these three baboons, pure bile salt (sodium taurocholate) was substituted for human gallbladder bile. Four grams of taurocholate were dissolved in 100 ml of saline and fed to the animals via the gastrostomy tube. Hepatic bile flow and lipid composition were measured as described in the previous experiments. Taurocholate produced a similar prompt increase in bile flow and bile salt excretion in hepatic bile. The effect of this increase in bile salt excretion on the solubility of cholesterol is depicted in figure 8. Cholesterol solubility is increased as evidenced by the changes in the relative proportions of the three major lipid substances in bile.

Radiologic Evaluation of Biliary Tract Function

In conjunction with our studies of the physical and biochemical properties of bile, studies of gallbladder function have been carried out using radiographic techniques. We have presumed that the changes in the lipid composition of bile may be important in the initial stages of stone formation, but such alterations may also be accompanied by disorders of gallbladder physiology, including motility and absorption. The evolution from a microscopic crystal to a macroscopic stone in the gallbladder implies some element of bile stasis within this structure. In order to study the concentrative ability and the contractility of the gallbladder, oral and intravenous cholecystography have been utilized.

The technique employed for oral cholecystography is relatively simple. On the evening of the day prior to study, the baboons are fed fruit into which

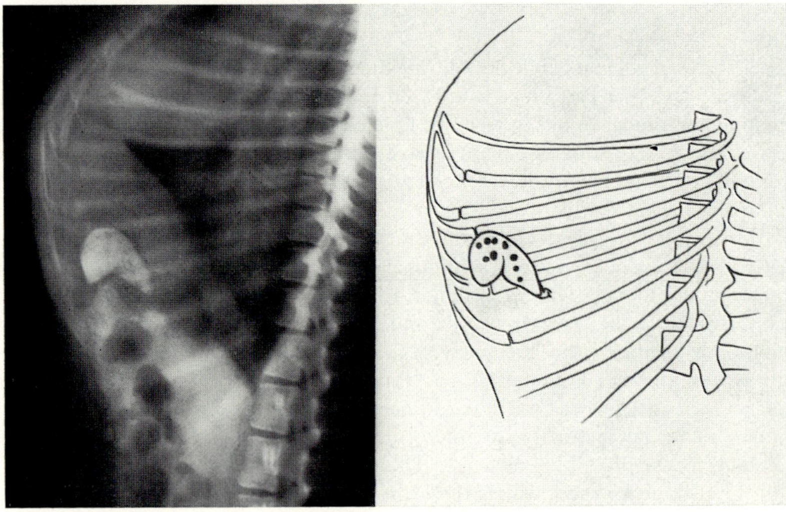

Fig. 9. The radiographic appearance of cholesterol cholelithiasis in the baboon.

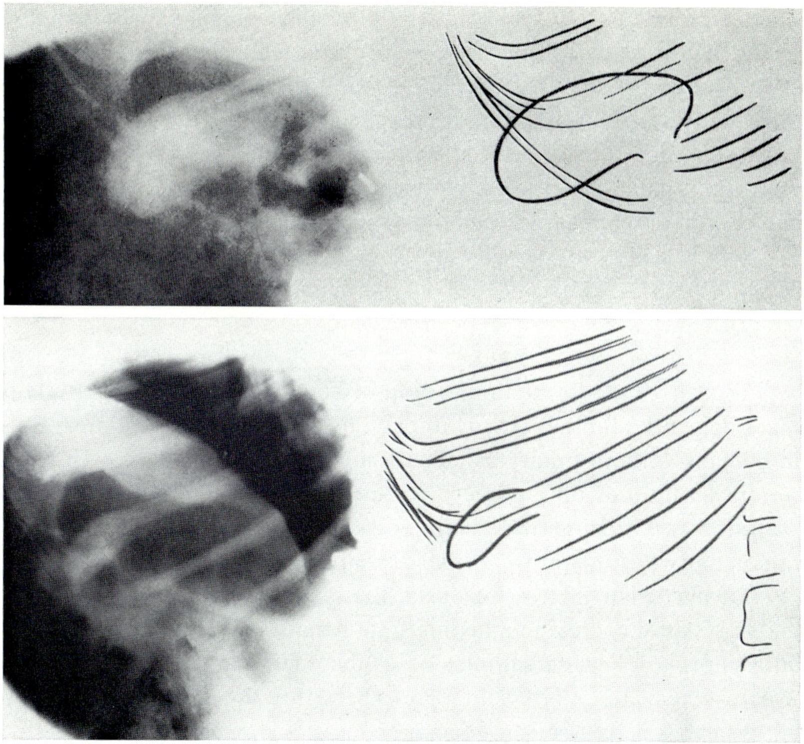

a

b

Fig. 10. Oral cholecystogram of a pregnant baboon *a* before and *b* 15 min after intravenous cholecystokinin. The contractile response of the gallbladder to cholecystokinin is less than that observed in the nonpregnant state.

the contrast agent has been impregnated. Three tablets (1.5 g) of iodopanoic acid have provided satisfactory opacification of the gallbladder in most of the animals. No additional food or fluid is then permitted until the X-ray examinations are completed on the following morning. The radiographs are obtained after sedating the animals with phencyclidine hydrochloride.[1] Figure 9 illustrates the radiographic appearance of cholesterol cholelithiasis in the baboon. To date oral cholecystograms have been obtained in 70 baboons and gallstones detected in three of the animals.

Experience with human cholesterol cholelithiasis has suggested that there is a close association between pregnancy and the onset of symptoms of biliary dysfunction. Because of the potential hazards of radiation to the fetus,

1 Sernylan, Parke Davis, Detroit, Mich.

opportunities to study gallbladder function in pregnant women are infrequent. We have, therefore, been particularly interested in studying the concentrative ability and contractility of the gallbladder in pregnant baboons. Such studies suggest that during pregnancy the gallbladder is capable of concentrating the contrast agent as well as during the nonpuerperal state. However, the capability of the gallbladder to contract in response to intravenous cholecystokinin appears to be diminished (fig. 10).

Discussion

Studies of the lipid composition of hepatic and gallbladder bile in humans have been limited largely to patients with gallstones. Comparable data in humans without biliary calculi are not as readily available. This imposes serious limitations on the interpretation of biliary lipid data obtained from such studies. Utilizing the baboon as an experimental model we have had the opportunity to study both normal as well as stone-forming animals.

Of particular interest has been the observation that most normal animals appear to excrete in a fasting state hepatic bile which is supersaturated with cholesterol while samples of gallbladder bile from the same animal usually reveal cholesterol solubility to be within the micellar zone. However, when fasting hepatic bile composition is studied in animals with permanent biliary fistulas, cholesterol solubility is almost always within the micellar zone. To explain these observations in the intact animal we have thought it reasonable to distinguish hepatic bile secreted under basal conditions from that occurring under stimulated conditions. Basal bile flow is defined as the type of bile secreted by the liver after 90% of the bile salt pool has returned to the gallbladder, as occurs for example in the fasting state. Stimulated bile flow occurs during the reabsorption from the intestines and secretion into bile of 90% of the bile salt pool. This phase of bile flow would be anticipated to occur following meals and contraction of the gallbladder with evacuation of gallbladder bile into the small intestine. From our studies in the baboon it appears that basal hepatic bile composition can be characterized by having more cholesterol in solution than can be accounted for by micellar solubility. Stimulated hepatic bile, in contrast, is less than saturated with cholesterol. Following cholecystectomy and removal therefore of a reservoir from the enterohepatic circulation, bile salt secretion by the liver continues at a more or less constant level resembling only stimulated bile flow. In the intact animal, the lipid composition of gallbladder bile may be considered to represent a combination of both basal and stimulated hepatic bile. Therefore, the observation that cholesterol solubility in simultaneously-obtained samples of gallbladder and hepatic bile is different may be readily explained.

Analysis of the bile and serum lipid data from the five animals with cholesterol cholelithiasis fails to reveal any significant distinctions from the normals. Four of these animals had been pregnant while in captivity. The remaining baboon had not been bred at the primate laboratory and because of her apparent youth had probably not been pregnant prior to capture. Of the nine baboons that we have studied during pregnancy, only one was found to have gallstones. The fact that cholesterol was supersaturated in the gallbladder bile of five of eight animals studied suggests that in comparison to the normal animals there may be some effect of pregnancy on the gallbladder. As yet, however, an insufficient number of pregnant baboons has been studied.

To evaluate further the effect of pregnancy on gallbladder function, radiographic studies of the concentrative ability and motility of this structure are under active investigation. Our preliminary findings suggest that during pregnancy the contractile response of the gallbladder to intravenous cholecystokinin appears to be slower and less vigorous.

References

1 Admirand, W. H. and Small, D. M.: The physicochemical basis of cholesterol gallstone formation in man. J. clin. Invest. *47:* 1043–1052 (1968).

2 Berger, J.; Redinger, R., and Small, D. M.: Instrument for sampling and measuring bile flow. Med. Electron. biol. Engng. *8:* 19–24 (1970).

3 Glenn, F. and McSherry, C. K.: Gallstones and pregnancy among 300 young women treated by cholecystectomy. Surg. Gynec. Obstet *127:* 1067–1072 (1968).

4 Glenn, F. and McSherry, C. K.: The baboon and experimental cholelithiasis. Arch. Surg. *100:* 105–108 (1970).

5 Hamerton, A. E.: Report on the deaths occurring in the Society's gardens during the year 1931. Proc. zool. Soc., Lond. *2:* 613–638 (1932).

6 Lapin, B. A. and Yakovleva, L. A.: Comparative pathology in monkeys, p. 265 (Thomas, Springfield 1963).

7 McSherry, C. K.; Javitt, N. B.; Carvalho, J. M. de, and Glenn, F.: Cholesterol gallstones and the chemical composition of bile in baboons. Ann. Surg. *173:* 569–577 (1971).

8 Ruch, T. C.: Diseases of laboratory primates, p. 265 (Saunders, Philadelphia 1959).

9 Svanborg, A. and Vikrot, D.: Plasma lipid fractions, including individual phospholipids at various stages of pregnancy. Acta. med. scand. *178:* 615–630 (1965).

10 Vagtborg, H.: The baboon in medical research, p. 212 (University of Texas Press, Austin 1965).

11 Zyl, A. van: Serum protein-bound iodine and serum lipid changes in the baboon *(Papio ursinus)*. II. During pregnancy, lactation and abortion. J. Endocrin. *14:* 317–324 (1957).

12 Wolf, R. C.; Temte, L., and Meyer, R. K.: Plasma cholesterol in pregnant rhesus monkeys. Proc. Soc. exp. Biol. Med. *125:* 1230–1231 (1967).

Authors' address: Charles K. McSherry, M.D., Kevin P. Morrissey, M.D., Norman B. Javitt, M.D., Ph.D., Patricia May, B.A. and Frank Glenn, M.D., New York Hospital-Cornell Medical Center, 525 East 68th Street, *New York, NY 10021* (USA)

Medical Primatology 1972. Proc. 3rd Conf. exp. Med. Surg. Primates, Lyon 1972, part II, pp. 29–38 (Karger, Basel 1972)

The Effect of Hemorrhagic Shock on Pulmonary Interstitial Sodium Distribution in the Primate Lung[1]

G. S. Moss, T. K. Das Gupta, B. Newson and L. M. Nyhus

Department of Surgery, Abraham Lincoln School of Medicine, University of Illinois College of Medicine and Hektoen Institute for Medical Research, Chicago, Ill.

In a previous study, the effect of hemorrhagic shock on primate lung structure was investigated [3]. Using special techniques, light microscopy demonstrated diffuse interstitial edema following two hours of profound hemorrhagic hypotension. Electron microscopy revealed that the edema was primarily located in the interstitial space of the alveolar capillary membrane. The ultrastructural findings included increased distance between individual collagen fibers and in some cases distortion of the collagen fibers. Both endothelial and epithelial layers appeared undamaged. In the presence of intact lining layers and low capillary hydrostatic pressure, the development of interstitial edema is puzzling. Among a number of possible explanations, one relates to the possible changes in collagen sodium concentration during prolonged hypoperfusion. Hemorrhagic shock is known to be associated with the development of metabolic acidosis, primarily as a result of lactic acid accumulation. A reduction in pH has been shown to heighten the affinity of collagen for sodium and water [2]. Thus it is possible that the acidosis existing during hemorrhagic shock may lead to the accumulation of sodium and water in areas occupied by the collagen fibers of the alveolar capillary membrane.

The object of this report is to describe the distribution of sodium in the primate lung interstitium in the control state and following profound hemorrhagic shock.

1 Supported by Office of Naval Research grant No. NR 105-539 and Institutional Funds from the Veterans Administration West Side Hospital.

Methods

Hemorrhage. After a baseline period of approximately 3 h, 8 baboons were bled into a sterile plastic bag containing 67.5 ml of acid citrate dextrose solution. The arterial pressure was reduced over a period of 15 min to 60 mm Hg, and maintained at this level for 1 h. It was then reduced to 40 mm Hg and maintained for a second hour.

Technique of in vivo *dual lung fixation.* At the conclusion of the 2-hour shock period, the animal was placed in the supine position and the extremities were restrained. A small dose of pentobarbital was given i.v. The thoracic cavity was entered through a midline sternal splitting incision. A purse string suture was then placed in the main trunk of the pulmonary artery and a siliconized plastic catheter was threaded into the left pulmonary artery. A similar catheter was threaded through the endotracheal tube to the left main bronchus. By cross-clamping the right lung hilum, the left lung was isolated. A dual method of perfusion fixation through both the vascular and endobronchial systems was then initiated. The perfusion pressure in the arterial system was maintained at a steady 40-cm of water. After a few minutes of perfusion through the vascular bed, the ascending aorta was clamped, thus reducing the quantity of the perfusate needed. Approximately 500–700 ml of the perfusate

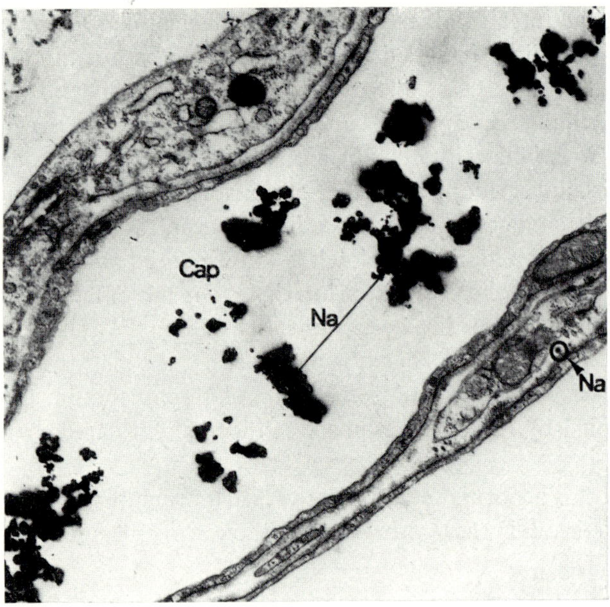

a

Fig. 1. a The sodium pyroantimonate precipitate in a pulmonary capillary. Note the alveolar capillary (Cap) membrane in either side of the capillary lumen. Sodium pyroantimony precipitate deposits (Na) can be seen distributed in the interstitial space. × 15,150. *b* The endothelial lining of a pulmonary capillary under high-power, showing the tight junc-

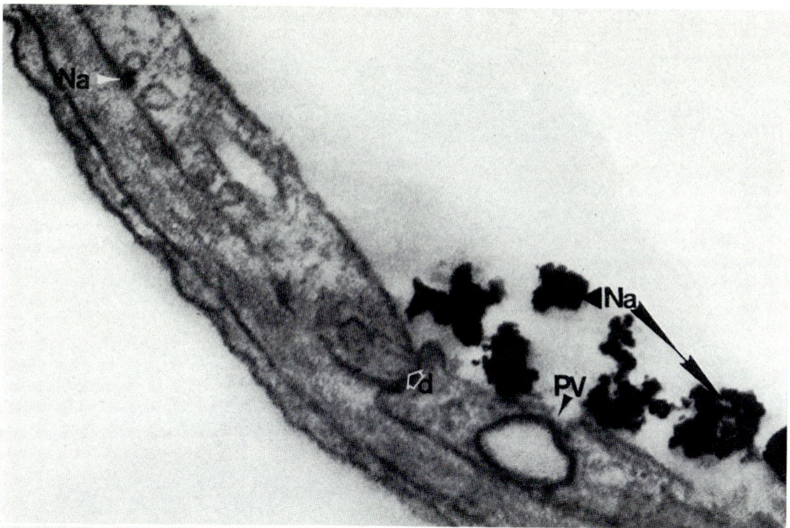

b

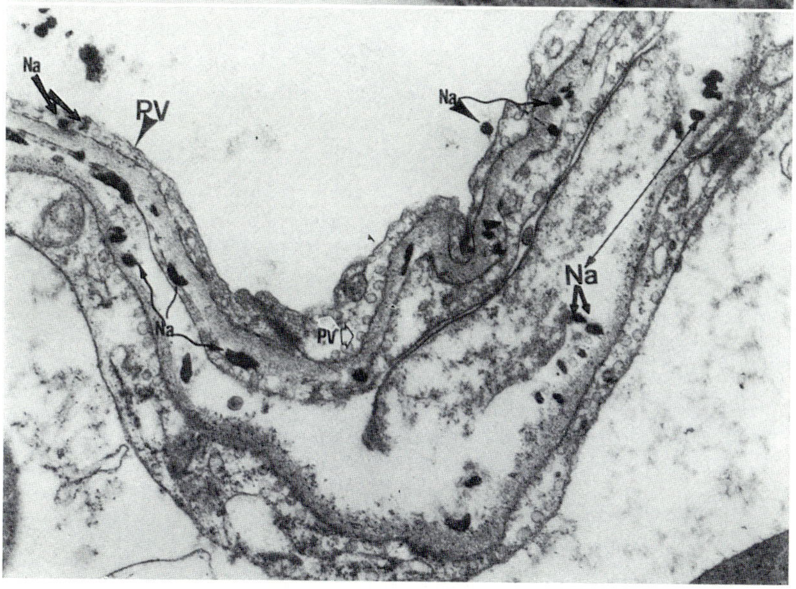

c

tion (d) and the surrounding sodium pyroantimony precipitate (Na). Note that no precipitate is passing through this desmosomal contact. A cluster of precipitates are adjacent to a pinocytotic vesicle (PV). ×20,900. *c* The distribution of sodium pyroantimony precipitate (Na) within the pinocytotic vesicles (PV) and the interstitial space in a control lung. Note the progression of the sodium pyroantimony precipitates from the pinocytotic vesicles to the collagen fibers. ×25,200.

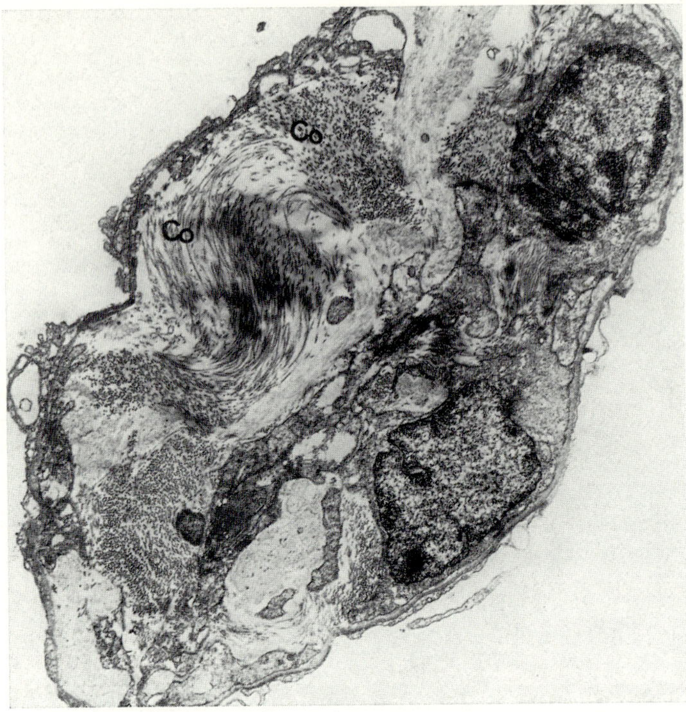

2a

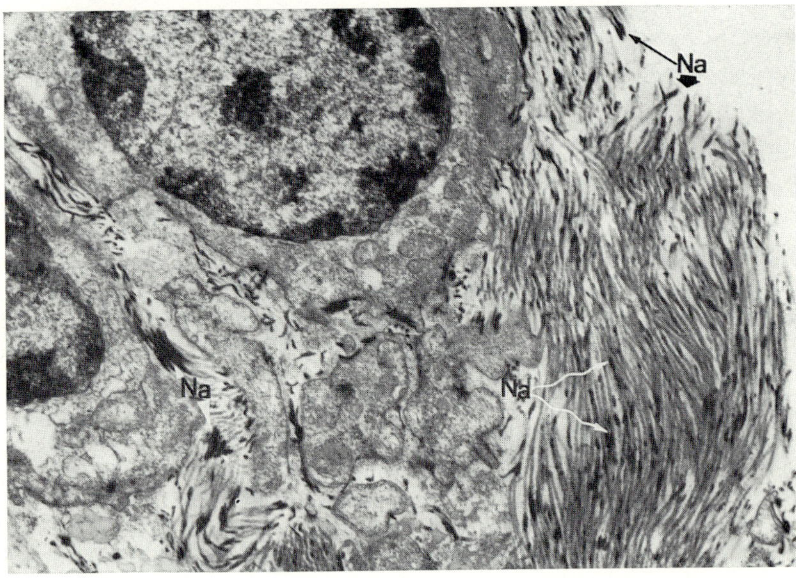

2b

was given in 30 min. Another 20–40 ml of the perfusate (the volume of fluid to fill the main stem bronchus and its branches) was manually instilled into the bronchus through the catheter. The catheter was then removed and the endotracheal tube clamped. At the end of 30 min of this dual fixation the lung was firm in consistency and resembled liver in gross appearance. The lung was then removed and tissue samples were taken from different areas. The baboon was then killed.

Following perfusion, slices of lung tissue were further fixed by immersion in the same perfusate for 2 h at 4 °C. Fixed tissues were quickly dehydrated in acetone and embedded in Araldite. The specificity of the chemical reaction was ascertained by adding the unbuffered pyroantimony solution to 0.9% sodium chloride solution. The resulting precipitate, sodium pyroantimonate, was sprayed in a grid and viewed.

Technique of electron microscopy. Ultrathin sections were cut and mounted on a copper grid. These were stained on the grid with an alcohol solution of uranylacetate and viewed in a RCA EMU 4 microscope.

Results

The mechanism of movement of sodium from the plasma to the pulmonary interstitium is not known. Theoretically, it might pass through intercellular clefts or it might be carried across the endothelial layer by way of pinocytotic vesicles. In figure 1a, sodium precipitates can be seen within the capillary lumen and also within the interstitial zone. In figure 1b, sodium is found adjacent to both an intercellular junction as well as a pinocytotic vesicle. However, in figure 1c, sodium precipitates can be located within pinocytotic vesicles. While this same relationship seen in figure 1c was frequently observed in other sections, sodium precipitates were never seen within intercellular junctions. This suggests a role for pinocytosis in sodium transport in the lung.

In figure 2a, a low-power view of the normal alveolar capillary membrane, the discrete distribution of the sodium precipitates can be seen. It should be noted that the major sodium concentration is in proximity to collagen fibers. At higher magnification, as demonstrated in figure 2b, a large fraction of the sodium is actually within individual fibers.

The effects of 2 h of hemorrhagic shock on sodium distribution are shown in figures 3a and b. In figure 3a, a low-power view of the alveolar cap-

Fig. 2. a This is a low-power view of the normal alveolar capillary membrane. The discrete distribution of sodium (Na) can be seen in and around collagen fibers (CO). × 5,940. *b* High-power view of an area of the alveolar capillary membrane in a control lung. Note the discrete localization of sodium pyroantimony deposits (Na) within the collagen fibers. × 23,800.

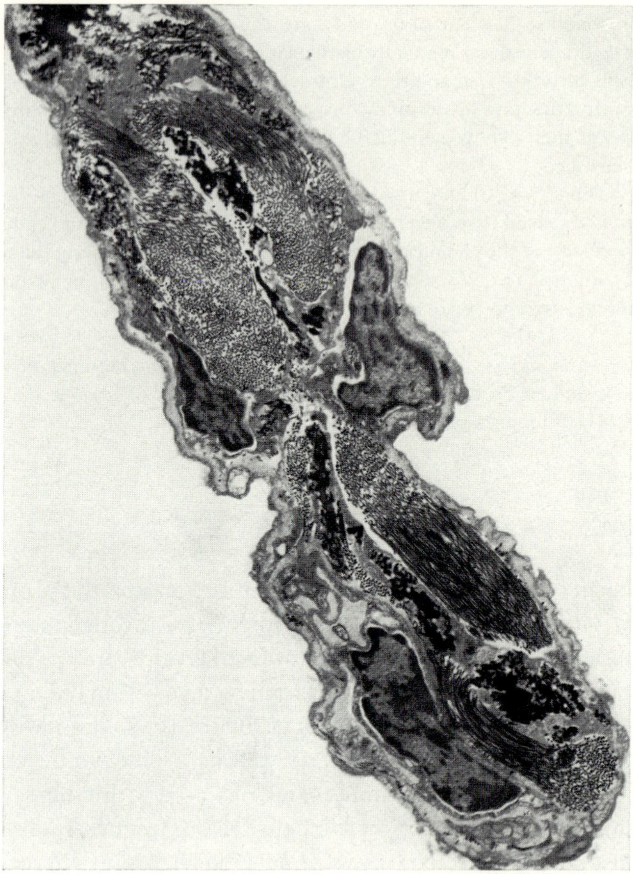

3a

Fig. 3. a Low-power view of the alveolar capillary membrane in shock. The appearance of large areas of interstitial intensely-stained smudging indicating sodium-rich edema fluid. Dispersion of collagen fibers is obvious. × 9,220. *b* High-power view of one area showing spectacular deposition of sodium pyroantimonate (Na) as well as smudging in shock lung. × 15,700.

illary membrane, profound differences can be seen as compared to the control state (see fig. 2a). These differences include the appearance of large areas of interstitial intensely stained smudging, indicating sodium-rich edema fluid; dispersion of individual collagen fibers, again suggesting edema; and intensely-stained individual collagen fibers, indicating increased take-up of sodium. Figure 3b shows in greater detail the smudging, the dispersion of individual collagen fibers, and the intense staining of the fibers.

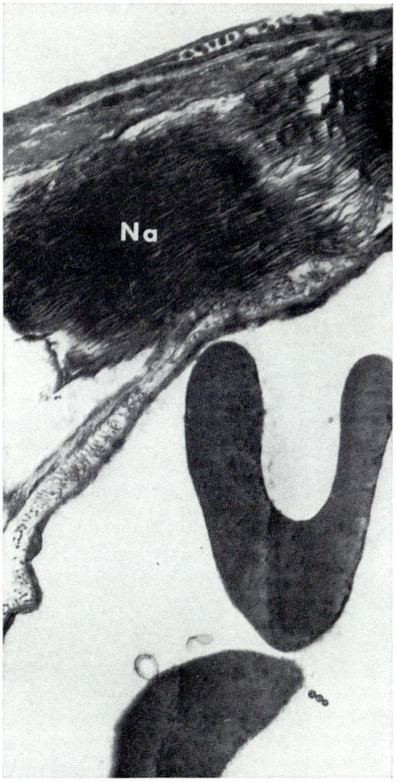

3b

Discussion

To our knowledge, this is the first histochemical study of sodium distribution in the lung. The movement of small and large molecules across intact membranes has been studied by several investigators [1]. Their results indicate that small molecules, such as sodium ions, are more likely to move across the membranes between intercellular junctions than by way of pinocytosis. However, we observed numerous instances in which sodium was visualized within pulmonary endothelial vesicles in both the control and shock state; and conversely we never observed sodium within an intercellular junction. In addition, we found no evidence of endothelial damage or disruption following shock. Another point in favor of pinocytosis as a bulk carrier of sodium is the observation made by us [3] in a previous report that the number of pinocytotic vesicles substantially increases during shock. Thus, one of the

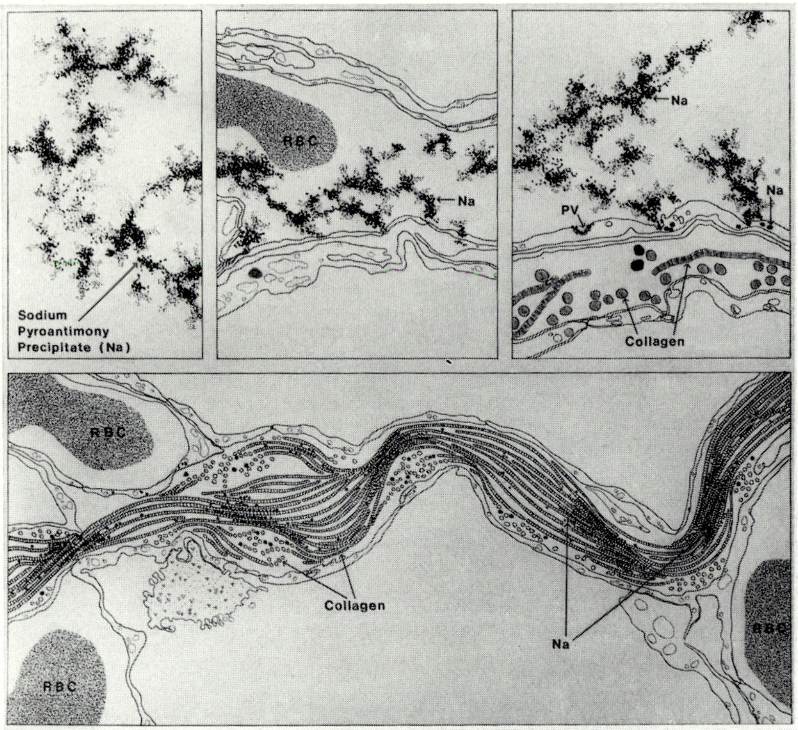

Fig. 4. Diagramatic rendition of the possible mode of endothelial transport of sodium and its ultimate distribution in the interstitial space of the alveolar capillary membrane of a control baboon lung. The upper left panel demonstrates the characteristics of sodium pyroantimony precipitate (Na). The upper middle panel shows sodium within the capillary lumen, as well as presence of the deposits within the pinocytotic vesicles (PV). The upper right panel shows the discrete localization within the vesicles. Several vesicles are close to the basement membrane. The lower panel shows the distribution of sodium precipitates in and around the collagen fibers. RBC

factors determining the amount of sodium deposited in the interstitium may be the rate of pinocytotic activity in the endothelial layer.

A second point of interest is the pattern of sodium distribution once it reaches the interstitial zone. Under normal circumstances sodium is distributed around groups of collagen fibers and within individual fibers. After a period of hemorrhagic hypotension, the concentration of sodium pyroantimonate in these same areas is substantially increased and pools of sodium-rich edema fluid develop within the interstitium. These observations sug-

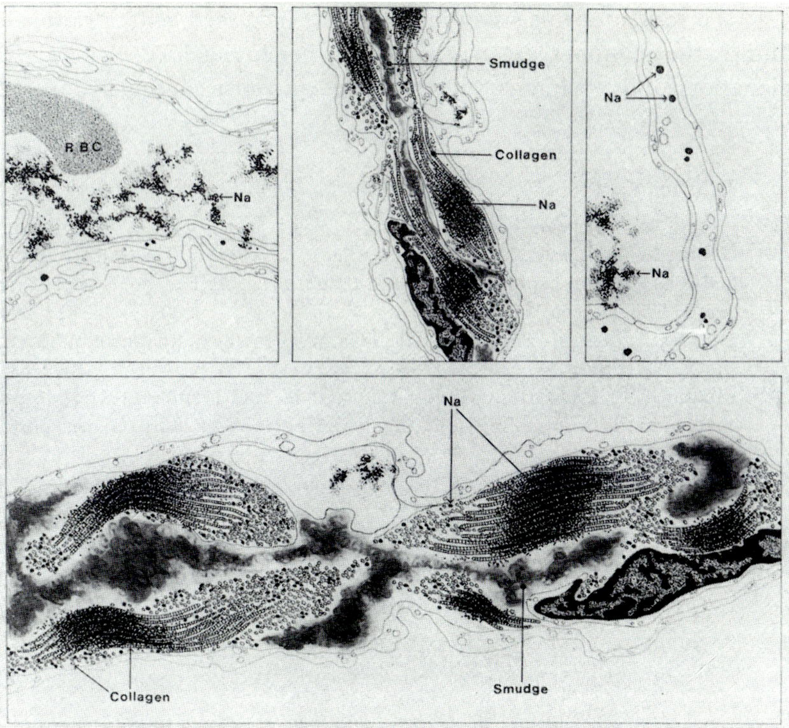

Fig. 5. Diagramatic rendition of the possible mode of endothelial transport of sodium and its final distribution in the interstitial space of the alveolar capillary membrane of a baboon lung in hemorrhagic hypotension. The upper panel shows the distribution of sodium pyroantimonate precipitate (Na) in the capillary. Note the precipitates are also within pinocytotic vesicles and progressing to the interstitial space. In the upper middle panel a low-power view shows the massive profile of sodium pyroantimony precipitate (Na) within the collagen fibers and the intense smudging representing edema. In the upper left panel the migration of sodium pyroantimony precipitate via pinocytotic vesicles are seen and no migration through the intercellular junctions are noted. In the lower panel the intense smudging, massive deposition of sodium pyroantimony precipitates are graphically shown. RBC

gest that collagen possesses a base line affinity for sodium and, furthermore, this affinity is heightened after a period of hemorrhagic hypotension. In addition the increased local concentration of interstitial sodium should favor the movement of plasma water into the interstitial zone. From these observations a conceptual rendition of sodium transport and distribution in control lung tissue and in lung tissue examined after 2 h of hemorrhagic shock is shown in figures 4 and 5.

The results of this study indicate that increased concentrations of sodium in the pulmonary interstitium during hemorrhagic shock may be the explanation for the pulmonary interstitial edema noted during hypoperfusion.

References

1 Crone, C. and Lassen, N. A. (ed.): Capillary permeability. The transfer of molecules and ions between capillary blood and tissue. Proc. A. Benzon. Symp. II (Munksgaard, Copenhagen 1970).

2 Dillon, J.; Lynch, L. J.; Myers, R., *et al.:* A bioassay of treatment of hemorrhagic shock. Arch. Surg. *93:* 537–555 (1966).

3 Moss, G. S.; Das Gupta, T. K.; Newson, B., and Nyhus, L. M.: Morphologic changes in the primate lung following hemorrhagic shock. Surg. Gynec. Obstet. *134:* 3–9 (1972).

Authors' address: Dr. G. S. Moss, M.D., Associate Professor of Surgery, Dr. T. K. Das Gupta, M.D., Dr. B. Newson, B.S. and Dr. L. M. Nyhus, M.D., The Abraham Lincoln School of Medicine, University of Illinois at the Medical Center, 840 South Wood Street, *Chicago, IL 60612* (USA)

Medical Primatology 1972. Proc. 3rd Conf. exp. Med. Surg. Primates, Lyon 1972, part II, pp. 39–43 (Karger, Basel 1972)

The Splanchnic Circulation in Shock

Species Variation Between Monkey and Dog

D. G. Reynolds and K. G. Swan

Walter Reed Army Institute of Research, Washington, D.C.

Among the many hypotheses concerning the pathogenesis of shock is the so-called 'target organ' or 'adrenergic' theory. Hypotension, regardless of its etiology, causes the release of endogenous catecholamines which are known vasoconstrictors in the splanchnic viscera. The degree of vasoconstriction sufficiently damages the splanchnic organs so as to permit enterotoxins to escape from within the lumen of the gut into the systemic circulation. Endotoxemia thus complicates the original hypotensive episode and a vicious cycle is established [7]. This theory has been derived almost exclusively from experimentation in canine shock models and yet has found clinical application in man. Adrenergic blocking drugs, such as phenoxybenzamine, have been advocated in human shock therapy to interrupt the vasoconstrictor effects of the endogenous catecholamines released during hypotensive syndromes [7]. However, two lines of evidence seriously question this theory. Although the catecholamines, norepinephrine and epinephrine are known vasoconstrictors in the gut of the anesthetized dog, whether injected [4] or infused [5], the amount of catecholamine required experimentally to induce significant vasoconstriction is significantly in excess of that amount found experimentally in shock [6]. Also in the presence of even high concentrations of catecholamines 'autoregulatory escape' [3] prevents sustained ischemia in the canine gut [5]. In addition, the subhuman primate has been shown to respond to endotoxic shock in a manner distinctly different from the dog, despite comparable degree of lethality and comparable elevations in circulating plasma catecholamines [6].

In the anesthetized dog, endotoxin (*E. coli*, LD_{80}, i.v.) causes a 60-% reduction in mean arterial pressure within two hours. Hypotension is associated with a 75-percent reduction in mesenteric blood flow, measured with an electromagnetic flowmeter. Hepatic and splenic arterial blood flows also fall. Blood flows and arterial pressure all exhibit a transient, partial recovery and this biphasic response of the splanchnic hemodynamic parameters in the dog may be due to the effects of endotoxin upon the portal venous circulation. Endotoxin causes a marked, but transient, increase in portal venous pressure, raising that value fourfold within minutes followed by a gradual return to control values. The increased portal venous pressure causes a significant fall in venous return to the heart and, presumably, accounts for the abrupt fall in arterial pressure and splanchnic blood flow. Following release of the hepatic venous sphincters from the transient effects of endotoxin, venous return to the heart is restored. Arterial pressure and blood flows are also restored but eventually undergo a second, significant and sustained fall, which persists until death. These hemodynamic findings are summarized in figure 1. Four hours following endotoxin injection the dog exhibits a hemorrhagic necrosis of the splanchnic viscera associated with bloody diarrhea.

In contrast to these studies in the dog are similar studies performed in the anesthetized baboon *(Papio anubis)*. Following injection of a comparable dose of *E. coli* endotoxin, there is a fall in arterial pressure to less than 50% of the control value within two hours. Hypotension is sustained at this low level for the duration of the four-hour observation period. During the entire experiment no significant change is observed in either portal venous pressure, mesenteric arterial blood flow or hepatic arterial blood flow. Splenic arterial blood flow does fall significantly and remains depressed during the four-hour observation period. These hemodynamic data are summarized in figure 2. In contrast to the canine shock model bloody diarrhea does not accompany lethal endotoxic shock in the baboon; and, in fact, there is neither gross nor microscopic pathologic change within the splanchnic viscera of the subhuman primate.

To further define the marked species differences between these two shock models, an investigation of the intestinal microvascular architecture of dog and baboon was performed using a silicone-rubber injection technique [2]. The results of this study provide morphologic evidence which may account for the different hemodynamic responses described. The blood supply to the small intestinal villus of the dog includes marginal arteries, supplying capillaries which are drained by a central vein. In contrast, the intestinal villus of the rhesus monkey and the baboon contain only capillaries drained by a similar, centrally-located vein. During endotoxic shock, the only significant change in intestinal microvascular architecture of either

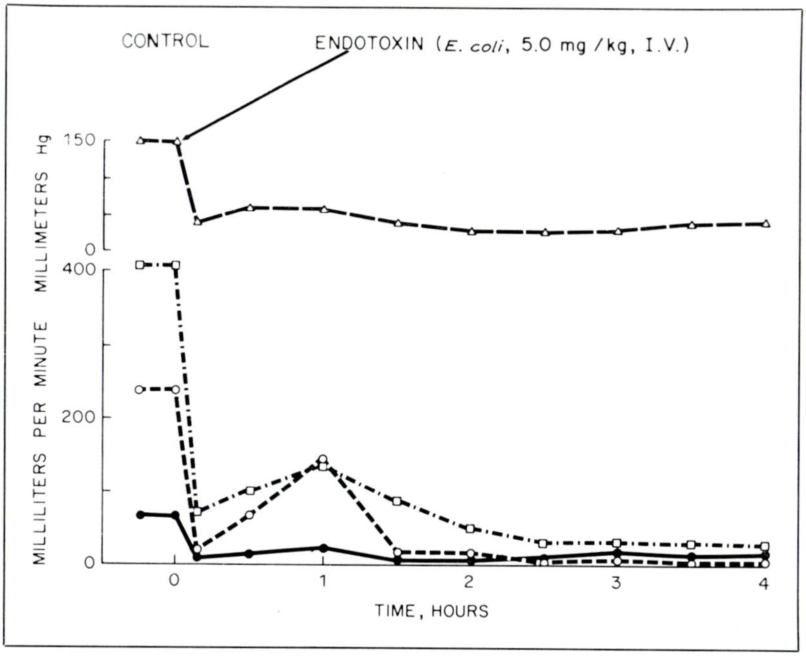

Fig. 1. The effects of endotoxic shock upon splanchnic hemodynamics in the anesthetized dog. Hepatic arterial blood flow (○), mesenteric arterial blood flow (□), and splenic arterial blood flow (●) are represented in ml/min along the ordinate along with arterial pressure (△), in mm Hg. The time course of the experiment is plotted along the abscissa in hours.

of the subhuman primates was a slight dilation of the subepithelial capillary plexus. In contrast, the dog exhibited four distinct alterations in the microcirculation of the villus: (1) The marginal arteries were coiled along with the central vein. (2) There was shortening of the villus vascular architecture by 50%. (3) The subepithelial capillary plexus was dilated and congested. (4) There was extravasation of silicone rubber into the surrounding parenchymal tissues. These findings in the dog suggest anatomic changes which could account for the rise in mesenteric vascular resistance observed in canine endotoxic shock as opposed to the fall in mesenteric vascular resistance accompanying endotoxic shock in the subhuman primate.

In hemorrhagic shock both dog and monkey exhibit similar hemodynamic changes in the gut [1]. These changes include a fall in arterial pressure and mesenteric arterial blood flow. Reinfusion of the shed blood,

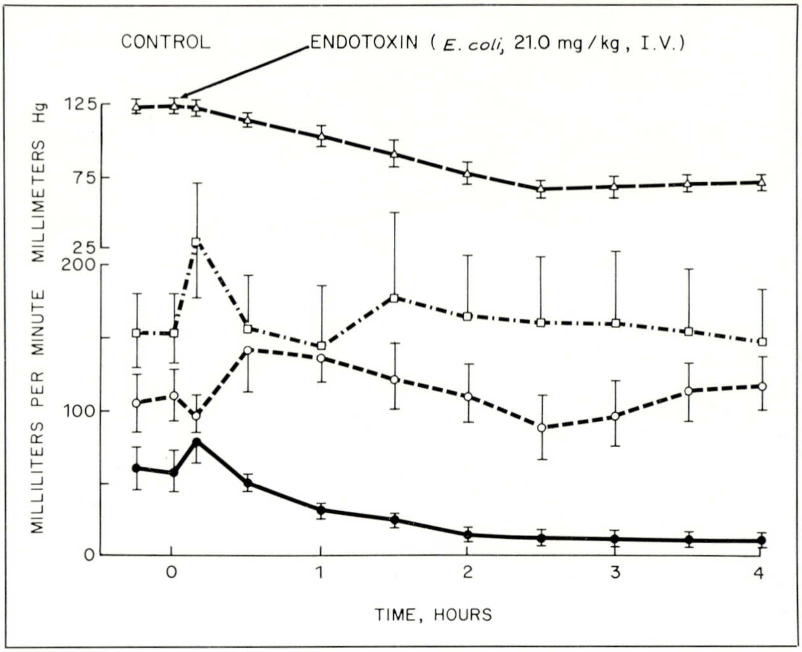

Fig. 2. The effects of endotoxic shock upon splanchnic hemodynamics in the anesthetized baboons. Hepatic arterial blood flow (○), mesenteric arterial blood flow (□), and splenic arterial blood flow (●) are represented in ml/min along the ordinate along with arterial pressure (△), in mm Hg. The time course of the experiment is plotted along the abscissa in hours. Each point in the graph represents the mean ± SE of five experiments in five animals.

after four hours of hypotension, restores all parameters to control values in both species; however, this degree of hemorrhagic shock in the dog is lethal, whereas the monkey survives. In addition, the subhuman primate does not demonstrate any significant change in intestinal microvascular architecture, either at the end of the fourth hour of hemorrhagic hypotension or following reinfusion of the shed blood. The canine shock model also does not exhibit any significant change in microvascular architecture at the end of the period of hypotension; however, reinfusion of the shed blood is accompanied by marked changes in the villous vasculature. These changes are as follows: (1) There is extravasation of silicone rubber into the core of the villus. If this extravasation is minimal, there is no distortion of the villus. However, if the extravasation is extensive, then the villus is distorted with ultimate rupture of the villus. (2) No change is noted in the subepithelial capillary plexus unless villus rupture occurs and then there is collapse of the entire villous vascular bed. (3). There is no shortening of the villus. (4) There is no coiling of the core vessels within the villus.

These studies have identified marked species differences between the splanchnic hemodynamic responses to experimental shock in the canine and subhuman primate models. Although the splanchnic viscera are clearly target organs of shock in the dog, they do not appear to be involved in the pathogenesis of shock in the monkey. The applicability of the target organ or adrenergic theory of shock to clinical therapy in man is thus seriously questioned since the subhuman primate is phylogenetically closer to man than the dog. The probability that man would respond to shock in a manner more like the monkey than the dog seems difficult to challenge.

References

1 Barton, R. W.; Reynolds, D. G., and Swan, K. G.: Mesenteric hemodynamics of hemorrhagic shock in the baboon. Ann. Surg. *175:* 204–209 (1972).

2 Reynolds, D. G.; Sheehy, T. W., and Brim, J.: The vasculature architecture of the small intestinal mucosa of the monkey *(Macacca mulatta)*. Anat. Rec. *159:* 211–218 (1967).

3 Richardson, D. R. and Johnson, P. C.: Comparison of autoregulatory escape and autoregulation in the intestinal vascular bed. Amer. J. Physiol. *217:* 586–590 (1969).

4 Swan, K. G. and Reynolds, D. G.: Adrenergic mechanisms in canine mesenteric circulation. Amer. J. Physiol. *220:* 1779–1785 (1971).

5 Swan, K. G. and Reynolds, D. G.: Effects of intraarterial catecholamine infusions on blood flow in the canine gut. Gastroenterology *61:* 863–871 (1971).

6 Swan, K. G. and Reynolds, D. G.: Mesenteric hemodynamics during endotoxemia in the baboon. Gastroenterology *61:* 872–876 (1971).

7 Wangenstein, S. L.; Geissinger, W. T.; Lovett, W. L.; Glenn, T. M., and Lefer, A. M.: Relationship between splanchnic blood flow and a myocardial depressant factor in endotoxin shock. Surgery *69:* 410–418 (1971).

Author's address: LTC Kenneth G. Swan, MC, Division of Surgery, Walter Reed Army Institute of Research, *Washington, DC 20012* (USA)

Medical Primatology 1972. Proc. 3rd Conf. exp. Med. Surg. Primates, Lyon 1972, part II, pp. 44–51 (Karger, Basel 1972)

A Comparison of Canine and Primate Circulatory and Metabolic Response to Graded Hemorrhagic Shock

W. SCHUMER, A. G. GREENBURG, R. L. NICHOLS and B. MILLER

Department of Surgery, University of Illinois, Abraham Lincoln School of Medicine and Surgical Service, West Side VA Hospital, Chicago, Ill.

Introduction

The use of the canine, particularly the dog, for investigations of metabolic and microcirculatory response to graded hemorrhagic shock has recently been criticized. As a result, many recent hemorrhagic shock studies have utilized primate animals ostensibly on the hypothesis that experimental results obtained from biological models in these animals will give results more readily extrapolated to humans. It is generally accepted fact that simians cost more, require a more elaborate environment and greater care. With research funding generally curtailed in the basic sciences and a necessity to justify all costs to granting agencies, one can rhetorically inquire: 'Are we justified in using nonhuman primates in the study of experimental shock?' Because of the plethora and variety of factors which must be considered if one were to answer this question, this paper does not attempt to place a value judgment on such an answer. This study in comparative physiology was undertaken to provide data to those investigators who must make decisions relative to their own work.

There is little doubt that the common denominator of all types of shock is a unified effect on the microcirculatory and cellular metabolism, resulting in metabolic anaerobiosis, intracellular acidosis and eventual cell death. This study was undertaken to compare the microcirculatory and metabolic response of selected biochemical parameters in a model of graded hemorrhagic shock in dog and in rhesus monkey *(Macaca mulatta)*.

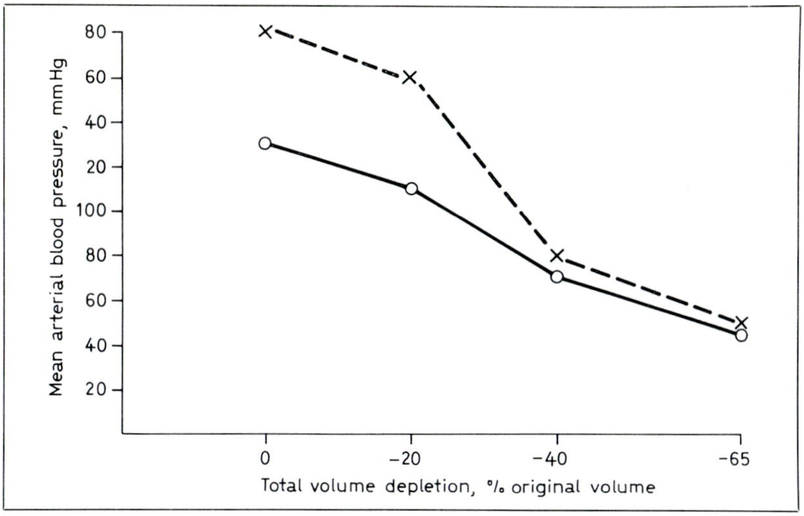

Fig. 1. Relationship of mean arterial pressure to volume depletion × = Dog; o = rhesus monkey.

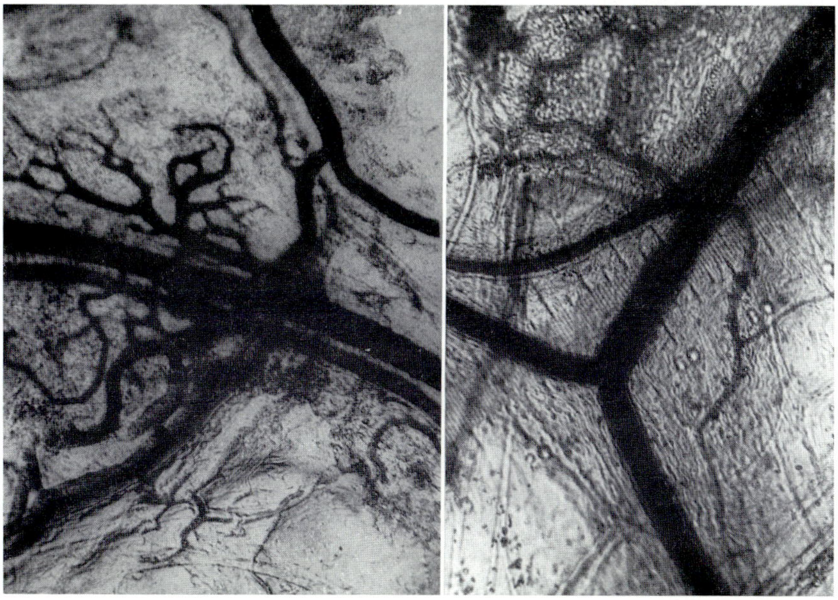

Fig. 2. Comparison of microcirculatory changes canine (right) and primate (left) after 20% volume depletion. Minimal changes are seen for either species.

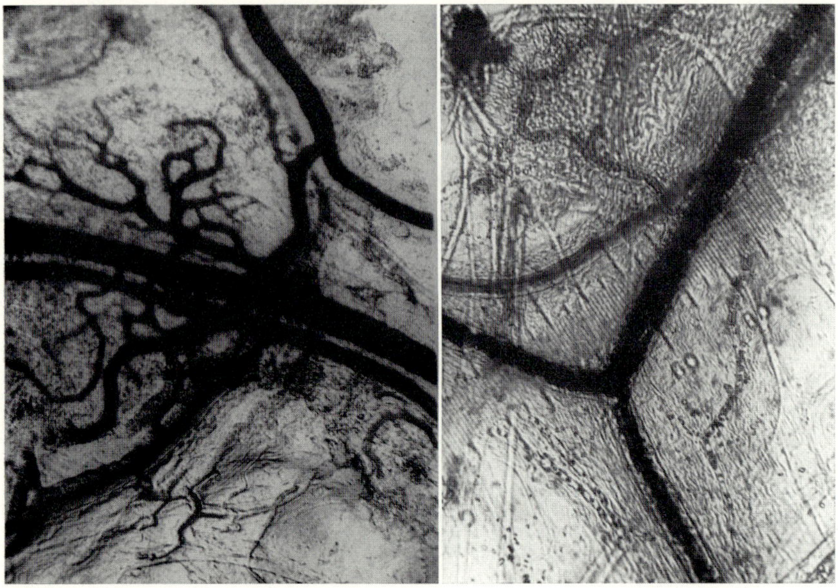

Fig. 3. Microcirculatory changes, canine (right) and primate (left) after 40% volume depletion. Early cellular aggregation is noted.

Experimental Design

Fifteen adult male rhesus monkeys of 3 kg average weight, and 15 young male splenectomized German shepard dogs, average weight 20 kg, were utilized for the study. All animals were anesthetized with sodium pentabarbital, 25 mg/kg body weight, and prepared in an identical manner. A femoral vessel cut-down was performed and polyethylene arterial and central venous catheters of appropriate size were placed. Through a small laparotomy the omentum was mobilized, exteriorized and fixed to the stage of a microscope. To prevent drying, saline moistened packs were placed around the omentum and the light source beneath the stage was passed through a heat filter (Zeiss C). Each animal served as his own control for hemodynamic and metabolic parameters.

Prior to the onset of graded hemorrhage, a modified Walcott [4], blood volume determinations were accomplished on all animals using the radioactive iodinated albumin method. After stabilization following the surgical procedure, animals were bled in sequence to 20, 40 and 65% of their actual blood volume. One hour after each bleed the microcirculation was photographed and blood and tissue samples obtained. Photographs were used to observe arteriolar, venular and capillary responses in the omentum.

Arterial and mean central venous pressures were continuously recorded. Arterial blood samples were anaerobically drawn into heparinized syringes for determination of pH, PO_2 and PCO_2. Determinations of circulating lactate and pyruvate were also per-

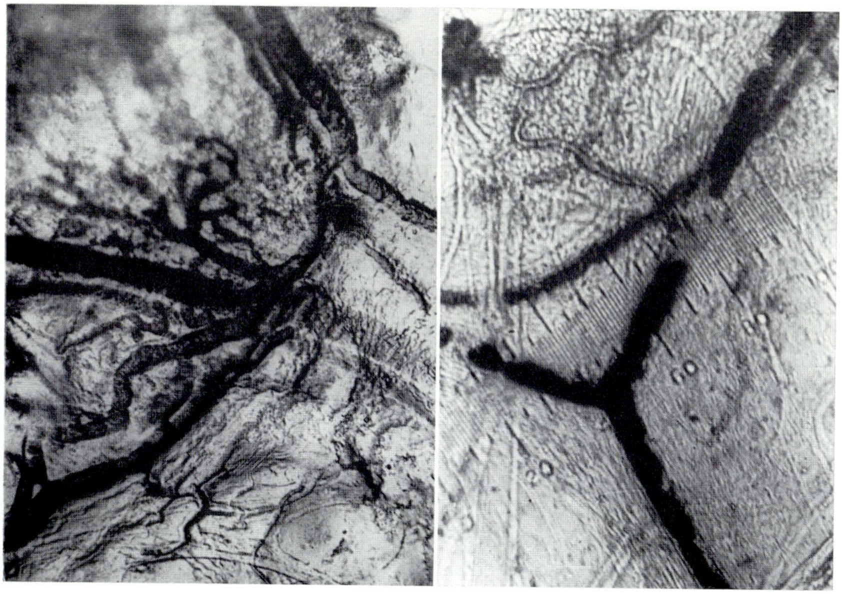

Fig. 4. Microcirculatory changes, canine (right) and primate (left) after 65% volume depletion. Marked cellular aggregation and lessened microcirculatory flow are evident.

Table I.

	% volume depletion			
	0 control	20	40	65
Monkey, intracellular lactate, mg/g wet tissue				
muscle x̄	0.74	1.41	1.77	2.57
SD	0.06	0.21	0.28	0.26
bowel x̄	1.90	2.34	2.77	3.42
SD	0.29	0.38	0.46	0.39
Dog, intracellular lactate, mg/g wet tissue				
muscle x̄	0.83	1.39	1.93	2.49
SD	0.13	0.34	0.32	0.31
bowel x̄	1.46	1.89	2.56	3.21
SD	0.17	0.33	0.24	0.24

Table II.

Dog	% volume depletion			
	0 control	20	40	65
Blood lactate, mm/l				
$\bar{x}$	1.09	2.50	7.02	10.72
SD	0.65	1.06	2.4	7.4
Blood pyruvate, mm/l				
$\bar{x}$	0.06	0.15	0.21	0.26
SD	0.025	0.012	0.11	0.09
Blood pH				
$\bar{x}$	7.40	7.34	7.21	7.04
SD	0.18	0.04	0.47	0.31
Blood PCO_2, mm Hg				
$\bar{x}$	42.2	38.6	29.1	22.7
SD	2.06	3.4	4.8	5.4
Blood PO_2, mm Hg				
$\bar{x}$	93.5	74.4	57.3	38.9
SD	3.2	13.0	16.1	14.7

formed on blood samples. Tissue samples from the adductor magnus muscle, the side opposite the cut-down and the terminal ileum were taken at each increment of volume depletion and analyzed for lactate and pyruvate [1].

Results

Initial blood volume determinations indicated the baseline blood volume in the monkey to be 52 ml/kg and in the dog, 90 ml/kg. These generally agree with published results of similar studies [5]. Figure 1 documents graphically the relationship between volume depletion and mean arterial blood pressure. The dogs response appears more labile. In each species, two of the 15 animals (13%) survived reinfusion after 65% volume depletion for 1 h.

Photomicrographs of the microcirculation for 20, 40 and 65% volume depletion are presented in figures 2–4. Obvious differences in the microcirculation are apparent with volume depletion. However, minimal interspecies differences are discernible with progressive volume depletion.

Table I shows the intracellular lactate levels of muscle and bowel for both species. Statistically, when corrected for tissue weight, there are no significant differences between the values with progressive volume depletion. In

Table III.

Monkey	% volume depletion			
	0 control	20	40	65
Blood lactate, mm/l				
$\bar{x}$	1.34	2.92	8.64	13.22
SD	0.49	0.31	2.53	2.24
Blood pyruvate, mm/l				
$\bar{x}$	0.16	0.14	0.21	0.30
SD	0.037	0.048	0.07	0.09
Blood pH				
$\bar{x}$	7.41	7.35	7.26	7.21
SD	0.20	0.10	0.20	0.08
Blood PCO_2, mm Hg				
$\bar{x}$	42.06	38.13	31.20	22.7
SD	4.8	3.9	8.5	4.7
Blood PO_2, mm Hg				
$\bar{x}$	107.4	90.73	67.53	50.8
SD	3.9	1.4	17.0	16.1

each species, there is a significant progressive increase in intracellular lactate, more pronounced in muscle than in bowel.

Table II contains blood lactate and pyruvate levels and the arterial pH, PCO_2 and PO_2 data for dog, and table III shows this information for monkeys. Circulating lactate increases markedly and significantly at essentially the same rate and to the same levels in both species. Minimal pyruvate increases occur to the same level in both species.

Both groups of animals show a progressive decrease in pH, the dog demonstrating a more rapid and greater decrease, which is significantly different from monkey at 65% volume depletion.

No statistically significant differences between the dog and the monkey with respect to either the PO_2 or PCO_2 is noted with graded hemorrhage.

Discussion

Zweifach [5] has reviewed the problem of comparative physiology of common laboratory species with respect to 'shock'. He has maintained that the effect of shock on the microcirculation resulting in a cellular metabolic anaerobiosis with acidema is the 'common denominator' of all shock models. A validation of this concept can be accomplished only if comparative studies of cellular alterations in shock models indicate no statistically significant differences between species. The results from this study have demonstrated no statistically significant differences in the biochemical parameters measured between dog and monkey.

In this study we observed an apparent decreased omental perfusion which was accompanied by biochemical evidence of increased intracellular anaerobic metabolism. The chemical reflection of tissue anoxia in circulatory collapse, with no significant difference in the species, supports previous work performed on the individual species. The use of metabolites of anaerobic metabolism to monitor the shock state both clinically and experimentally has been reported by us and other investigators. The studies presented here support this concept and indicate the minimal species differences that exist.

Acute hemorrhagic enteritis, reportedly associated with hemorrhagic shock in the dog and not the monkey, was not observed in these experiments. Increasing anaerobic metabolism of the bowel of equal magnitude in both species could theoretically explain the production of such an enteritis, particularly in nonacute experiments. Its occurrence in one species and not the other cannot, however, be explained.

Conclusion

This study has demonstrated that the microcirculatory and biochemical response to graded hemorrhage in the dog and monkey are essentially the same. We were unable to discern any significant differences. These studies also support the concept of biochemical changes occurring at the cellular level and that these changes may play a significant role in altered hemodynamics. In experimental work, it should be possible to use either animal and obtain similar metabolic and microcirculatory responses in the graded shock.

References

1 ANTONIS, A.; CLARK, M., and PILKINGTON, T. R. E.: A semiautomated fluorimetric method for the enzymatic determination of pyruvate, lactate, acetoacetate, and β-hydroxybutyrate levels in plasma. J. Lab. clin. Med. *68:* 340–355 (1966).

2 BOUNOUS, G.; SUTHERLAND, N. G.; MCARDLE, A. H., and GURD, F. N.: The pro-

phylactic use of an 'elemental' diet in experimental hemorrhagic shock and intestinal ischemia. Ann. Surg. *166:* 312–343 (1967).

3 CHIU, C.; MCARDLE, A. H.; BROWN, R.; SCOTT, H. J., and GURD, F. N.: Intestinal mucosal lesion in low-flow states. I. A morphological, hemodynamic, and metabolic reappraisal. Arch. Surg. *101:* 478–488 (1970).

4 WIGGERS, C. J.: Physiology of shock, p. 135 (The Commonwealth Fund, New York 1950).

5 ZWEIFACH, B. J.: Aspects of comparative physiology of laboratory animals relative to the problem of experimental shock. In SEELEY and WEISIGER Recent progress and present problems in the field of shock, pp, 18–29 (Federation of American Societies for Experimental Biology, Bethesda, 1961).

Authors' address: Prof. WILLIAM SCHUMER and Dr. A. GERSON GREENBURG, Depart ment of Surgery, University of Illinois, the Abraham Lincoln School of Medicine and Surgical Service, West Side VA Hospital, 820 South Damen Ave., *Chicago, IL 60612* (USA)

Medical Primatology 1972. Proc. 3rd Conf. exp. Med. Surg. Primates, Lyon 1972, part II, pp. 52–57 (Karger, Basel 1972)

A Technique for Sampling Gallbladder Bile and Some Observations on Bile Lipids in the Baboon During Pregnancy

B. Thorbjarnarson, H. Gilder and K. Morrissey

Department of Surgery, The New York Hospital Cornell Medical Center, New York City, N.Y.

Surgeons have always believed that some connection exists between gallstones and pregnancy. Although statistical evidence is not available it is well known that gallstones occur four times more commonly amongst women than men and surprisingly often the onset of gallbladder symptoms can be traced to a pregnancy period.

The baboon is a suitable model for gallstone experiments since the composition of its bile and anatomical arrangements are similar to the human beings. To carry out valid experiments on bile composition during pregnancy it was necessary to have a preparation which allowed, firstly, easy and repeated sampling of the gallbladder bile, secondly, would not interfere with the animals normal living habits nor interfere with impregnation or progress of pregnancy and, lastly, would not cause damage in any significant way to the gallbladder itself or interfere with its function.

Method

Anesthesia is induced by phencyclidine (1 mg/kg) (Sernylan, Parke Davis) and maintained by i.v. Nembutal-Sodium (Abbott). Under aseptic conditions a midline incision is made in the upper abdomen (fig. 1). After exploring the abdomen a 2.5 cm circular area is excised from the peritoneum and posterior rectus fascia where the fundus of the gallbladder is most closely approximated. The fundus of the gallbladder is then sutured circumferentially into this defect with interrupted sutures of silk. While the abdomen is open a small incision is made in the skin corresponding to the center of the circular area already excised. A No. 28 steel wire suture is placed in the anterior rectus fascia at this point. The small skin incision is closed and a circle tattooed in the skin around it. The abdominal incision is closed in layers and the animal returned to its cage.

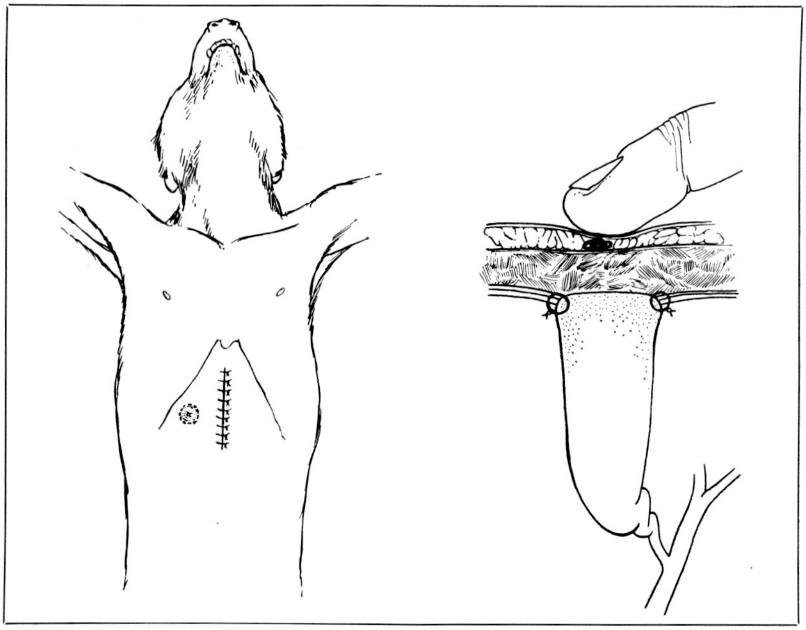

Fig. 1. The cholecystopexy is done through a midline incision. The gallbladder is sutured into the defect following removal of a button of peritoneum and posterior rectus fascia. The steel wire suture in the anterior rectus fascia can be palpated for direction.

Sampling of Bile

The baboon is given a tranquilizer (phencyclidine, 1 mg/kg) and the abdominal area cleansed with an antiseptic solution. Inside the tattooed circle the steel wire suture in the anterior fascia is palpated and a No. 21 needle introduced into the gallbladder and the desired amount of bile aspirated (fig. 2 and 3). In this study samples were taken in the fasting state. Usually 1 ml of gallbladder bile suffices to permit analysis of biliary lipids and bile salts. Specimens are frozen immediately and analyzed, by the methods of PEARSON *et al.* [2] and ZILVERSMIT and ARI [3] for bile salt, cholesterol and phospholipids. A preparation similar to this was used by us on other animals in the past. We have found that simple suturing of the gallbladder to the posterior adbominal wall usually is not adequate and the organ becomes detached in time. By excising a button of peritoneum and fascia and suturing the gallbladder into this defect, detachment is rare and has not happened in the baboon. The tattoo marks and wire suture

2 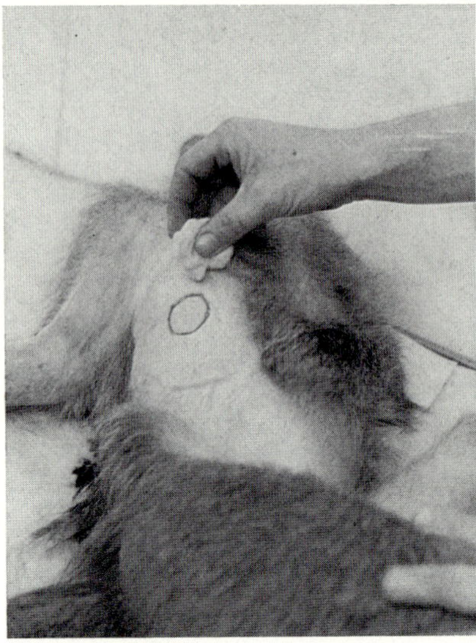 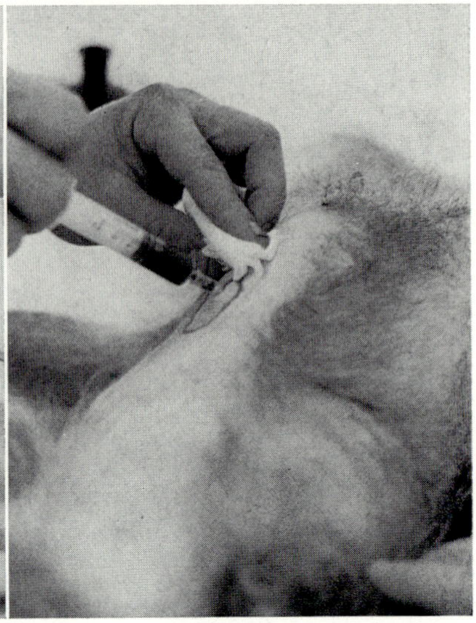 3

Fig. 2. The tattoo circle indicates the location of the gallbladder.

Fig. 3. After palpating the wire suture inside the tattoo a needle is inserted and the bile sample withdrawn.

aid in accurately identifying the center of the gallbladder fundus. Histological examinations of gallbladders prepared this way have failed to show significant structural changes. Cholecystograms on the baboon prepared in this way have shown a normally-functioning gallbladder (fig. 4).

For the experiment described here four adult female baboons were prepared in the above-described manner. Bile samples were obtained weekly, at different times of the menstrual cycle for a period of two months. The animals were then bred and two months later three were pregnant. The fourth did not conceive and was eliminated from the study. The three pregnant baboons had bile samples drawn at two-week intervals through their pregnancy and at two-week intervals after delivery. All delivered healthy babies six months after conception.

Results of Bile Lipid Studies

The aim of the bile lipid studies was primarily to determine whether changes occurred in the solubility status of the cholesterol during pregnancy, which would lead to precipitation and possibly gallstone formation. For

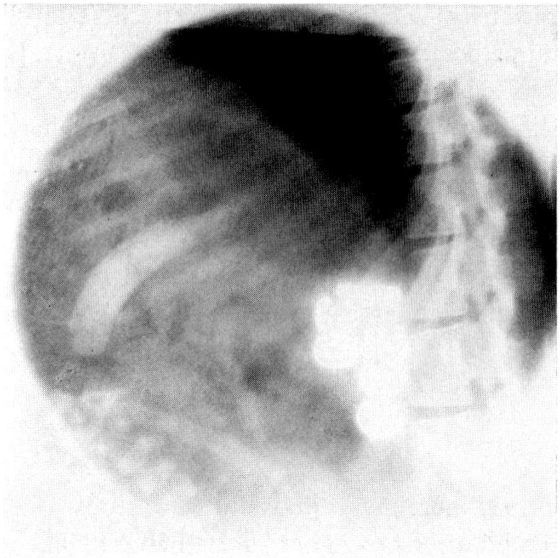

Fig. 4. A cholecystogram on a pregnant baboon.

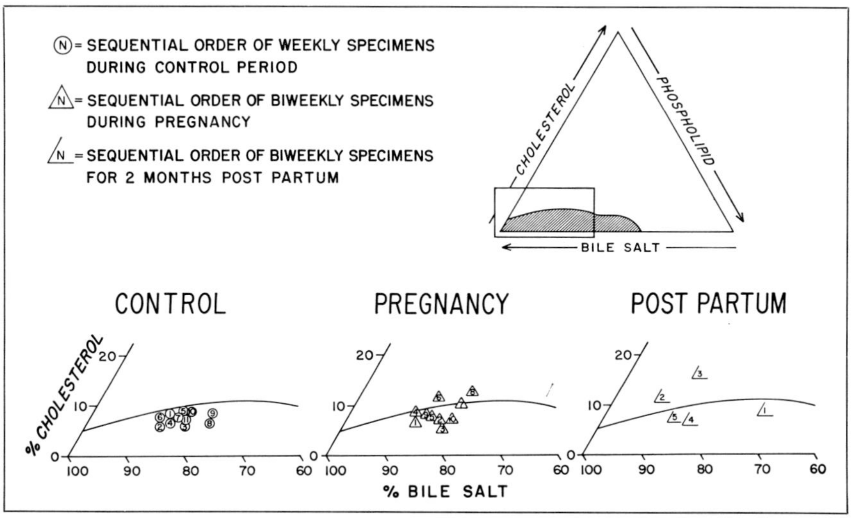

Fig. 5. Phase diagram of bile lipid in a baboon (No. 246) before, during and after preparing. ○ Sequential order of weekly specimens during control period. △ Sequential order of biweekly specimens during pregnancy. □ Sequential order of biweekly specimens of two months *post partum.*

this purpose the concentrations of lecitin, bile salts and cholesterol were plotted on a phase diagram as described by ADMIRAND and SMALL [1] and the zone of abnormal bile plotted on the triangle (fig. 5). Two of the baboons remained within normal limits before, during and after pregnancy. The cholecystograms on these animals also remained normal. The third animal moved completely out of the micellar zone on two occasions and almost so on two other occasions during pregnancy, and continued the same pattern in the *post partum* period. This third animal has on two occasions since had a nonfunctioning gallbladder as tested by the cholecystogram.

Discussion

The preparation described here has allowed us to study bile lipid composition before, during and after pregnancy in the baboon. The results indicate that one of three animals produced a cholesterol precipitable bile on several occasions during pregnancy and in the *post partum* period. The reasons for this have not yet been fully elucidated but it may indicate that pregnancy can indeed produce conditions favorable to precipitation of cholesterol in bile and predispose to gallstone formation. Cholecystograms have been performed in all animals during and after their pregnancy. As compared with other nonoperated, nonpregnant and pregnant females, they show good concentration of contrast media and emptying of the gallbladder in response the cholecystolchinin. The one animal which did not become pregnant was reoperated upon six months after tacking the gallbladder up to the abdominal wall. The gallbladder appeared grossly normal and did not distort the remainder of the biliary tree, which also appeared normal. This preparation is clearly far superior to multiple laparotomies as a means of monitoring gallbladder bile composition in experimental animals. It allows truly long-term studies of gallbladder function and bile composition to be carried out, without producing iatrogenic pathologic change in the biliary tract. We are presently extending this model system to the study of total bile salt pool size before and during pregnancy.

Acknowledgements

The help of Mrs. PATRICIA MAY in this investigation is gratefully acknowledged. The technical assistance of the staff at LEMSIP is gratefully acknowledged.

This study was made possible by financial support from the Irving Strasburger Memorial Medical Foundation, New York, N.Y.

References

1 Admirand, N. H. and Small, D. M.: The physicochemical basis of cholesterol gallstone formation in man. J. clin. Invest. *47:* 1043–1052 (1968).
2 Pearson, S.; Stern, S., and McGavac, T. H.: A rapid accurate method for determination of total cholesterol in serum. Analyt. Chem. *25:* 813 (1953).
3 Zilversmit, D. B. and Ari, A. K.: Microdetermination of plasma phospholipids by trichloracetic acid precipitation. J. Lab. clin. Med. *35:* 155 (1950).

Author's address: Dr. B. Thorbjarnarson, Department of Surgery, The New York Hospital Cornell Medical Center, *New York, N.Y.* (USA)

Medical Primatology 1972. Proc. 3rd Conf. exp. Med. Surg. Primates, Lyon 1972, part II, pp. 58–67 (Karger, Basel 1972)

Plasma Heparin Concentration following Intravenous Administration to Baboons[1]

M. E. TODD, E. McDEVITT, T. GOODMAN and E. I. GOLDSMITH

New York Hospital-Cornell Medical Center, New York, N.Y. and Laboratory for Experimental Medicine and Surgery in Primates (LEMSIP), Sterling Forest, N.Y.

The clotting time has been the accepted test to measure the coagulation defect produced in a patient's blood ever since heparin has been used in the treatment of thromboembolic disorders. There is a well-recognized need for a test to more accurately assay patients' requirements for heparin and to measure the effect of therapy; this test would also be useful in the standardization of different lots of heparin.

The *in vitro* methods for the bioassay of heparin have recently been reexamined by BANGHAM and WOODWARD [1] because the standard heparin preparations were formerly prepared from bovine lung, while recently the intestinal mucosa of oxen, sheep and pigs have been used as the starting material. Heparins prepared from intestinal mucosa gave a higher biologic activity *in vitro* by the two standard bioassay procedures, but the differences were not considered significant enough to warrant preparing a new standard.

There were also anecdotal reports to pharmaceutical manufacturers [6] that the potency of heparin made from lung and that from intestinal mucosa differed when administered to patients undergoing clinical anticoagulation.

We have shown [12] that the coagulation mechanism of the baboon (*Papio* sp.) closely resembles that of man enabling one to use baboons as models for study of human blood clotting. We, therefore, studied in baboons the *in vivo* potency of five of the commercially-available heparin preparations comparing clotting times (CT) and the activated partial thromboplastin times (APTT) following single dose full heparinization.

1 Supported by grants from the Upjohn Company and by the Vascular Research Fund, Department of Medicine, Cornell University Medical College.

Materials and Methods

Healthy baboons were tranquillized with phencyclidine hydrochloride (Sernylan, Bioceutic Laboratories, Inc., St. Joseph, M.) 1 mg/kg and were subsequently given barbiturates (Nembutal, Abbott Laboratories, Abbott Park, North Chicago, Ill.) in small amounts as needed. Femoral vein cut downs were performed using siliconized tubing and were kept patent with 5% dextrose and water. Test materials were injected and the blood samples withdrawn from this site.

The five heparin preparations were coded by the pharmacy of the New York Hospital and diluted so that 1 ml contained 1,000 international units (IU) and the unknown preparations were given i.v. to each of 16 animals in a standard dose of 400 IU/kg body weight. The APTT results were calculated from a standard curve and recorded in units of heparin per milliliter of plasma. The glass clotting times were reported in minutes. Each of the coded preparations was given in duplicate.

The two standard preparations – (1) the second international standard of the World Health Organization (WHO) and (2) that from the Connaught Laboratories – were given as knowns and the animals were followed according to an identical protocol.

Blood samples for clotting times and the APTT (4.5 ml of blood mixed with 0.5 ml of 3.8% sodium citrate) were withdrawn (1) before injection for control studies, (2) at 30 min, and (3) hourly thereafter until the clotting time returned to normal. The clotting times were determined by the Lee White method [7]. The APTT was performed according to the method of MARDER [8]. The standard curves for the heparin concentrations were prepared before the *in vivo* studies were attempted. These were done as described by MARDER [8] except that Liquamin sodium (Liquamin Sodium, Organon, Inc., West Orange, N.J.) was used as the stock heparin. The anticoagulant activity of this was 100 IU/mg as compared to a standard preparation (Connaught Laboratory, Toronto, Canada). Two milliliters of citrated whole blood was added to less than 0.1 ml of sodium heparin from a concentration of 0.1–10.0 IU. These were spun in a refrigerated centrifuge and the plasma removed. The test was performed on 0.2 ml of plasma to which was added 0.1 ml of a solution containing Kaolin (0.25% in 0.05 M Veronal buffer, pH 7.3) plus $^{1}/_{100}$ volume of 3.7% Inosithin. This was recalcified with 0.1 ml of 0.025 M $CaCl_2$. The test was performed in a water bath at 37 °C. The plasma heparin concentrations were calculated by correction for the hematocrit.

The test was performed on undiluted plasma and plasma which was diluted 1 :4, 1 :10 and 1 :20 with normal citrated human plasma. When used as the diluent baboon plasma gave essentially the same results as human. Statistical analysis of the data was limited to calculations of the regression lines by the method of least squares which demonstrated that the regression lines for all dilutions were essentially the same. The concentration of heparin in the diluted test plasmas was read from the standard curves.

Blood volumes were determined in four animals using Evans Blue Dye [4] (Evans Blue Dye (T-184), Warner Lambert, Morris Plains, N.J.).

Results

Table I shows the comparison between the glass clotting times and the APTT (heparin units/ml plasma) in all experiments. The CT were at infinity

Table I. Glass CT vs. APTT

Time (min) after i.v. injection	Heparin code										Standard			
	A		B		C		D		E		Connaught		WHO	
	CT	hep. U/ml	CT	hep. U/ml	CT	hep. U/ml	CT	hep. U/ml	CT	hep. U/ml	CT	hep. U/ml	CT	hep. U/ml
Control	5	0	5^{50}	0	5^{50}	0	5^{30}	0	7^{30}	0	5^{10}	0	5^{10}	0
	6^{10}	0	5^{50}	0	5^{20}	0	4^{50}	0	4^{30}	0	6^{20}	0	6	0
Immediately	∞	8.5	∞	7.0	∞	10.8	∞	9.9	∞	6.6	∞	10.0	∞	12.1
	–	10.5		8.1	–	11.3	–	9.3	–	9.8	–	10.4	–	10.4
30	∞	8.3	∞	5.3	∞	7.3	∞	7.3	∞	4.3	∞	5.4	∞	7.8
	–	6.7	–	6.5	–	8.7	–	7.6	–	4.8	–	4.2	–	6.0
60	∞	4.7	∞	2.8	∞	6.5	∞	4.6	∞	2.2	∞	4.3	∞	3.2
	–	3.3	∞	3.2	–	5.4		5.4		5.0		3.6		4.5
120	∞	2.7	∞	1.4	∞	3.7	∞	2.9	∞	1.5	∞	2.0	∞	1.6
	–	2.9	–	3.0	–	4.2	–	2.1	–	3.3	–	1.3	–	1.6
180	85	2.4	50	0.6	∞	1.8	91	1.5	60	0.9	42	1.1	∞	0.5
	–	2.2	79	1.5	–	1.5	–	1.0	69	1.6	46	1.6	–	0.5
240	50	1.1	31	0.2	37	0.7	48	1.1	31	0.4	25	1.0	∞	0.4
	–	1.3	48	0.7	20^{20}	1.0	55	0.7	29	1.5	31^{30}	1.2	–	0.3
300	25	0.5	13	0.1	18^{30}	0.3	24^{50}	0.6	19	0.0	11^{40}	0.2	53	0.05
	27^{20}	0.5	28	0.3	13^{40}	0.7	39	0.1	19^{40}	0.5	22^{10}	0.7	43^{40}	0.1
360	17	0.2	5^{40}	0.1	6^{30}	0.0	13^{30}	0.06	8^{10}	0.0	11^{40}	0.2	53	0.05
	23^{30}	0.4	16^{30}	0.2	6^{30}	0.6	18^{50}	0.04	17^{50}	0.4	10^{50}	0.3	17^{40}	0.0
420	8^{10}	0.10	–	–	–	–	10^{30}	0.01	5	–	5	0.06	16^{10}	0.0
	8^{30}	0.2	8^{40}	0.03	–	0.3	9	0.01	11^{10}	0.06	6^{40}	0.2	9^{40}	0.0
480						–							5^{50}	0.0

Table II. Origin of heparin starting material

Code	Name	Species origin	Organ
WHO	standard	bovine	lung
Connaught	standard	porcine	intestinal mucosa
A	lipoheparin (Riker)	porcine	intestinal mucosa
B	Liquamin (Organon)	porcine	intestinal mucosa
C	Heparin (Upjohn)	porcine	intestinal mucosa
D	heparin (Upjohn)	bovine	lung
E	Panheparin (Abbott)	porcine	intestinal mucosa

(more than 60 min) until 180 min after heparinization at which time there was about 1.5 units of heparin/ml of plasma. The CT were sensitive to very small amounts of heparin and in some instances did not return to normal until seven or eight hours. By this time there was only 0.1–0.3 units of heparin present in a milliliter of plasma.

Figure 1 shows the regression line for one of the heparin preparations (unknown B) which is a mean of two determinations. The points scattered around the line are the actual units of heparin as determined from the standard curve of the APTT. A good correlation exists between the actual data and that calculated ($r = 0.8–0.9$).

Figure 2 depicts the unknown heparins compared to the Connaught standard heparin which is prepared from hog intestinal mucosa. The mean regression lines of all except unknown D are essentially parallel to the standard. Figure 3 is the same as figure 2 except that the comparison is with the WHO standard which is a beef lung preparation. Here the unknown heparins are more active than the standard. These results demonstrate that all of the unknown heparins are similar to the standard made from hog intestinal mucosa except for unknown D which appears to lose activity faster than the other unknowns and has a slope similar to the standard prepared from bovine lung (WHO).

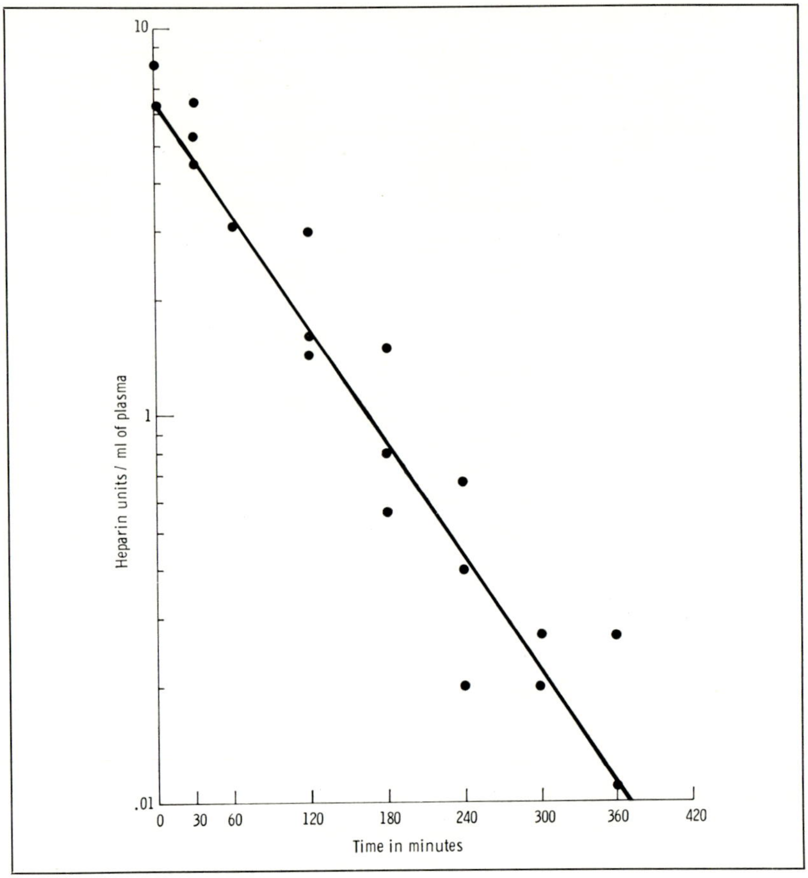

Fig. 1. The regression line for one of the unknown heparin preparations (unknown B).

When the code was broken and the results compared, the only lung preparation was unknown D; all the rest were prepared from intestinal mucosa (table II).

The half life of all the preparations was calculated in two ways; first ½ the initial heparin units was placed in the formula determined by the least squares analysis ($\log y = mx + b$) and the time calculated, and second the heparin units were read from the original data of the APTT and the time noted. Table III shows the comparison of the two methods. The mean T ½ for the two methods differs by 5 min, being 55 min for the APTT and 60 min for the calculated method.

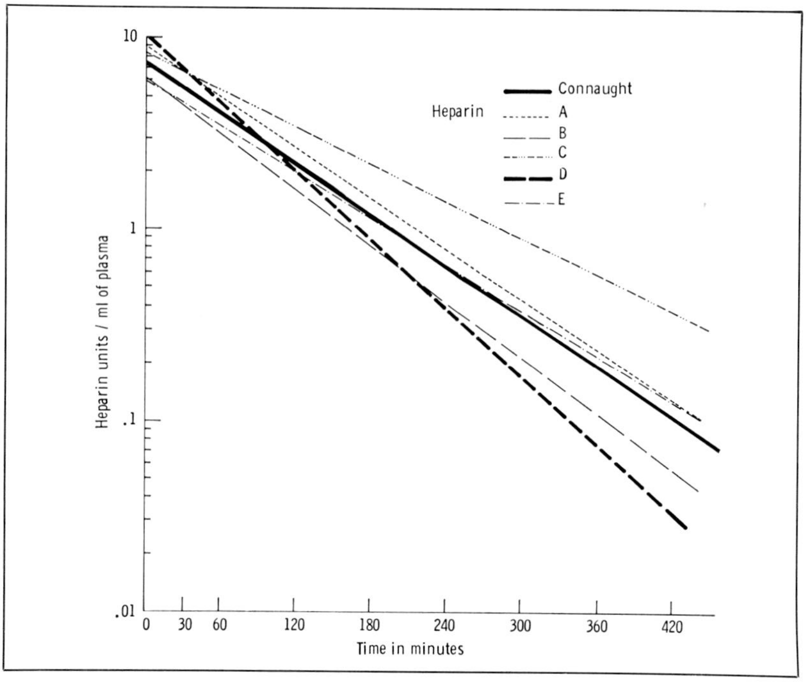

Fig. 2. The regression lines of all the unknown heparin preparations compared to the standard prepared from porcine intestinal mucosa (Connaught).

The initial concentration of heparin in the plasma was calculated in several different ways for comparison (table IV). (A) Each animal was given heparin in a standard dose of 400 IU/kg body weight in grams. Therefore, using the formula: plasma volume = 5% body weight in grams, the initial heparin concentration was 8 units/ml. (B). The concentration of heparin can be more accurately calculated by using: (1) the APTT done on the initial blood sample at time zero; (2) the value for heparin concentration on the regression line; (3) the value for plasma volume determined directly by the Evans Blue Dye test which can then be used to calculate heparin concentration.

Discussion

The co-operative study reported by BANGHAM and WOODWARD [1] compared heparins prepared from ovine and porcine intestinal mucosa to three

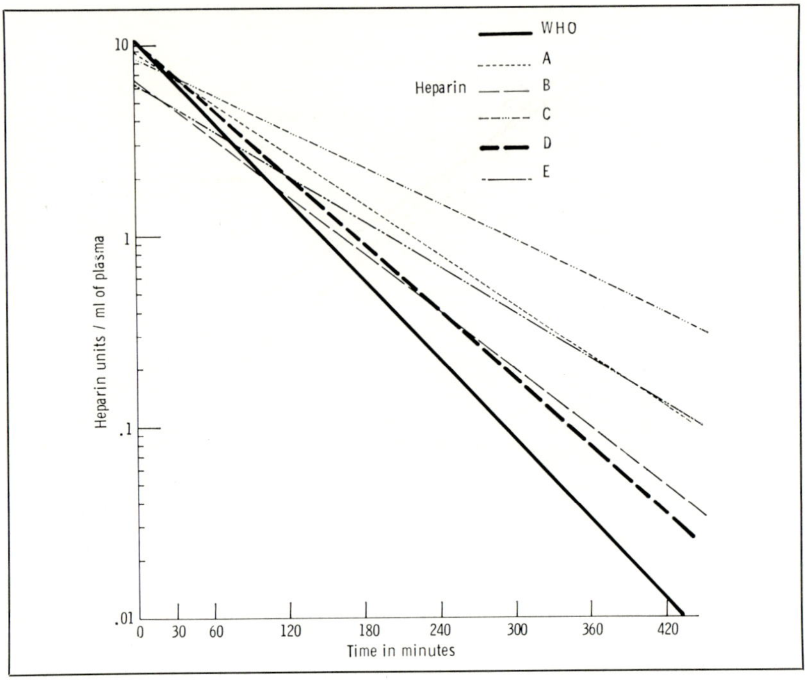

Fig. 3. The regression lines of all the unknown heparin preparations compared to the standard prepared from bovine lung (WHO).

existing standards. These were WHO international standard prepared from bovine lung, the USP reference standard made from bovine intestinal mucosa and the British Standard from hog intestinal mucosa. The laboratories participating in that study used many *in vitro* techniques to assay the relative potencies. Only two used an *in vivo* assay using the dog and measuring CT, APTT and lipoprotein lipase.

The study reported here was designed to determine if heparin prepared from bovine lung differed *in vivo* from that made from intestinal mucosa when given i.v. to baboons. The conclusions are based on the results of the APTT, which as well as the glass CT, is sensitive to small amounts of heparin. Marder's [8] modification of the APTT provides a rapid and accurate measurement of high as well as low plasma heparin concentrations by diluting the plasma before the assay, thus bringing the heparin concentration within the sensitivity of the traditional test. Once the standard curves are prepared, the assays of the test plasma requires very little time.

Table III. The $T^1/_2$ values for all heparin preparations

Heparin code	Heparin units APTT	$T^1/_2$	Heparin units calc.	$T^1/_2$
A1	8.5	75	9.6	67
A2	10.8	60	9.9	68
B1	7.0	52	7.7	57
B2	8.1	48	5.2	40
C1	10.8	80	11.8	76
C2	10.8	92	6.2	48
D1	9.9	65	13.0	80
D2	9.3	60	8.3	60
E1	6.6	45	4.5	42
E2	9.8	30	8.0	59
WHO	1 12.1	43	12.6	78
	2 10.4	45	8.3	60
Connaught	1 10.0	42	6.6	51
	2 10.4	30	8.9	63

Table IV. Heparin concentration in plasma

Method	Animal			
	B352	B256	B60	B124
Evans blue calculated	10.2	13.0	8.4	6.8
APTT standard curve	11.1	10.0	8.5	6.5
Regression line (y calc.)	11.8	13.0	9.6	4.5
5% body weight[1] (plasma volume)	8.0	8.0	8.0	8.0

1 Plasma heparin conc. IU/ml $= \frac{\text{dose heparin injected (IU)}}{\text{body weight, g} \times 0.05}$.

When the regression lines were calculated for all the preparations, duplicate determinations were parallel. The mean regression lines for the different preparations were similar except for 'unknown D' which had a steeper slope. When the results were plotted on similog paper and compared to each standard preparation, certain differences became apparent (fig. 2 and 3). All the heparins with the exception of 'D' were parallel to the Connaught standard which is prepared from hog intestinal mucosa. When they were compared to the WHO standard (beef lung), all except D showed more heparin per milliliter of plasma in each time period. Thus, heparin prepared from intestinal mucosa maintains a higher concentration in the plasma for a longer period of time. This difference, though not great, may explain the clinical observations that heparin activity is not always predictable.

STRUVER and BITTNER [10] were the first to observe that the partial thromboplastin time correlated well with the amount of heparin added *in vitro*. Since then there have been numerous studies correlating the APTT with the glass CT [3, 5, 9, 13] and the consensus is that there is a linear correlation between the tests when the concentration of heparin is below 1.5 units of heparin/ml of plasma. If the therapeutic range of heparin were above 1.5 units/ml, the glass clotting time would reach infinity and no longer have any value in clinical management. The APTT can measure these higher concentrations and thereby will help determine the amount of heparin to give to individual patients.

Heparin's interference with the thrombin fibrinogen reaction is well known. WESSLER and YIN [12] recently demonstrated that heparin potentiates the natural inhibitor of factor X_a and, therefore, retards the intrinsic blood coagulation system. Thrombin also catalyses the activation of factor IX by activated factor XI. BIGGS *et al.* [2] also noted that heparin in low doses, with a plasma cofactor, promoted the progressive destruction of factor X_a and suggested that perhaps the assay for heparin should be on its ability to destroy factor X_a. The APTT measures the intrinsic pathway of blood coagulation and is, therefore, the logical test for this function of heparin.

Acknowledgement

The authors wish to acknowledge with appreciation the technical assistance of W. NIEMANN, D. V. M., MARIE GESSERT and D. ARMSTRONG.

References

1 BANGHAM, D. R. and WOODWARD, P. M.: A collaborative study of heparin from different sources. Bull. Wld Hlth Org. *42:* 129–149 (1970).

2 BIGGS, R.; DENSON, K. W., and BARRETT, A. R.: Antithrombin III, antifactor X_a and heparin. Brit. J. Haemat. *19:* 283 (1970).
3 COLMAN, R. W.; OXLEY, L., and GIANNUSA, P.: Statistical comparison of the automated active partial thromboplastin time and the clotting time in the regulation of heparin therapy. Amer. J. clin. Path. *53:* 904–907 (1970).
4 GREGERSEN, M. I.; GIBSON, J. G., and STEAD, E. A.: Plasma volume determinations with dyes, errors in colorimetry. Use of the blue Dye T-2824. Amer. J. Physiol. *113:* 54 (1935).
5 HIRSH, J.; O'SULLIVAN, E. F.; GALLUS, A. S., and MARTIN, M.: The activated partial thromboplastin time in the control of heparin therapy. Austr. Ann. Med. *19:* 334–337 (1970).
6 KRATOCHVIL, C.: Personal commun.
7 LEE, R. I. and WHITE, P. D.: A clinical study of the coagulation time of blood. Amer. J. med. Sci. *145:* 495 (1913).
8 MARDER, V.: A simple technique for the measurement of plasma heparin concentration during anticoagulant therapy. Thromb. Diath. haemorrh. *10:* 230–239 (1970).
9 PITNEY, W. R.; PETTIT, J. E., and ARMSTRONG, L.: Control of heparin therapy. Brit. med. J. *iv:* 139–141 (1970).
10 STRUVER, G. P. and BITTNER, B. L.: The partial thromboplastin time (cephalin time) in anticoagulant therapy. Amer. J. clin. Path. *38:* 473–481 (1962).
11 TODD, M. E.; MCDEVITT, E., and GOLDSMITH, E.: Blood clotting mechanisms of nonhuman primates. Choice of the baboon model to simulate man. J. med. Prim. *1:* 132–141 (1972).
12 WESSLER, S. and YIN, E. T.: Heparin accelerated inhibition of activated factor X by its natural plasma inhibitor. Biochem. Biophys. *201:* 387–390 (1970).
13 ZUCKER, S. and CATHEY, M. H.: Control of heparin therapy. Sensitivity of the activated thromboplastin time for monitoring the anticoagulant effect of heparin. J. Lab. clin. Med. *73:* 320–326 (1969).

Authors' address: Dr. M.E. TODD, Dr. E. MCDEVITT, Dr. T. GOODMAN and Dr. E.I. GOLDSMITH, New York Hospital-Cornell Medical Center, *New York, N. Y.* (USA)

Medical Primatology 1972. Proc. 3rd Conf. exp. Med. Surg. Primates, Lyon 1972, part II, pp. 68–76 (Karger, Basel 1972)

Prosthetic Cardiac Valve Placement in the Baboon *(Papio anubis)* with Circulatory Arrest Utilizing Cerebral Hypothermia[1, 2]

S. K. WOLFSON, jr., T. P. COMER, C. S. THOMAS and R. G. SELKER

Division of Neurological Surgery, University of Pittsburgh School of Medicine and Surgical Research Laboratories, Montefiore Hospital, Pittsburgh, Pa.

Preferential cerebral hypothermia has been previously described as a method of safely producing upwards of one hour total circulatory arrest without the aid of cardiopulmonary bypass. The investigation of this method was begun in dogs [12, 13] and continued in various primates including *Cercopithecus aethiops*, *Macaca nemestrina*, *Macaca speciosa* and *Papio anubis* [3, 8, 11, 14]. The method depends upon the intentional establishment of differential hypothermia in which various organ systems are cooled to a degree determined by their sensitivity to total ischemia. This is accomplished by a 5-min headward perfusion of one common carotid artery with 0 °C buffered electrolyte solution. Thus, the central nervous system (brain and upper spinal cord) is between 10 and 15 °C, heart, liver and other core organs are at a midrange temperature (22–26 °C) and the periphery least cool at 29–30 °C. Resuscitation requires a short period of manual or intrapericardial balloon massage [4], surface rewarming and administration of cardiocirculatory drugs and sodium bicarbonate. Only a small amount of blood need be mobilized through the lungs to oxygenate the myocardium and permit adequate cardiac function for rewarming. Cardiopulmonary bypass is thus avoided. It is anticipated that the method could find clinical usefulness in neurosurgery and certain types of intracardiac surgery, especially in the very young. It also could find a place in transplantation and major vascular surgery. To this end, clinical trials in human patients with recurrent gliomas have been instituted with successful resuscitation [7].

1 This research was performed at the Michael Reese Hospital and Medical Center, Chicago, Ill.

2 Supported in part by USPHS Grant GM-16654 and General Research Support Grant 5-SO1-RR-05416-11.

In order to test the feasibility of using this procedure for open heart surgery, we decided to implant a prosthetic tricuspid valve. This is a procedure that could reasonably be accomplished with uniform results within the 60-min period arbitrarily chosen as the limit of circulatory arrest time. This report is of 14 consecutive tricuspid valve implants in the baboon *(Papio anubis)*.

Methods

The baboons were of both sexes and weighed 9–11 kg. They were residents of our primate colony for 30–60 days prior to experiment and were gaining weight and were in apparent good health. The night before surgery, water was withheld and the animal received half (0.015 mg/kg) of an i.m. digatilizing dose of lanatoside C (Cedilanid®). On the morning of surgery they received the second half of the lanatoside C, lincomycin, 10 mg/kg, (Lincocin®) and streptomycin (10 mg/kg), all i.m. Phencyclidine HCl, 0.5 mg/kg i.m. (Sernylan®), was given to facilitate removal from colony to animal operating room. Anesthesia was induced with 2% sodium thiopental (Pentothal®) i.v. and maintained with 0.5–2% halothane (Fluothane®) in 100% O_2 by closed circuit endotracheal administration. Central arterial and venous blood pressure, esophageal, rectal, cerebral and cardiac (after thoracotomy) temperature as well as arterial pH, P_{CO_2} and P_{O_2} were all monitored throughout the procedure.

After the establishment of stable anesthesia, cooling was begun by means of hypothermia blankets. Initially the blankets were above and below the animal. After 30–45 min, when the esophageal probe indicated a definite core cooling trend, the upper blanket was removed and the chest was prepared, draped and thoracotomy (median sternotomy) was performed. The pericardium was opened and tapes placed around the inflow and outflow vessels in preparation for circulatory arrest. A catheter containing heparinized saline was inserted well into the right common carotid artery. The carotid was approached via the mediastinal incision. By the time the preliminary surgery was complete (90–120 min after the onset of cooling), the midesophageal temperature had usually reached 30 °C which was the desired level for onset of circulatory arrest and carotid perfusion.

Heparin, 300 units/kg was administered i.v. and while it was allowed to circulate for 1 min, stimulating electrodes were attached near the apex of the left ventricle. Ice bags were placed around the head to prevent upward drift of brain temperature after perfusion ended. The ventilator was switched to 1 cycle/5 min period and circulatory arrest (ventricular fibrillation) begun by direct stimulation of the ventricle with 60 Hz AC, 1–3 V. Common carotid artery perfusion was begun within 20 sec of the last recorded aortic pulse. This headward, pulsatile, perfusion with 0 °C buffered electrolyte[1] was continued at a rate of 10 ml/kg/min for 5 min. Thus, a total volume of 50 ml/kg was given. During the perfusion, the heart was seen to dilate somewhat and the coronary vessels were washed free of visible blood. At the end of perfusion, the tapes were drawn up occluding the aorta and both vena cavae, as well as the azygos vein. The ventricular stimulation was discontinued and the right atrium opened. About 50 ml of a mixture of blood and perfusion fluid were aspirated.

1 The components of the perfusion fluid are listed in table I.

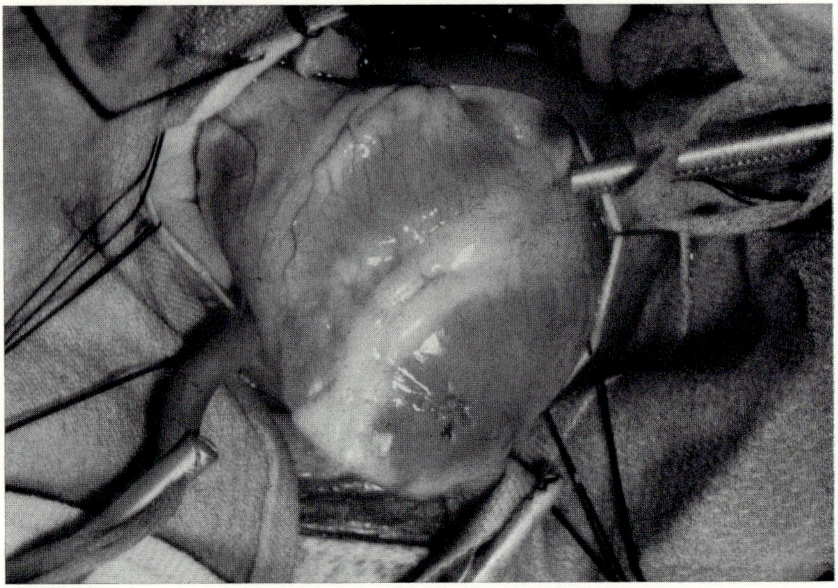

Fig. 1. Baboon heart after placement of the tricuspid prosthesis and closure of the atriotomy. Note the virtual absence of blood in the anterior descending coronary vessels. The site of stimulating (fibrillatory) electrode attachment is visible near the apex. The electrode has been removed in preparation for resuscitation.

The prosthesis, a Starr-Edwards animal valve (00, silastic poppet), was sutured in place after excision of the existing tricuspid leaflets. During the valve placement, the operative field was intermittently irrigated with 2–6 °C saline and this fluid together with a small amount of blood seepage was aspirated periodically. The valve was in place and the atriotomy closed before the expiration of 60 min from the time of initial ventricular fibrillation. The inferior vena cava tape was partially relaxed just prior to final closure of the atriotomy to facilitate aspiration of air. Figure 1 is a photograph of the heart just before resuscitation. The absence of blood in the anterior descending coronary vessels is apparent. The site of stimulatory electrode attachment is also visible near the apex. This electrode has been removed in readiness for resuscitation.

Resuscitation was carried out similarly to the standard preferential cerebral hypothermia protocol cited above. Inasmuch as the chest was open, two notable variations were made, both of which added to the ease and rapidity of resuscitation. Cardiac massage was by direct gentle manual pressure and warm (40 °C) saline was alowed to irrigate the pericardium and pleura. Both of these contributed to the ease of defibrillation and re-establishment of effective cardiac action within a few minutes. At the onset of cardiac massage, all tapes were removed and epinephrine 100 μg was given i.v. followed by a slow phenylephrine (Neosynephrine®) drip. The ventilator was switched to the original rate of about 20 breaths/min. Ice bags were removed from the head about 10 min prior to this time. Electri-

Table I. Composition of carotid perfusion fluid

Substance	Concentration mEq or mM/l[1]
Na^+	107.0
K^+	3.6
Ca^{++}	3.5
Cl^-	155.8
Tromethamine$^+$	41.6
Tromethamine (undissociated)	10.4
Total	321.9
Osmolality[2]	305.0

1 Undissociated tromethamine is expressed in mM/l.
2 Freezing point depression method.

cal defibrillation was employed if necessary as soon as active ventricular fibrillation was observed on the ECG. The ECG was isoelectric until the onset of massage and warm saline irrigation. Spontaneous onset of heart beat without the need of electrical defibrillation was a frequent occurrence. As cardiac action improved, bleeders were detected and ligated. Usually the chest was closed within 30 min of the beginning of resuscitation. The arterial pH was monitored and corrected to >7.3 by administration of i.v. $NaHCO_3$. Zero to 40 mEq were required.

Beginning 1 h after the onset of resuscitation but always after the chest incision was closed, a neutralizing i.v. dose of protamine sulfate was given (4.5 mg/kg in three equal doses). Operative blood loss was countered by the administration of about 100 ml of human hemoglobin solution (7.5% Hgb in balanced electrolyte solution). This was never administered until after the full neutralizing dose of protamine was administered. Rewarming was by external blankets and required about 1 h after chest closure. During rewarming, the animal was ventilated with 100% O_2, semiclosed. Additional halothane frequently had to be given to maintain a satisfactory surgical level of anesthesia for chest closure. At the end of the procedure the initial antibiotic dose was repeated and continued twice daily for ten days. The animal was returned to its cage in a semiconscious condition and usually was sitting up by the next morning. These animals were ordinarily able to nourish themselves beginning with the first postoperative day and i.v. fluids were rarely given. Except for two, animals which survived eight to ten weeks were killed at that time and the valves examined at autopsy. Two animals were permitted to survive longer.

Results

All 14 animals of this series survived and fully recovered from the procedure. They were able to maintain posture and take a usual diet beginning the

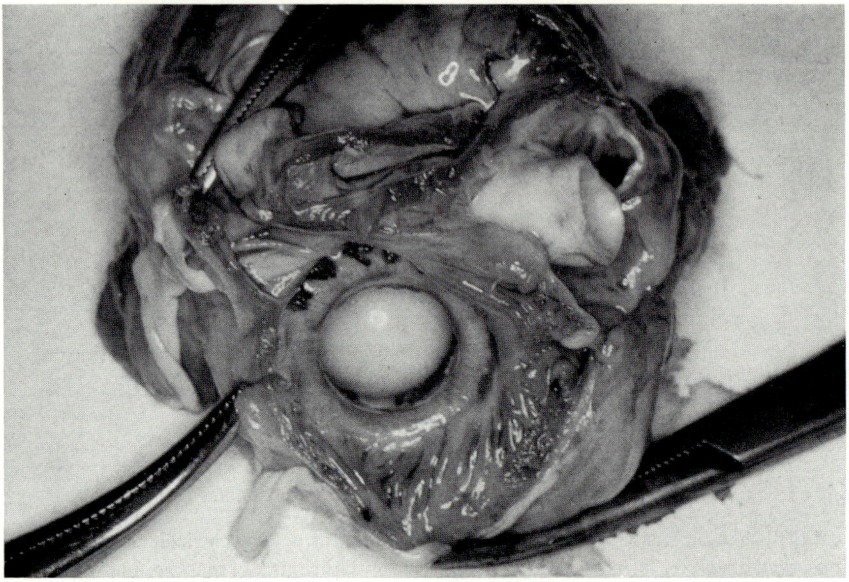

Fig. 2. Atrial aspect of valve ten weeks after implantation. All surfaces have been endothelialized. The orifice is fully open.

first day after the operation. They had no signs of circulatory embarrassment during the first week. No neurologic deficits were observed. Preliminary experiments indicated that the animal needed anticoagulation for 1 h after the intracardiac procedure and until a reasonable volume of circulation was effected. In the absence of this short period of heparinization, acute thrombosis of the valve was seen. Table II is a summary of the results. Figures 2 and 3 are typical photographs of valves ten weeks after insertion.

One animal developed chest wound drainage and fever. Cultures were positive on several occasions for *Pseudomonas aeruginosa* and *Enterobacter cloacae*. He was switched to the systemic antibiotics indicated by sensitivity tests and had irrigation of the mediastinal fistula with polymyxin B and neomycin (Neosporin®) solution on alternate days for a week. When killed eight weeks later, no evidence of this infection remained, the wound was well healed and the valve was functioning. One of the two animals which were permitted long-term survival died suddenly at 17 weeks. His valve had sufficient organized thrombosis to immobilize the ball and he had signs of congestive heart failure (300 ml of pale yellow fluid in pleura and peritoneal

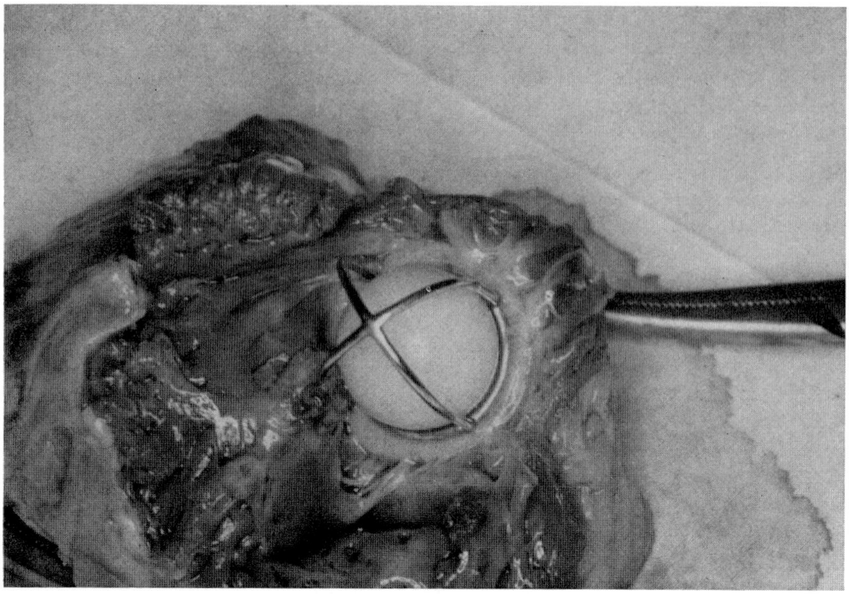

Fig. 3. Ventricular aspect of same valve seen in figure 2. Some organized thrombus is present on the septal cage wire. The poppet has no restriction and the valve is fully competent by water test.

cavities). He was otherwise normal. The other animal died similarly after 12 weeks.

Discussion

A number of points about the study warrant brief comment. The method of resuscitation is such that the circulation is gradually restored and thus even large bleeders begin to ooze and increase flow slowly until they are easily attended to during closure without excessive blood loss. The method of circulatory arrest does not of itself require anticoagulation. For neurosurgery, or any procedure not involving the insertion of an intravascular prosthesis, heparin can be omitted. In clinical intrathoracic work, the carotid catheter could be inserted through a small stab of the aortic arch and closed by a purse string or 'U' suture. The carotid itself need not be incised. Digitalis was included in the protocol since it seemed to provide a smoother return of cardiac action. We do not have positive evidence for its need. In any case, it was

Table II. Results of tricuspid ball-valve implants in baboons *(Papio anubis)*

Baboon No.	Survival weeks	Cause of death	Condition of valve	Remarks
251	4	spontaneous	obstruction thrombosis	
253	8	sacrificed	functioning no thrombosis	
256	10	sacrificed	functioning no thrombosis	
257	8	sacrificed	functioning no thrombosis	mediastinitis completely healed
259	9	sacrificed	functioning no thrombosis	
260	1	spontaneous	obstructed thrombosis	
263	8	sacrificed	functioning no thrombosis	
264	9	sacrificed	partial function thrombosis	poppet excursion limited 50%
270	1	spontaneous	obstructed thrombosis	
278	2	spontaneous	obstructed thrombosis	
283	8	sacrificed	functioning minimal thrombosis	
285	8	spontaneous	functioning no thrombosis	cause of death unknown
248	17	spontaneous	obstructed thrombosis	permitted to survive beyond 10 weeks
279	12	spontaneous	obstructed thrombosis	permitted to survive beyond 10 weeks

not continued postoperatively with these presumably normal hearts. Ice bags were placed about the head which is the only part of the body well below room temperature and which would tend to drift upward rapidly. An insulating helmet is also adequate but it must fit well. Ice was easily available in the laboratory. Since we did not have a baboon blood bank, replacement of operative blood loss was a problem. This was nicely solved by the use of human or sheep hemoglobin solution which was present in the laboratory for another study. No adverse effects of this treatment could be detected [6]. We believe that the subhuman primate is an important experimental animal for work such as this. We are interested in observing personality and higher cerebral function after a circulatory arrest protocol. More importantly, the cerebral circulation (a key element in this procedure) of lower mammals, including the dog, is different in important aspects from that of man.

Severely malformed hearts in the newborn are frequently incompatible with life much beyond the neonatal period and yet, open heart surgery in the very young has been frought with numerous difficulties. Earlier attempts at providing periods of circulatory arrest in children and infants by rapid blood cooling techniques have had an alarming incidence of neurological sequellae which have usually been attributed to the effects of cold itself upon the brain [2, 5]. In our laboratory it has been possible to cool the brain of the dog and subhuman primate to temperatures well below 10 °C, followed by 60 min of cardiac arrest without neurological sequellae [10]. We have also demonstrated that perfusion of the carotid artery with 0 °C blood does cause cerebral damage after circulatory arrest with brain temperatures of 8–12 °C, while a similar perfusion with 0 °C saline and similar brain temperatures during circulatory arrest is without neurological deficits [3]. Thus, we attribute the brain damage in the earlier reports of rapid blood cooling procedures to the effects of increased viscosity and sludging of the very cold blood with consequent impairment of the cerebral microcirculation. More recent efforts, elsewhere, with total body immersion hypothermia of infants have also had less than desired success, probably related to the incurring of heavy metabolic deficits [1, 9]. We believe there is a place, in intracardiac surgery, for a procedure such as that described. In addition to the suggested application of the method to the very young where conventional bypass is difficult, it could serve for adults when the procedure is of short duration and might be especially useful in medical centers of countries where technology has not progressed to the point of supporting a modern by-pass program but where surgeons and anesthesiologists can receive adequate training by visiting centers which have well-established cardiac surgery programs.

References

1 Baum, D.; Dillard, D. H.; Mohri, H., and Crawford, E. W.: Metabolic aspects of deep surgical hypothermia in infancy. Pediatrics *42:* 93 (1968).

2 BJORK, V. O. and HULTQUIST, G.: Contraindications to profound hypothermia in open heart surgery. J. thorac. cardiovasc. Surg. *44:* 1–13 (1962).
3 ICÖZ, M. V. and WOLFSON, S. K., jr.: Tolerance to total ischemia in dogs after cold blood or cold physiologic saline perfusion of the brain. Surg. Forum *17:* 403–405 (1966).
4 JOHNSON, A. S.; PRISK, B., and WHITNEY, D. R.: Substernal cardiac massage and assistance. Surgery *63:* 800–805 (1968).
5 PARKINS, W. M.; JENSEN, J. M., and VARS, H. M.: Brain cooling in the prevention of brain damage during periods of circulatory occlusion in dogs. Ann. Surg. *140:* 284–289 (1954).
6 PESKIN, G. W.; O'BRIEN, K., and RABINER, S. F.: Stroma-free hemoglobin solution: The 'ideal' blood substitute? Surgery *66:* 185–193 (1969).
7 SELKER, R. G.; WOLFSON, S. K., jr., and DEUTSCH, S. B.: Preferential cerebral hypothermia with elective cardiac arrest. II. Clinical experience. Proc. 4th Europ. Congr. Neurosurgery, 1971 (in press).
8 TYERS, G. F. O. and WOLFSON, S. K., jr.: Automatic external cardiac massage as an adjunct to recovery from moderate general hypothermia, profound cerebral hypothermia and prolonged cardiopulmonary arrest. Arch. Surg. *98:* 771–775 (1969).
9 WOLFSON, S. K., jr.: Rewarming following surgical hypothermia. Pediatrics *43:* 628–629 (1969).
10 WOLFSON, S. K., jr.: Unpubl. observations.
11 WOLFSON, S. K., jr.; ICÖZ, M. V., and LUBER, S.: Preservation of conditioned responses in primates after total circulatory arrest with preferential cerebral hypothermia. Surg. Forum *16:* 409–411 (1965).
12 WOLFSON, S. K., jr.; INOUYE, W. Y.; KAVIANIAN, A. R.; ICÖZ, M. V., and PARKINS, W. M.: Preferential cerebral hypothermia for circulatory arrest. Surgery *57:* 846–855 (1965).
13 WOLFSON, S. K., jr.; INOUYE, W. Y.; KAVIANIAN, A. R., and PARKINS, W. M.: Circulatory arrest of 30–90 minutes utilizing preferential cerebral hypothermia without extracorporeal circulation (abstract). Physiologist *6:* 298 (1963).
14 WOLFSON, S. K., jr. and SELKER, R. G.: Preferential cerebral hypothermia with elective cardiac arrest. I. A sub-human primate model for clinical application. Proc. 4th Europ. Congr. Neurosurgery, 1971 (in press).

Authors' addresses: SIDNEY K. WOLFSON, jr., M.D. and ROBERT G. SELKER, M.D., Department of Neurosurgery, Montefiore Hospital, 3459 Fifth Avenue, *Pittsburgh, PA 15213;* T. P. COMER, M. D., Department of Surgery, Kern General Hospital, *Bakersfield, Calif.;* C. S. THOMAS, M. D., Department of Surgery, Vanderbilt University School of Medicine, *Nashville, Tenn.* (USA)

Transplantation Immunology

Medical Primatology 1972. Proc. 3rd Conf. exp. Med. Surg. Primates, Lyon 1972, part II, pp. 78–91 (Karger, Basel 1972)

Major Histocompatibility System of Rhesus Monkeys, RhL-A

Relations between RhL-A and the Locus Determining Reactivity in Mixed Lymphocyte Cultures[1]

H. BALNER[2], J. D'AMARO, E. K. TOTH, H. DERSJANT and W. VAN VREESWIJK

Primate Centre TNO, Rijswijk

Tissue typing of rhesus monkeys was initiated around 1964 when the transplantation program of our Institute required methods to determine the degree of histocompatibility between individual animals. At that time the significance of leukocyte antigens for histocompatibility had been proven in several laboratories [7, 14, 21] and the serological identification of a few human leukocyte antigens had just been reported. Since monkeys were intended to serve as an experimental model for clinical transplantation, we decided to make a major effort to identify the rhesus monkey's own system of histocompatibility antigens.

This communication describes our current knowledge of the major histocompatibility system (RhL-A) of monkeys and briefly reviews the results of an international workshop on transplantation genetics of primates held at this Institute in September, 1971. Data obtained with mixed lymphocyte cultures (MLC) are also reported and the relation between MLC reactivity, RhL-A matching and histocompatibility is discussed.

1 Supported in part by a contract from the Dutch Government and Contract No. 6243-22/6/001 of the Commission of European Communities (EEG), Luxembourg. This publication is contribution No. 714 of the Euratom Biology Division.

2 Member of the Biology Division of Euratom.

Table I. Stages in identifying a system of leukocyte antigens

Leukocyte grouping:

produce antisera, test with cell panel;
group sera with similar reactivity pattern by computer analysis;
assign provisional specificities to sera;
check and improve specificities by absorption, if necessary.

Population analysis (unrelated individuals):

test groups of sera with many cell samples;
look for possible allelic relationships between specificities (negative associations and χ^2 for allelism low, no 'triple positives' for alleles of a series);
check for Hardy-Weinberg fit (comparison of observed with expected phenotypes).

Family analysis:

check for inheritance of antigens as alleles of one or more segregant series;
antigens of linked series inherited in coupling (haplotypes).

Serology and Genetics of the RhL-A System

Methods

Our methods of producing and screening alloantisera have been described extensively [8]. Optimal reagents are obtained if skin grafts are exchanged between siblings sharing one RhL-A haplotype (see later). Sera are tested with a one-stage microcytotoxicity method and the international microcomplement fixation technique. Absorption methods used have been described elsewhere [5].

Table I schematically depicts the successive steps which are taken to identify a system of leukocyte antigens. A more detailed description of that approach has been published elsewhere [8].

RhL-A Specificities

In 1970, 14 specificities were recognized with groups of alloantisera; six of those specificities were rather well defined by operationally monospecific reagents, the others were broad 'supertypic' specificities. Population analysis and family studies showed that those specificities belonged to one complex system probably controlled by two closely-linked series of alleles [10, 17]. The system was called RhL-A and its relevance to histocompatibility proven by prolonged survival of skin grafts exchanged between genotypic identical sibs [9].

Table II. Results of First Primate Workshop, Rijswijk 1971. Leukocyte antigens of rhesus monkeys

Workshop nomenclature	Original Rijswijk nomenclature	Original Bethesda nomenclature	Gene frequencies[1]
RhL-A 6	RhL-A 6	Group 2	0.219
RhL-A 9	RhL-A 9	Group 1	0.197
RhL-A 10	RhL-A 10	Group 5	0.149
RhW 2	RhL-A 2		0.149
RhW 7	RhL-A 7		0.284
RhW 11	RhL-A 11		0.257
RhW 12	RhL-A 12		0.169
RhW 13	RhL-A 13		0.090
RhW 14	RhL-A 14		0.065
RhW 15		Group 8	0.442
RhW 16		Group 3	0.545

1 Calculated from sample of 87 unrelated animals.

Since several other laboratories were then also involved in tissue typing of rhesus monkeys, the time seemed ripe to compare serum reactivities and reach agreement about definition of antigens and nomenclature. Thus the first primate histocompatibility workshop was held in September, 1971, in our Institute [23]. A large number of animals, including many rhesus families, were typed by the various teams using their own reagents and techniques. During the symposium following the workshop, the concept of RhL-A as the rhesus monkey's major histocompatibility system was unanimously accepted; some of the well-defined allelic specificities kept the original RhL-A nomenclature, others were given provisional workshop designations (table II) [4]. A second primate workshop is planned for September, 1973, to be held in Atlanta, USA.

More recent developments have led to an improved understanding of the RhL-A system. Well-planned intersib immunizations yielded several highly specific reagents recognizing at least two, possibly five new specificities of the RhL-A system. Figure 1 shows the distribution of six currently-accepted specificities of the first segregant series of RhL-A in 127 unrelated rhesus monkeys. As can be seen, there were no triple positive individuals. Table III shows that the fit between observed and expected phenotypes is

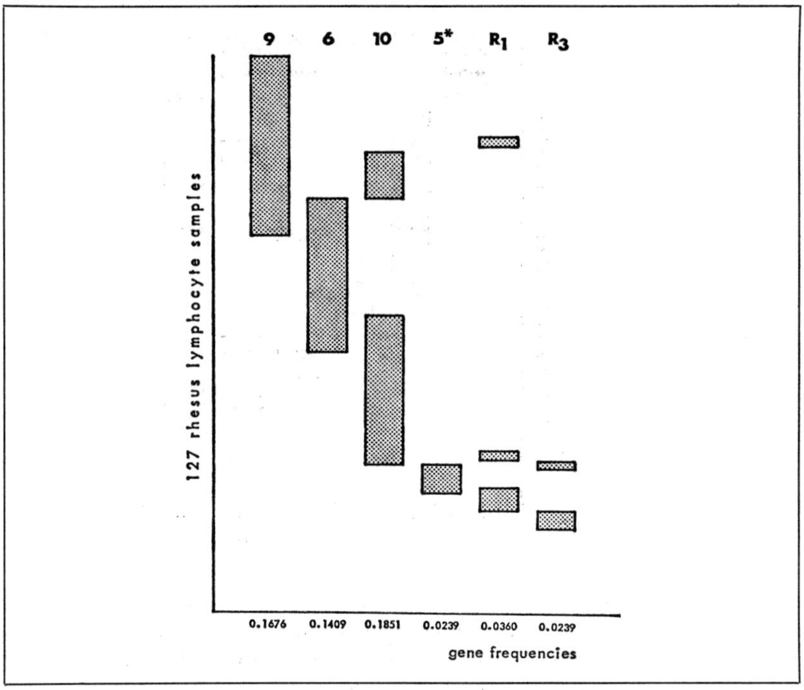

Fig. 1. Phenotype distribution in 127 unrelated rhesus monkeys (first segregant series of RhL-A). Cross-hatched bars indicate that the majority of sera defining the group was strongly positive. No triple positive individuals were found.

good. Figure 2 and table IV show the same for the second segregant series. If specificities 13,11*, 2* and 14 are considered (table IV), the fit between observed and expected phenotypes is excellent and no triple positive individuals are encountered. Two triple positive individuals were found when specificities DA_1 and R_2 were provisionally added to this segregant series (fig. 2).

Allelic segregation of the antigens of both series was confirmed in family studies. When assigning the antigens to phenotypic combinations which are inherited in coupling, 19 different RhL-A haplotypes were found (25 unrelated parents). In studying the parental haplotypes in more than a hundred offspring, crossing-over at the RhL-A locus was found to have occurred in one case, and possibly in one other; this confirmed our concept of two separate segregant series within the RhL-A system. Analysis of the segregation of the antigens in backcross matings (+/— × —/—) with at least

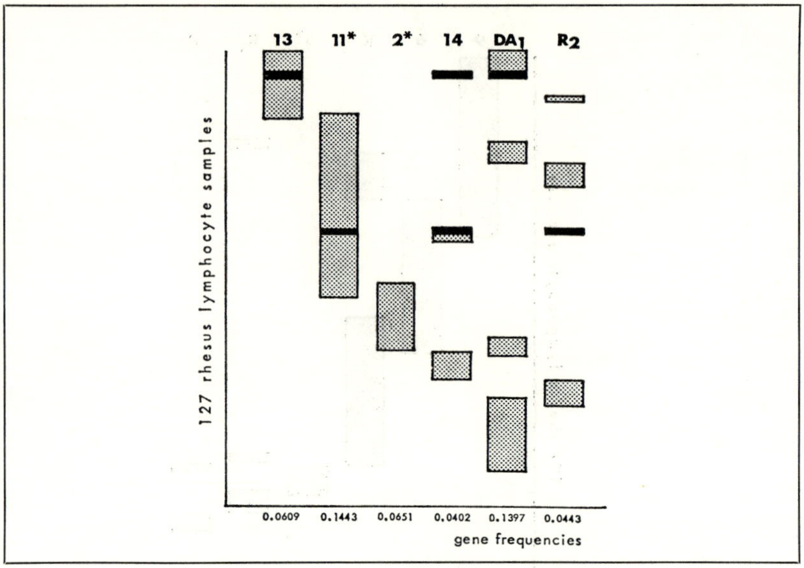

Fig. 2. Phenotype distribution in 127 unrelated rhesus monkeys (second segregant series of RhL-A). Cross-hatched bars indicate that the majority of sera defining the group was strongly positive. Specificities DA_1 and R_2 are tentative additions to this series. The only two triple positives were produced by adding those specificities; they are indicated by black bars.

one negative offspring, showed that the ratio of +/— children had a near-perfect fit with the Mendelian expectation of 0.5 (table V).

Relevance to Histocompatibility

As has been briefly mentioned, the relevance of RhL-A to histocompatibility was proven in families; similar to the situation in man, grafts exchanged between sibs sharing both parental RhL-A haplotypes had a significantly-prolonged survival time [9]. This proved that the RhL-A genetic region is of major importance for histocompatibility in rhesus monkeys. However, from studies in mice and several other species, it is known that loci for other genetic traits (MLC reactivity, immune response, etc.) are located in the same genetic region and may also exert an influence on graft rejection. This is best demonstrated in experiments performed with un-

Table III. Distribution of 6 leukocyte specificities in 127 unrelated rhesus monkeys (first segregant series of RhL-A)

RhL-A 9	RhL-A 6	RhL-A 10	5*	R1	R3	Number observed	Number expected	χ^2
+	+	–	–	–	–	8	6.00	0.67
+	–	+	–	–	–	10	7.87	0.63
+	–	–	+	–	–	0	1.02[a]	
+	–	–	–	+	–	2	1.52[a]	2.38[a]
+	–	–	–	–	+	0	1.02[a]	
–	+	+	–	–	–	8	6.60	0.30
–	+	–	+	–	–	0	0.84[a]	
–	+	–	–	+	–	0	1.27[a]	
–	+	–	–	–	+	0	0.84[b]	
–	–	+	+	–	–	0	1.12[b]	0.02[b]
–	–	+	–	+	–	2	1.68[b]	
–	–	+	–	–	+	1	1.12[b]	
–	–	–	+	+	–	1	0.20[b]	
–	–	–	+	–	+	1	0.13[b]	
–	–	–	–	+	+	0	0.20[b]	
+	–	–	–	–	–	19	21.54	0.30
–	+	–	–	–	–	17	17.63	0.03
–	–	+	–	–	–	21	24.21	0.43
–	–	–	+	–	–	4	2.63	0.71
–	–	–	–	+	–	4	4.01	0.00
–	–	–	–	–	+	4	2.63	0.71
–	–	–	–	–	–	25	22.67	0.24

a, b Indicate that values were pooled to obtain expectations greater than 5 for the χ^2 calculation.

related phenotypic identical host-donor combinations. These have the same histocompatibility antigens but are not necessarily identical for other relevant traits. In such experiments, phenotypic identity did not always lead to a significant prolongation of graft survival. This has been shown for man [18] and rhesus monkeys [6]. Furthermore, there is evidence that in human cases of phenotypic identity for all four HL-A antigens ('full-house' identity), negative or low MLC reactivity between host and donors correlates with prolonged graft survival [19]. It would seem, therefore, that the degree of stimulation in the MLC test is related to graft survival although its relative importance is not yet known. A crucial question is obviously

Table IV. Distribution of four leukocyte specificities in 127 unrelated rhesus monkeys (second segreant series of RhL-A)

13	11*	2*	14	Number observed	Number expected	χ^2
+	+	−	−	1	2.23[a]	
+	−	+	−	0	1.00[a]	
+	−	−	+	1	0.62[a]	0.02[a]
−	+	+	−	0	2.38[a]	
−	+	−	+	3	1.47[a]	
−	−	+	+	0	0.66[a]	
+	−	−	−	13	11.14	0.31
−	+	−	−	33	27.91	0.93
−	−	+	−	12	11.94	0.00
−	−	−	+	6	7.24	0.21
−	−	−	−	55	60.39	0.39

a Indicates that values were pooled to obtain expectations greater than 5 for the χ^2 calculation.

whether host-donor combinations, which are not identical for the major histocompatibility antigens but display low MLC reactivity (a rare event), will show prolonged graft survival. Like the possible influence of the immune response loci, this is one of the controversial issues and the subject of current investigations in many laboratories.

In any case, it is apparent that typing and matching for HL-A (in man) or RhL-A (in monkeys) is not sufficient to select the most compatible host-donor combinations in unrelated individuals and that other selection methods should be taken into account as well. Yet, the antigens of the major histocompatibility systems are still the only convenient markers to serologically identify the chromosomal region, which is of primary importance for graft survival as well as for other essential biological functions.

Mixed Lymphocyte Cultures

As indicated above, reactivity in MLC is regarded as an important additional method to define histocompatibility. In fact, the MLC reaction is considered to be one of the *in vitro* models of certain cellular events taking

Table V. Test on segregation ratios for RhL-A specificities in back-cross matings (+/− × −/−)

	RhL-A specificity	Number of families	Children +	Children −
First series	9	4	7	9
	6	8	13	17
	10	15	30	28
	5*	2	4	4
	R_1	5	7	9
	R_3	1	2	2
Second series	13	10	19	17
	11*	5	9	11
	2*	5	11	8
	14	4	10	6
	R_2	6	12	11

Specificity R_2 is provisionally assigned to the second segregant series.

place during allograft rejection[3]. For this reason we have started a systematic study of MLC reactivity in rhesus monkeys. Only some of the available results will be presented here; a more detailed report of our MLC studies in rhesus monkeys is in preparation.

Technique

A conventional MLC technique was used in the initial experiments [2] but the more economical and convenient micromethod of EIJSVOOGEL *et al.* [16] is currently used.

Only 3×10^5 lymphocytes (1.5×10^5 per donor) are cultured in a volume of 1 ml. To measure reactivity, ^{14}C-thymidine is added on day 5. The culture is terminated 24 h later. Cells are collected and washed on fiberglass filters; after drying and addition of scintillation fluid activity is counted for 10 min. One-way MLC are performed by treating stimulator cells with mytomycin as described by BACH and VOYNOW [3].

All MLC determinations are performed in triplicate. Reproducibility of experiments done on the same day is excellent; when repeated on another day, absolute values are not identical, but the 'trend' is the same (high and low values for the same combinations). In the tables, only the means of the triple values and standard errors are given. Appropriate controls were included in each experiment but are not presented here.

3 Another very promising *in vitro* model is the CML test (cell-mediated lympholysis) as described by MIGGIANO *et al.* [20]. Experiments using this method in rhesus monkeys are in progress.

Mixed Lymphocyte Cultures in Unrelated Animals

In several hundred two-way MLC performed with cells from nonmatched combinations, values ranged between 8 and 25×10^3 counts with an occasional lower or higher stimulation. One-way MLC nearly always gave lower counts, usually about half the two-way value.

Using unrelated animals matched for defined RhL-A antigens has so far always produced stimulation in MLC. However, until now no unrelated 'full-house' identical combinations (all four RhL-A antigens identified) have been tested.

Mixed Lymphocyte Cultures in Related Animals

Like in humans [1], close linkage has been found between a locus defining MLC reactivity and the genetic region controlling the expression of the major histocompatibility antigens in rhesus monkeys. Early experiments showed that genotypic identical sibs do not stimulate in MLC [2] and this has been confirmed in numerous recent experiments. Until now, we have not come across genotypic identical sibs which *do* stimulate in MLC (recombination between the loci for RhL-A and for MLC). However, a pair of sibs was found whose cells are not identical for the RhL-A antigens and *did not* stimulate in MLC (fig. 3). One of these sibs was a paternal recombinant for the antigens of the second series of RhL-A. A likely explanation for this nonreactivity would be that the crossing-over did not

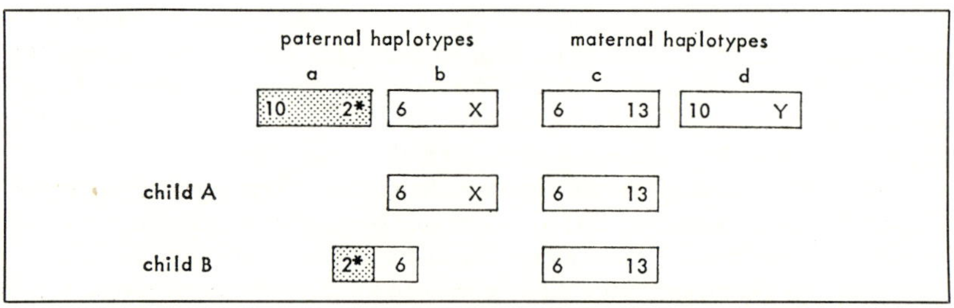

Fig. 3. MLC reactivity and crossing-over at the RhL-A locus of rhesus monkeys. A+B showed no stimulation in mixed lymphocyte cultures in spite of nonidentity for antigen 2* (for conclusion see text). X and Y stand for unidentified RhL-A specificities in the second segregant series.

Table VI. MLC in rhesus monkeys. Stimulation between phenotypic identical half-sibs

	MLC stimulation		RhL-A genotypes	
AE			a_1	6,13
			d_1	6,13
BS	7,668		a_1	6,13
	(239)		d_2	6,13
ET	8,732	914	a_1	6,13
	(105)	(41)	d_2	6,13
	AE	BS		

Body of table presents mean values of triple determinations of two-way MLC (cpm) and standard errors (in parentheses). The genotypes show that all animals had inherited haplotype a_1 from the same father; AE inherited haplotype d_1 from mother 1 while BS and ET inherited the same haptolype d_2 from mother 2. All individuals are homozygous for RhL-A 6 and 13. Note that there was stimulation between phenotypic identical half-sibs but not between genotypic identical full-sibs.

affect the chromosomal region determining MLC reactivity. Consequently, the genetic control of MLC in rhesus monkeys may well be located near the area coding for the antigens of the first series of RhL-A. Analogous linkage situations exist for the mouse (MLC adjacent to the K-end of H-2 [22]) and for man (MLC adjacent to the four series of HL-A [11, 15, 16, 24]).

Interesting data can be obtained if sibs or half-sibs sharing only one parental haplotype are used for MLC. If they share one haplotype but are not (phenotypic) identical for RhL-A, the MLC values are sometimes lower than when they differ for both haplotypes. But what happens if they share only one haplotype but carry the same RhL-A antigens on the other chromosome? On one such occasion, normal stimulation was found even when all four antigens were identified (table VI). On the other hand, several other combinations of this type failed to show stimulation in MLC (for instance, G+AX in table VII). This suggests that the MLC locus is not as polymorphic as for instance the two loci of the RhL-A system. Alternatively, there could be a rather strong linkage disequilibrium between RhL-A and MLC genes so that haplotypes (from different parents) with identical RhL-A specificities may have identical 'MLC phenotypes' more often than would be expected by chance.

Table VII. Inhibition of MLC reaction by preincubation of stimulator cells with alloantibody (anti-R_1 serum)

Stimulator cells	Responder cells								RhL-A genotype	
	G								b_1	10,13
									c_1	9,11*
					CV				b_1	10,13
									d_5	6, –
	two-way MLC	+antibody	one-way MLC	+antibody	two-way MLC	+antibody	one-way MLC	+antibody		
AG	23,661	*1,665*	11,153	*1,586*	10,723	*2,734*	6,262	*2,386*	a_1	R_1, –
	(280)	(130)	(295)	(136)	(367)	(94)	(113)	(81)	c_1	9,11*
ED	15,317	*1,767*	7,611	*1,442*	14,639	*1,520*	9,653	*1,434*	a_1/b_1	R_1,13
	(130)	(82)	(1193)	(119)	(628)	(122)	(202)	(172)	d_4	10, –
BY	17,693	*2,707*	8,626	*2,812*	12,311	*2,894*	6,051	*3,516*	a_2	10, –
	(603)	(113)	(266)	(55)	(156)	(563)	(59)	(200)	c_7	R_1, –
AX	1,404	1,551	1,203	1,311	9,680	8,623	5,989	6,475	b_1	10,13
	(206)	(134)	(102)	(54)	(61)	(396)	(400)	(259)	c_3	9,11*
DC	8,442	7,939	5,186	5,457	8,159	8,304	5,495	4,511	b_1	10,13
	(123)	(428)	(145)	(180)	(554)	(534)	(566)	(150)	c_6	–, –

Body of table presents mean values of triple determinations (cpm) and standard errors (in parantheses). Under 'RhL-A genotype', a and b stand for paternal, c and d for maternal haplotypes; subscripts indicate individual families (a_1/c_1 and b_1/c_1 are thus full sibs, sharing the maternal haplotype c_1).

Responder cells G and CV were tested against five stimulator cells in two- and one-way MLC; 'antibody' under column means that stimulator cells were preincubated with monospecific anti-R_1 serum. Values showing significant reduction (in comparison with nonincubated values) are in italics. Note very low stimulation of G with phenotypic identical AX (see further text).

Anti-R_1 serum was produced by immunizing monkey G with cells from monkey AG.

Inhibition of Mixed Lymphocyte Cultures with Alloantisera

Does the coating of histocompatibility antigens on lymphocytes (with alloantisera) inhibit MLC reactivity? For man, CEPPELLINI *et al.* [12, 13] had shown that stimulator as well as responder cells of MLC can be inhibited by alloantibodies. We were interested to confirm this phenomenon for rhesus monkeys and to find out how specific the inhibition would be. The antibodies were raised in monkeys immunized with cells from sibs or half-sibs differing for one parental haplotype. They were directed against an RhL-A antigen but possibly also against hypothetical products of an MLC locus. If inhibition can be caused by such hypothetical antibodies against MLC determinants, then it should be feasible to inhibit an occasional cell which carries that MLC determinant but not the RhL-A antigen in question.

Table VII shows an example of an inhibition experiment. Monkey G had been immunized with cells from sib AG; the antibody was directed against antigen R_1 on the paternal haplotype a_1. Cells from G (and from another animal which had also produced a monospecific anti-R_1 serum) were reacted against cells from other monkeys, some carrying the paternal a_1 chromosome. Testing was done in two- and one-way MLC with and without preincubation of the stimulator cells with the antibody (dilution $^1/_5$; incubation ½ h at 37°C; three washings). It can be seen that preincubation with antibody reduced the MLC values only when the R_1 antigen was on the stimulator cells, regardless of whether these carried haplotype a_1 or not. The design was such that the antibody used could only affect the stimulator cells. But inhibition of reactor cells is also possible; preincubation of cell AG with anti-R_1 serum reduced reactivity against mytomycin-treated DC cells from 6,000 to 2,000 counts.

Several inhibition experiments have now been carried out. Inhibition of stimulator cells was always highly specific; it occurred only when the cells carried the antigen in question. Inhibition of cells negative for the antigen, but possibly carrying the relevant 'MLC determinant', has not been found. It seems likely, therefore, that inhibition is caused by specific binding of immunoglobulin to the RhL-A antigens on the cell surface. As suggested by CEPPELLINI [11], this might prevent close contact of responder cells with the MLC determinants on stimulator cells (also if the antibody is used to coat the antigens of the responder cells). It should be interesting to see whether sera specifically raised against hypothetical MLC determinants, for instance by cross-immunization of phenotypic identical individuals which do stimulate in MLC (table VI), will inhibit MLC reactivity; such experiments are currently in progress.

References

1 AMOS, D. B. and BACH, F. H.: Phenotypic expressions of the major histocompatibility locus in man (HL-A). Leukocyte antigens and mixed leukocyte culture reactivity. J. exp. Med. *128:* 623–638 (1968).

2 APPELMAN, A. W. M. and BALNER, H.: Mixed leukocyte cultures in rhesus monkeys. Transplant. Proc. *4:* 17–20 (1972).

3 BACH, F. H. and VOYNOW, N. K.: One-way stimulation in mixed leukocyte cultures. Science *153:* 545–547 (1966).

4 BALNER, H.; BARNES, A. D.; D'AMARO, J.; DARROW, C. C.; DÉMANT, P.; GABB, B. W.; PIAZZA, A.; METZGAR, R. S.; SEIGLER, H. F.; ROGENTINE, G. N.; ROOD, J. J. VAN; WARD, F., and ZWEERUS, R. L.: Joint report of first histocompatibility workshop on primates. Transplant. Proc. *4:* 141–147 (1972).

5 BALNER, H.; DERSJANT, H.; LEEUWEN, A. VAN, and ROOD, J. J. VAN: Identification of two major leukocyte antigens of rhesus monkeys and their relation to histocompatibility; in Histocompatibility testing 1967, pp. 267–276 (Munksgaard, Copenhagen 1967).

6 BALNER, H.; DERSJANT, H.; VREESWIJK, W. VAN, and GABB, B. W.: Tissue typing of rhesus monkeys. Application in transplantation research. Transplant. Proc. *4:* 3–9 (1972).

7 BALNER, H.; DERSJANT, H., and ROOD, J. J. VAN: A method to relate leukocyte antigens and transplantation antigens in monkeys. Transplantation *3:* 230–234 (1965).

8 BALNER, H.; GABB, B. W.; DERSJANT, H.; VREESWIJK, W. VAN; LEEUWEN, A. VAN, and ROOD, J. J. VAN: Leukocyte groups in nonhuman primates; in Comparative genetics in monkeys, apes and man, pp. 97–128 (Academic Press, London 1971).

9 BALNER, H.; GABB, B. W.; DERSJANT, H.; VREESWIJK, W. VAN, and ROOD, J. J. VAN: Major histocompatibility locus of rhesus monkeys (RhL-A). Nature *230:* 177–180 (1971).

10 BALNER, H.; LEEUWEN, A. VAN; VREESWIJK, W. VAN; DERSJANT, H., and ROOD, J. J. VAN: Leukocyte antigens of rhesus monkeys (RhL-A) and chimpanzees (ChL-A). Similarities with the human HL-A system. Tissue Antigens *1:* 229–238 (1971).

11 CEPPELLINI, R.: Old and new facts and speculations about transplantation antigens of man; in Progress in immunology (Academic Press, New York, in press).

12 CEPPELLINI, R.; BIGLIANI, E. S.; CURTONI, E. S., and LEIGHEB, G.: Experimental allotransplantation in man. Transplant. Proc. *1:* 390–394 (1969).

13 CEPPELLINI, R.; BONNARD, C. D.; COPPO, F.; MIGGIANO, V. C.; POSPISIL, M.; CURTONI, E. S., and PELLEGRINO, M.: Mixed leukocyte cultures and HL-A antigens. Transplant. Proc. *3:* 58–70 (1971).

14 DAUSSET, J.; RAPAPORT, F. T.; IVÁNYI, P., and COLOMBANI, J.: Tissue alloantigens and transplantation; in Histocompatibility testing 1965, pp. 63–69 (Munksgaard, Copenhagen 1965).

15 DUPONT, B.; STAUB-NIELSEN, L., and SVEJGAARD, A.: Relative importance of four LA loci in determining mixed lymphocyte reaction. Lancet *ii:* 1336–1340 (1971).

16 EIJSVOOGEL, V. P.; DUTOIT, E. D.; ROOD, J. J. VAN, and SCHELLEKENS, P. TH. A.: Position of the locus determining mixed lymphocyte reaction apart from the HL-A region. Transplant. Proc. (in press).

17 GABB, B. W.; PIAZZA, A.; D'AMARO, J., and BALNER, H.: Genetics of RhL-A system of rhesus monkeys. Transpl. Proc. *4:* 11–16 (1972).

18 KOCH, C. T.; FREDERIKS, E.; EIJSVOOGEL, V. P., and ROOD, J. J. VAN: Mixed-lymphocyte-culture and skin-graft data in unrelated HL-A identical individuals. Lancet 1334–1336 (1971).

19 KOCH, C. T.; LEEUWEN, A. VAN; FREDERIKS, E., and ROOD, J. J. VAN: Skin graft survival between HL-A identical unrelated individuals. Correlation with MLC reactivity; in Histocompatibility testing 1972 (Munksgaard, Copenhagen, in press).
20 MIGGIANO, V. C.; BERNOCO, D.; LIGHTBODY, J.; TRINCHIERI, G., and CEPPELLINI, R.: Cell mediated lympholysis (CML) *in vitro* with normal lymphocytes as target. Transplant. Proc. (in press).
21 ROOD, J. J. VAN; LEEUWEN, A. VAN; SCHIPPERS, A. M. J.; VOOYS, W. H.; FREDERIKS, E.; BALNER, H., and EERNISSE, J. G.: Leukocyte groups, the normal lymphocyte transfer test and homograft sensitivity; in Histocompatibility testing 1965, pp. 37–50 (Munksgaard, Copenhagen 1965).
22 RYCHLÍKOVÁ, M.; DÉMANT, P., and IVÁNYI, P.: Further studies on mixed lymphocyte reaction in H-2 incompatibility. Proc. Symp. Immunogenetics of the H-2 System, Liblice-Prague 1970, pp. 178–181.
23 BALNER, H. and ROOD, J. J. VAN (ed.): Transplantation genetics of primates. Proceedings of a Workshop and Symposium. Transplant. Proc. *4* (1972).
24 YUNIS, E. J.; PLATE, J. M.; WARD, F. E.; SEIGLER, H. F., and AMOS, D. B.: Anomalous MLR responsiveness among siblings. Transplant. Proc. *3:* 118–120 (1971).

Authors' address: H. BALNER, J. D'AMARO, E. K. TOTH, H. DERSJANT and W. VAN VREESWIJK, Primate Centre TNO, 151 Lange Kleiweg, *Rijswijk Z.H.* (The Netherlands)

Medical Primatology 1972. Proc. 3rd Conf. exp. Med. Surg. Primates, Lyon 1972, part II, pp. 92–99 (Karger, Basel 1972)

The Production of Leucocyte Typing Sera in Monkeys and Baboons

A. D. BARNES and R. J. HAWKER

Transplantation Immunology Research Unit, Department of Surgery, Queen Elizabeth Hospital, University of Birmingham, Birmingham

Introduction

Methods of improving the specificity and titre of leucocyte antisera in man are limited by ethical considerations, although antisera have been raised by skin grafting [12] and leucocyte [20] or whole blood innoculations [11] of selected donors and recipients. In subhuman primates a variety of methods has been used to produce high-titre leucocyte and erythrocyte alloantisera [3, 22]. We describe here an attempt to raise antisera to monkey and human transplantation antigens in a limited period of time, in three species of subhuman primate. The antisera, raised to human antigens detect lymphocyte and erythrocyte polymorphisms in a human cell panel.

Materials and Method

Three primate species, *Macacca mulatta*, *Papio anubis* and *Papio cynocephalus*, from the Department of Anatomy Primate Facility, have been used in these experiments. Many of the antisera were raised in animals which had been used in other experiments and were not available for long-term study. Thirteen rhesus monkeys and four baboons received subcutaneous skin implants from single selected donors at weekly intervals. Ten monkeys received antigen preparations in multiple sites, aimed at stimulating major lymphoid groups, as suggested by BINNS and his colleagues [10] (fig. 1). The first injection was emulsified with Freund's complete adjuvant (FCA) and repeated 14 days later without adjuvant. All animals were bled seven days after the second antigen injection.

A lymphoid antigen eluate [13] was prepared from human and monkey mesenteric, iliac and inguinal lymph nodes from a single tissue typed donor and stored at –20 °C. After thawing the nodes were pushed through a stainless steel sieve (pore size 320 μm) in ice-cold 0.8% NaCl and the stroma discarded. Following incubation for 60 min at 4 °C the super-

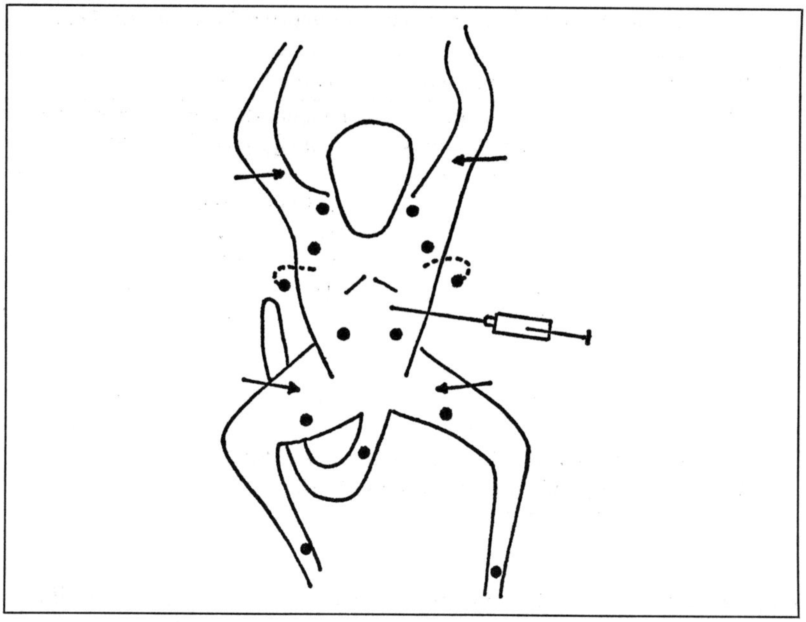

Fig. 1. 'Antigen' injection sites. ● = s.c. ×13; ⤙ = i.m. ×4; syringe = i.p.

natant was removed and the sediment incubated in 0.8% NaCl for 60 min. The two supernatants were pooled and centrifuged at 100,000 *g* for 60 min at 4 °C. The eluate pellet was washed in 0.9% saline and freeze-dried.

Lymphocytotoxicity using fresh cells was performed by a one-stage procedure on 'Compatype' plates (G.D. Searle and Co.). The reproducibility of this test has been reported previously [16]. Titres were measured using twofold serial dilutions of heat-inactivated serum and expressed as the 100-percent cytotoxic end-point.

Adsorptions were carried out by incubating antisera with packed erythrocytes or purified peripheral blood lymphocytes for 30 min at room temperature and repeating until negative against the adsorbing cells.

Results

Lymphocyte Alloantisera

We have reported previously [5, 16] the results of raising lymphoid alloantisera in three species of subhuman primates. These antisera were produced by sensitization with more than five repeated skin implants from single

Table I. Alloantibody response to s.c. skin implants in rhesus monkeys

Recipient animal typing		Typing RhW 13 +ve (2168)[1] +ve number of immunizations	lymphocytotoxic antibody titre		haemagglutination titre
RhW 13	(2168)		RhW 13	(2168)	
–	+	8	192	0	0
–	–	8	96	192	[2]
–	+	8	24	0	0
+	–	7	0	96	0
Frozen skin					
–	–	9	0	0	0

1 Serum (2168) known to be oligospecific and includes RhLA 6.
2 Animal subsequently immunized with blood.

serologically selected donors and proved to be of narrow specificity and mostly of high titre, suitable for adsorption studies. Antibody titres have been traced during and after sensitization in rhesus monkeys and baboons [16]. The appropriate donor was not always available, so we used whole lymphoid cells and an antigenic eluate as a source of antigen to raise antisera rapidly. Initial experiments resulted in low titre cytotoxic antibodies and weak haemagglutinins. A supply of easily available fresh lymphoid cells from one donor at regular intervals was difficult and the method of using such cells proved to be unsatisfactory.

The availability of a high titre RhW 13 antiserum (international nomenclature) raised by skin grafting and defined at the Workshop in Transplantation Genetics of Primates 1971 [4] has enabled us to select animals for this antigen. RhW 13 has a gene frequency of 0.090. Donor material containing this antigen was innoculated into various recipients combinations to study the efficacy of the antigen in sensitization schemes. Live skin (table I) was used to raise specific antisera of titres (24–192) after more than five weekly implants. Serum (2168) is oligospecific and includes RhLA-6. Attempts to store skin at –20 °C without any special treatment, to form an 'antigen bank' failed as it did not elicit an antibody response even after nine weekly implants. This failure was confirmed by the transplantation of frozen skin across H2 differences in mice [unpubl.). Freeze-dried lymphoid eluates produce a good heterologous antibody response (ALS) [6]. We, therefore, produced antigen from

Table II. Alloantibody response to 'lymphoid eluate antigen' in rhesus monkeys

Recipient animal typing		Typing RhW 13 +ve RhLA 6 −ve			
		number of sensitizations	lymphocytotoxic antibody titre		haemagglutination titre
RhW 13	RhLA 6		RhW 13	RhLA 6	
+	−	2	0	0	0
−	−	2	24	0	0
−	+	2	3	0	0
−	+	2	±[1]	0	0

1 Weak reaction >40% <100% cytotoxicity.

lymphoid tissues for alloantibody production. The yield of rhesus lymph node eluate (37.6 mg from 2.7 g of tissue) was lower than that derived from human lymphoid tissue [7]. Using an RhW 13 containing antigen and a two-pulse injection regime (3.75-mg eluate injection) specific cytotoxic antibodies were produced (table II). Titres to RhW 13 cells, by day 21, in RhW 13 negative recipients were between weak (40% cytotoxocity) and strong (1/24 100% cytotoxicity). The presence of non-specific lymphocytotoxins was excluded by reactions with RhLA 6 positive cells and by the negative antibody response in a monkey with identical typing for the two specificities. All antisera raised by this and the skin-implanting technique have proved to be negative for saline haemagglutinins.

Monkey Antihuman Sera

Three rhesus monkeys received 100 ml human group B rhesus positive blood at two-week intervals (table III). Although these sera had a high titre erythrocyte activity none was lymphocytotoxic.

Monkey 4 (table III) received 1 ml of red cell ghost antigen and 3.58 mg of lymphoid eluate from a single donor on three occasions. Antispecies antibodies were raised and were adsorbed with red cells and lymphocytes from a single donor source. After adsorption the antiserum was capable of detecting polymorphisms in both human erythrocytes (50%) and human lymphocytes (80%). Monkey 5 received only lymphoid material. The initial sensitization was achieved with a s.c. human lymph node implant, followed one week later

Table III. Production of monkey antihuman sera

Animal No.	Sensitization	Antibody titre		After adsorption			
		B rhesus positive RBC	lymphocytes	% positive to red cell panel (10)	titre	% positive to lymphocyte panel (10)	titre
1	whole blood (days 0, 14, 28)	512	0	60	2	0	0
2	whole blood (days 0, 14, 28)	256	0	80	2	0	0
3	whole blood (days 0, 14, 28)	8	0	60	2	0	0
4	erythrocyte + lymphoid antigen + FCA (day 0) and without adjuvant on days 14 and 21	512	3	50	4	80	1
5	lymph node implant (day 0) FCA + lymphoid antigen on day 7 and without adjuvant on day 21	48	12	80	8	70	2

1 = Weak reaction of this antiserum was >40% and <100% cytotoxicity.
2 = Only undiluted antiserum gave 100% cytotoxicity.

by lymphoid antigen eluate in FCA and 14 days later without FCA. This antiserum after adsorption was capable of distinguishing red and white cell polymorphisms in human populations, being positive with 80 and 70% of the cells tested, respectively. The titres of antibody to red cells were markedly reduced by using only lymphoid material. After adsorption the titres were comparable with those of the animal that also received red cell ghosts.

Discussion

Repeated immunization with skin allografts or with blood infusions produce potent alloantisera in rhesus monkeys [1, 3, 22]. The specificity of these may be narrowed by either adsorption or serological selection of the donor recipient pairs, using heteroantisera and alloantisera [2, 16]. Skin or blood donors of the appropriate type are not always available as an antigen source, so that selection is not always possible. This may be overcome by using an antigen eluted from lymphoid tissue and stored freeze-dried. This retains its ability to elicit an antibody response over long periods. The dose of antigen required is relatively small [14] so that one monkey donor can be used to immunize many monkey recipients over a long period. The eluate can be used retrospectively in cytotoxic inhibition tests to re-evaluate serological reactions.

Attempts to raise human lymphocyte typing reagents in rabbits, guinea pigs, rats and dogs, by SHULMAN and his colleagues [21] failed but both chimpanzees [17, 18] and rabbit [8, 9] have been successfully immunized with human material. The antisera produced in rabbits when semi-purified HLA antigens were injected detected antispecies antibodies. EINSTEIN and his colleagues [14, 15], using small quantities of purified papain, solubilized human and rhesus monkey lymphoid antigens produced antisera in rabbits which after adsorption of the antispecies antibody yielded low titre antibodies with HLA specificity. They claimed that species specific and HLA antigens were in close proximity or on the same molecule, but were distinct antigenic determinants. In a four-week immunizing period we have produced antihuman lymphocyte sera in rhesus monkeys with similar titres to those produced in rabbits. These are capable of detecting polymorphisms in a human lymphocyte cell panel. The reaction patterns of none of the antisera have been correlated with known HLA or red cells antigens but this work is in progress. The monkey appears to offer a useful model to raise antilymphocyte-typing reagents in closely related species [19].

Acknowledgements

We wish to thank Prof. J. T. EAYRS for the use of the Primate Facilities and all primate users and colony staff for their co-operation in these experiments. We had excellent technical assistance from Miss LINDA KING. The work was supported financially by G. D. Searle Research, UK and the United Birmingham Hospitals Endowment Research Fund.

References

1 BALNER, H.; DERSJANT, H., and ROOD, J. J. VAN: Leukoagglutinating isoantibodies in rhesus monkeys. Transplantation *3:* 402–422 (1965).

2 Balner, H.; Leeuwen, A. von; Dersjant, H., and Rood, J. J. van: Defined leucocytoantigens of chimpanzees. Use of chimpanzees isoantisera for leucocyte typing in man. Transplantation *5:* 624–642 (1967).

3 Balner, H.; Dersjant, H.; Leeuwen, A. van, and Rood, J. J. van: Identification of two major leukocyte antigens of rhesus monkeys and their relation to histocompatibility; in Histocompatibility testing 1967, pp. 257–276 (Munksgard, Copenhagen 1967).

4 Balner, H.; Barnes, A. D.; D'Amaro, J.; Darrow, C. C.; Demant, P.; Gabb, B. W.; Piazza, A.; Metzgar, S.; Seigler, H. F.; Rogentine, G. N.; Rood, J. J. van; Ward, F., and Zweerus, R. L.: Joint report on first histocompatibility workshop on primates. Transplant. Proc. *4:* 141–147 (1972).

5 Barnes, A. D. and Hawker, R. J.: Leukocyte antigens in baboons. A preliminary to tissue typing for organ grafting. Transplant Proc. *4:* 37–42 (1972).

6 Barnes, A. D.; Hawker, R. J.; Cowdall, E. I.; Dukes, H.; Fejfar, J., and Griffin, J.: The production of anti-human thymocyte antiserum in various species. Behring Inst. Mitt. 51 (in press).

7 Barnes, A. D.; Hawker, R. J.; Cowdall, E. I.; Dickerson, C.; Dukes, H.; Griffin, J.; Crozier, C.; Fejfar, J.; Robinson, B. H. B., and Blainey, J. D.: The production of immunosuppressive antisera. Proc. 9th EDTA Congr., Florence 1972.

8 Batchelor, J. R.: The relationship between HLA specificity and human species antigen. Transplantation *7:* 554–561 (1969).

9 Batchelor, J. R. and Sanderson, A. R.: HLA substances. Properties and immunization of rabbits. Transplant. Proc. *1:* 489–490 (1969).

10 Binns, R. M.; Simpson, E.; Nehlsen, S. L., and Ruszkiewicz, M.: A method for producing immunosuppressive antilymphocyte serum in pigs and ruminants. Transplant. Proc. *3:* 784–787 (1971).

11 Cepellini, R.; Celada, F.; Mattiuz, P. L., and Zanalda, A.: Study of the possible correlation between blood antigens and histocompatibility in man. I. Production of leukoagglutinins by repeated transfusions from one donor. Ann. N.Y. Acad. Sci. *120:* 335–347 (1964).

12 Colombani, J.; Colombani, M., and Dausset, J.: Leukocyte antigens and skin homograft in man. Demonstration of humoral antibodies after homografting by the antiglobulin consumption test. Ann. N.Y. Acad. Sci. *120:* 307–312 (1964).

13 Davies, D. A. L.; Manstone, A. J.; Viza, D. C.; Colombani, J., and Dausset, J.: Human transplantation antigens. The HLA (Hui) system and its homology with the mouse H-2 system. Transplantation *6:* 571–586 (1968).

14 Einstein, A. B., jr.; Mann, D. L.; Gordon, H. G., and Fahey, J. L.: The immune reactivity of heterologous antisera against solubilized human lymphoid membrane component. Tissue Antigens *1:* 209–218 (1971).

15 Einstein, A. B. jr.; Mann, D. L.; Gordon, H. G.; Trapani, R. J., and Fahey, J. L.: Heterologous antisera against specific HLA antigen. Transplantation *12:* 299–304 (1971).

16 Hawker, R. J.; Barnes, A. D., and King, L. M.: Raising leucocyte alloantisera in limited populations of subhuman primates. Tissue Antigens (in press).

17 Metzgar, R. S. and Seigler, H. F.: Immunological studies on the human response to chimpanzees tissue antigens; in Goldsmith and Moor-Jankowski Medical primatology 1970, pp. 2–11 (Karger, Basel 1972).

18 METZGAR, R. S.; ZMIJEWSKI, C. M., and AMOS, D. B.: In AMOS Histocompatibility testing 1965, p. 45 (National Academy of Sciences, Washington, D.C. 1965).
19 SANDERSON, A. R.: Personal commun. (1972).
20 SHANBROM, E.; FEINGOLD, E.; SHERPHERD, L., and WALFORD, R. L.: Antisera for tissue typing. The production of cytotoxic and agglutinating antibodies following intradermal leukocyte injections in man. Blood *32:* 402–411 (1968).
21 SHULMAN, N. R.; MARDER, V. J.; HILLER, M. C., and COLLIER, E. M.: Platelet and leukocyte isoantigens and their antibodies. Serologic, physiologic and clinical studies. Progr. Haemat., vol. 4, pp. 222–304 (Grune & Stratton, New York 1964).
22 WIENER, A. S.; MOOR-JANKOWSKI, J.; BALNER, H., and GORDON, E. B.: Blood groups in monkeys, demonstrated with antisera produced to combined skin transplantation and isoimmunization with red cells. Int. Arch. Allergy *34:* 386–391 (1968).

Author's address: Dr. A. D. BARNES, Transplantation Immunology Research Unit, Department of Surgery, Queen Elizabeth Hospital, University of Birmingham, *Birmingham B15 2TH* (England)

Medical Primatology 1972. Proc. 3rd Conf. exp. Med. Surg. Primates, Lyon 1972, part II, pp. 100–106 (Karger, Basel 1972)

Immunological Aspects of Kidney Cell Fractions with Immunosuppressive Properties

J. H. GROENEWALD, H. D. BREDE and H. W. WEBER

Departments of Surgery, Microbiology and Anatomical Pathology, Faculty of Medicine, University of Stellenbosch and Tygerberg Hospital, Bellville

The literature contains many reports on 'enhancement', 'tolerance' or 'prolonged survival' of allografts induced by differing regimes involving the pretreatment of recipients with prospective donor antigens. OWEN *et al.* [7] reported at the First International Congress of the Transplantation Society, Paris, June 1967, on the Prolongation of Rabbit Kidney Allograft Survival by 'Desensitisation. They found that small repeated doses of subcellular donor antigens given over a prolonged pretreatment period significantly prolonged survival of a functioning orthotopically-transplanted rabbit kidney allograft. Delayed onset or the rejection process was observed biochemically and histologically. We have followed the same lines using the South African Cape chacma baboon *(Papio ursinus)* as experimental animal. In 1968 we reported that ultrasonically-disintegrated subcellular kidney cell fractions in small doses (2 ml) prolonged renal allograft survival, and that large doses (20 ml) gave an accelerated rejection phenomenon, hallmarked by a high incidence of intrarenal vascular thromboses [3].

Subsequently we have further investigated baboon kidney cell fractions by modifying the antigenicity of these cell fractions in the manner described by LOISELEUR [5] to create what he termed contra-antigens. In 1938 LOISELEUR [6] at the Pasteur Institute in Paris, changed immunogen-like serum proteins. snake-venom and diphtheria toxin by treatment with ultrasonic energy and by change of the electric charge in such a way that the product of this treatment combined with the original antigen in the form of precipitates. Neutralization of toxins occurred at the same time. A year later, LOISELEUR's experiments were repeated and confirmed by SCHULZE [8] of the Behring Werke in Marburg. We applied LOISELEUR's method on pooled baboon kidneys and achieved prolonged renal allograft survival up to 40 days and longer in baboons pretreated for four weeks prior to transplantation with 2 ml 48-hourly injections. The mean survival of untreated control animals was 9.6 days.

Material and Methods

The altered kidney cell fractions or so-called 'contra-antigens' used in these experiments were prepared from pooled baboon kidneys removed under sterile conditions according to the methods described by LOISELEUR [5, 6].

The capsules and connective tissue from the kidneys were removed. The pulps were homogenized and filtered through a single layer of gauze. The pH was set at between 7, 8 and 8,4.

Disintegration in the ultrasonic disintegrator was carried out for one hour, avoiding excessive heat by keeping the suspension in an ice bath. The suspension temperature never exceeded 35 °C. The mixture was centrifuged at 10,000 rpm for 15 min and the pH of the supernatant (which should be clear) was set at 3 using N/1 HCl. It was then dialized for 24 h against N/5,000 HCl containing 0.7 g NaCl/100 ml. The pH was then set at 7.4 and Streptomycin added to a final concentration of 5,000 g/ml, following which the product was dispensed and sterility tests performed.

The evaluation was carried out in a baboon transplantation model, where kidneys were exchanged between healthy, conditioned ABO compatible baboons. Baboons from remote troops were paired for transplantation to avoid the possibility of inbreeding as this would prolong kidney graft survival [4]. The transplantation model and methods of evaluation have been previously described [2]. Only animals acceptable by the protocol stipulations were included in the series.

The sterile product as described above was administered in the following doses:

(1) 0.2 ml contra-antigen i.m. every second day *post op:* 11 baboons;
(2) 1.0 ml contra-antigen i.m. every second day *post op:* 8 baboons;
(3) 2.0 ml contra-antigen i.m. every second day *post op:* 16 baboons;
(4) 5.0 ml contra-antigen i.m. every second day *post op:* 4 baboons;
(5) 20 ml contra-antigen i.m. every second day *post op:* 12 baboons;
(6) 2.0 ml contra-antigen i.m. every second day for four weeks before, and similarly after transplantation until death due to rejection occurred: 15 baboons.

Results

Survival

Figure 1 illustrates graft survival of the animals receiving 0.2 and 20 ml 'contra-antigen' every second day as compared to control animals receiving

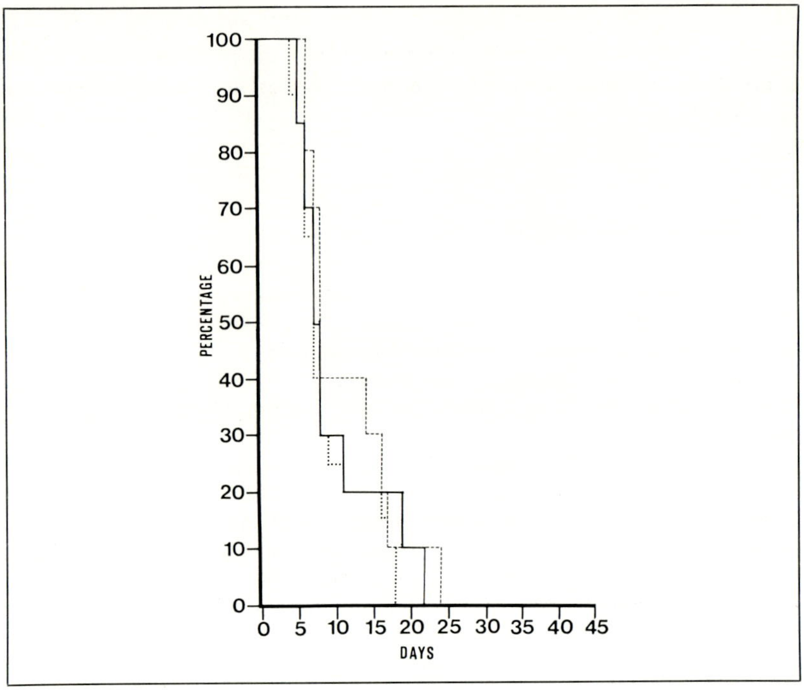

Fig. 1. Percentage survival of 0.2 (---, n = 10) and 20 ml (—, n = 10) contra-antigen-treated allotransplanted baboons. .... Controls (n = 20).

no treatment, and demonstrates that these two extreme dosages employed did not alter graft survival.

Figure 2 similarly illustrates the status with 1-ml and 5-ml doses, and although some prolongation of graft survival is noted it was not statistically significant.

In Figure 3 the 2-ml dosage every second day postoperatively and 2-ml dosage every second day pre- and postoperatively are shown. In both instances prolonged survival is noted. In the pre- and postoperative treatment group the results were statistically significant.

Most of the animals maintained reasonable kidney function until acute rejection induced renal failure. Immediately before rejection a decline in creatinine clearance and renal plasma flow was usually observed.

While total lymphocyte levels usually fell in the initial postoperative periods they usually recovered prior to rejection.

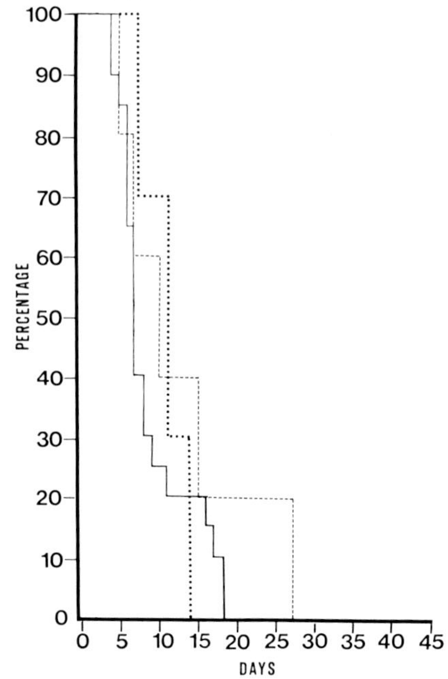

Fig. 2. Percentage survival of 1 (---, n=5) and 5 ml (..., n=5) contra-antigen-treated allotransplanted baboons. —— Controls (n=20).

Histology

As compared to untreated controls a much milder histological rejection reaction was observed in the contra-antigen-treated animals. There was little endothelial reaction with mild fibroid necrosis and less interstitial infiltration than usually observed in untreated controls. Few basophilic cells were seen in these infiltrations, but some eosinophilic cells were observed and occasional mast cells were also noted.

Serology Results

Figure 4 illustrated in graft form the titers of heterohemolysins, heterohemagglutinins and free complement levels of one of the longer surviving animals (E 106) in the pre- and posttransplant treatment groups reflecting

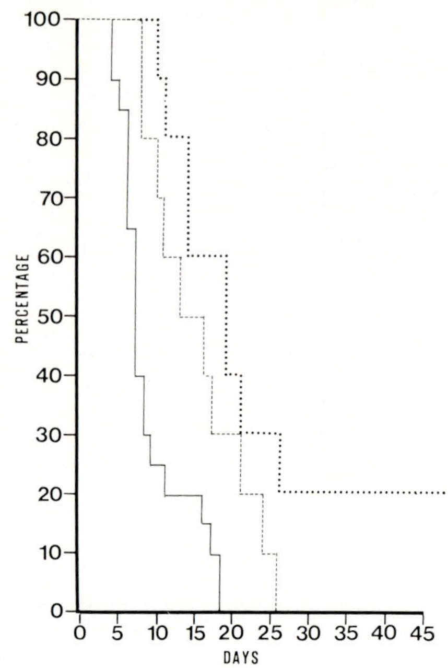

Fig. 3. Percentage survival of 2 ml pre- and postoperative (. . ., n = 15) and postoperative (---, n = 10) only contra-antigen-treated allotransplanted baboons. — Controls (n = 20).

a pattern of serological changes. Figure 4 shows that the baboon (E 106) before treatment had a normal free hemolytic complement titer varying between 1:32 and 1:64. During the preoperative treatment with 2ml contra-antigen i.m. every second day, the complement titer dropped to 1:16. We expect that contra-antigen absorbs complement. Under surgical operation the complement titer nearly diappears. This is typical and happens regularly during major surgery in baboons as also in human beings. Shortly after the kidney transplantation the complement titer returns to its preoperative level and slowly increases, showing that no complement binding rejection is under way. In the final rejection period the complement drops again as can be expected. The baboon (E 106) had a low titer of heterohemagglutinins (IgG) against sheep red cell. This titer is primarily not altered, it disappears in the rejection phase and reappears shortly before death. IgM hetero-hemolysins were primarily not present in the circulation of baboon E 106.

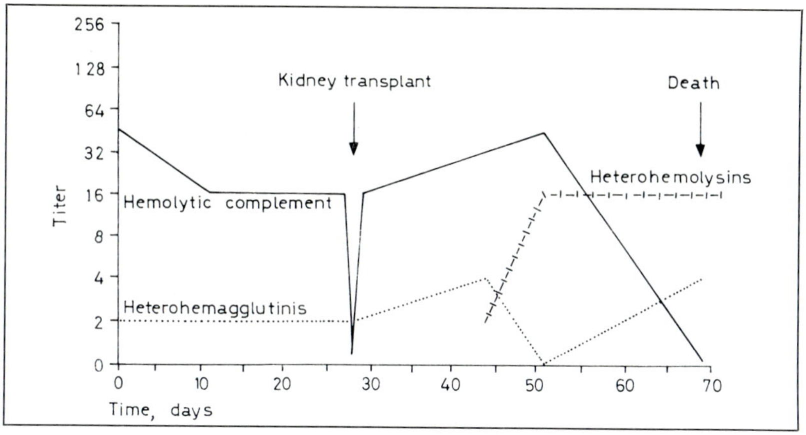

Fig. 4. Serological values of allotransplanted baboon (E 106) treated with contra-antigen, 2 ml every second day four weeks pre- and postoperatively.

As IgM antibodies are the first ones to develop, it may be assumed that their production is induced by the preoperatively-administered contra-antigen. We believe that they are absorbed to the contra-antigen. After four weeks of sensitization the IgM antibodies should be replaced by IgG antibodies which trigger a negative feed-back effect on the IgM production. This in our opinion may produce a moderate enhancement of the incompatible graft. About 40 days after the transplantation, IgM antibodies appear and coincide with the final rejection episode.

Serologically all groups showed reduced titers of heteroantibodies following transplantation except in the group receiving the 20-ml contra-antigen every second day, who showed some significant increases of heterohemolysins (IgM). In the longer surviving animals, low hemolytic complement titers were constantly observed.

Discussion

Contra-antigen enhanced the survival time of transplants considerably whenever it was administered in small amounts prior to transplantation. There was a decrease of free hemolytic complement, of heterohemolysins and of heterohemagglutinins. The question how contra-antigen works is not easy to answer. The following ways of working mechanisms have to be considered: contra-antigen is a compound of different chemically-modified antigens. We presume that contra-antigen injected in small amounts over a time of four weeks before transplantation may:

(1) Neutralize the bivalent antibody as postulated by EISEN and KARUSH [1].

(2) Form precipitates with the original antigens, covering the transplant for some time. We have not been able to prove this.

(3) Evokes blocking antibodies. As such antibodies are directed against chemically-modified antigens, they may only temporarily protect the graft.

(4) Create a delay of rejection by depleting the pool of 'designated lymphocytes' by continuous exposure to contra-antigen.

All these explanations are hypothetical. The fact remains that contra-antigen is able to prolong the graft survival. The exact mechanisms of this action has still to be elaborated.

References

1 EISEN, H. N. and KARUSH, F.: Immune tolerance and extracellular regulatory role for bivalent antibody. Nature, Lond. *202:* 677–682 (1972).

2 GROENEWALD, J. H.; ZYL, J. A. VAN; SCHOONEES, R.; ZYL, J. J. W. VAN, and MURPHY, G. P.: Methods of vascular anastomoses in baboon renal allotransplantations. Sth. afr. med. J. *17:* 3–4 (1968).

3 MURPHY, G. P.; BREDE, H. D.; WEBER, H. W.; GROENEWALD, J. H.; SCHOONEES, R.; ZYL, J. J. W. VAN; KLERK, J. N. DE, and ZYL, J. A. VAN: Treatment of baboon renal allografts with subcellular kidney cell fractions. Surg. Forum *19:* 215–217 (1968).

4 MURPHY, G. P.; WEBER, H. W.; BREDE, H. D.; RETIEF, F. P.; RETIEF, C. P.; ZYL, J. A. VAN; GROENEWALD, J. H., and ZYL, J. J. W. VAN: The significance of human ABO blood groups in the survival of untreated baboon renal allotransplants. Amer. Surg. *35:* 292–300 (1969).

5 LOISELEUR, J.: Sur la transformation d'un antigène en un contra-antigène spécifique. C. R. Soc. Biol. *129:* 172–176 and 250–253 (1948).

6 LOISELEUR, J.: Formation et propriétés de la fonction anticorps correspondant aux molécules organiques de faible poids moléculaire. Ann. Inst. Pasteur *78:* 151–176 (1950).

7 OWEN, E. R.; SLOME, D., and WATERSTON, D. J.: Prolongation or rabbit kindey allograft survival by 'desensitisation'; in Adv. Transplant. p. 385 (Munksgaard, Copenhagen 1968).

8 SCHMIDT, H.: Die künstliche *In vitro*-Herstellung von Antikörpern aus dem Antigen; in Fortschr. Serol., vol. 2, pp. 524–525 (Steinkopft, Marburg 1955).

Author's address: Dr. J. H. GROENEWALD, Department of Surgery University of Stellenbosch, P. O. Box 63, *Tiervlei C.P.* (South Africa)

Medical Primatology 1972. Proc. 3rd Conf. exp. Med. Surg. Primates, Lyon 1972, part II, pp. 107–116 (Karger, Basel 1972)

Pathology of Cardiac and Renal Allograft Rejection in the Rhesus Monkey

C. F. HOLLANDER and C. ZURCHER

Institute for Experimental Gerontology TNO, Rijswijk Z.H.

In man, obliterative arteritis seems to emerge as the critical factor in cardiac allograft survival [BIEBER *et al.*, 1970]. Obliterative arteritis also seems to be the limiting factor in renal allograft survival in man [KENNEDY and WEISSMAN, 1971].

In order to gain more insight into the mechanisms and morphology of chronic organ rejection and the influence of immunosuppressive treatment on this process, orthotopic, allogeneic, cardiac and renal transplantations were performed in randomly-selected rhesus monkeys of about 3 kg. The following experimental groups were established: a group of monkeys receiving no immunosuppressive treatment; a group receiving imuran, 4 mg/kg/day, i.m. for 36 days combined with ALS s.c., 2 ml/kg/3 times/week, for 60 days; furthermore, a cardiac allograft group was studied which received ALS as mentioned above, imuran i.m., 4 mg/kg/day for 28 days followed by 1 mg/kg/day until death, and from day 20 prednisone, 2 mg/kg/day, i.m., also until death.

All animals were completely autopsied when killed moribund or found dead. Tissues were fixed in 10% buffered formalin, embedded in paraffin, and routinely stained with hematoxylin-phloxin-saffron (HPS). Special stains were made when necessary.

Cardiac Allograft Survival and Morphology

The survival time of five *untreated animals* ranged from six to 27 days (mean survival, 13 days). At autopsy the hearts were dark red, edematous and enlarged. Microscopic examination of the hearts showed severe conges-

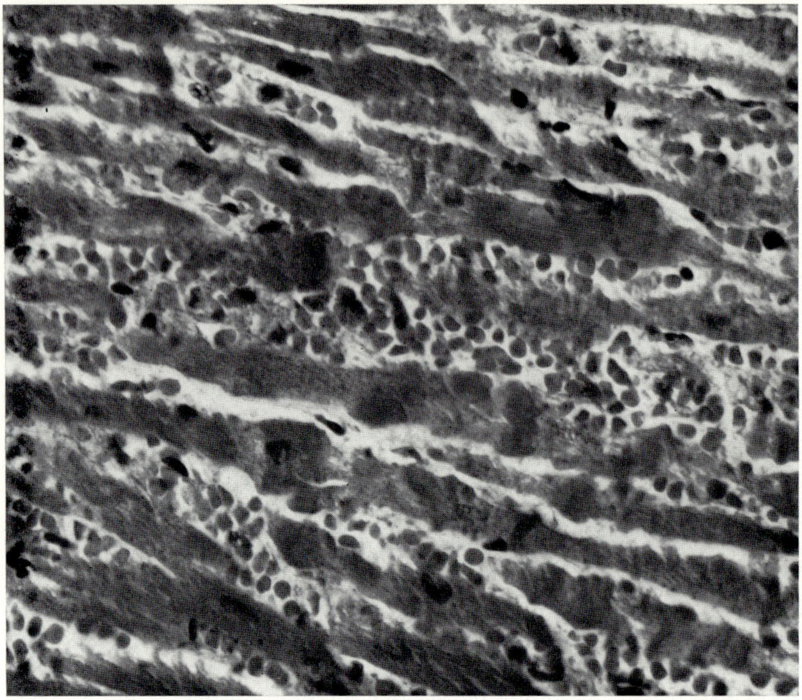

Fig. 1. Untreated monkey. Acute cardiac rejection at day 6 showing hemorrhages and muscle degeneration. HPS. ×450.

tion of small vessels with platelet thrombi and rupture of the wall. Extensive hemorrhages and degeneration of myocardial muscle were observed in conjunction with these vascular lesions (fig. 1). Also, a beginning mononuclear infiltration was seen in which the cells were mainly perivascularly localized. Only in the longest survivor did the main branches of the coronary arteries show signs of arteritis. These consisted of endothelial cell swelling, deposition of fibrinoid-like material on the intimal side, slight intimal proliferation and infiltration of the wall with poly- and mononuclear cells (fig. 2).

The survival time of 11 animals *treated with imuran and antilymphocyte serum (ALS)* ranged from 11 to 128 days (mean survival, 45 days). At autopsy the hearts were also enlarged, firm of consistence, with fibrinous pericarditis or its fibrous residue and showed a mottled brown-yellow myocardium. Histological examination revealed a diffuse infiltration of the

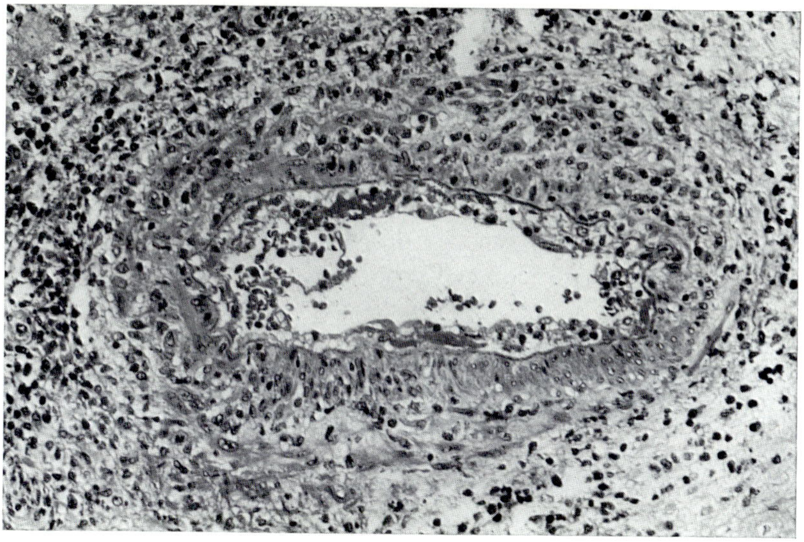

Fig. 2. Untreated monkey. Cardiac rejection at day 27 showing main branch of coronary artery with focal lesion (cf. text). HPS. ×180.

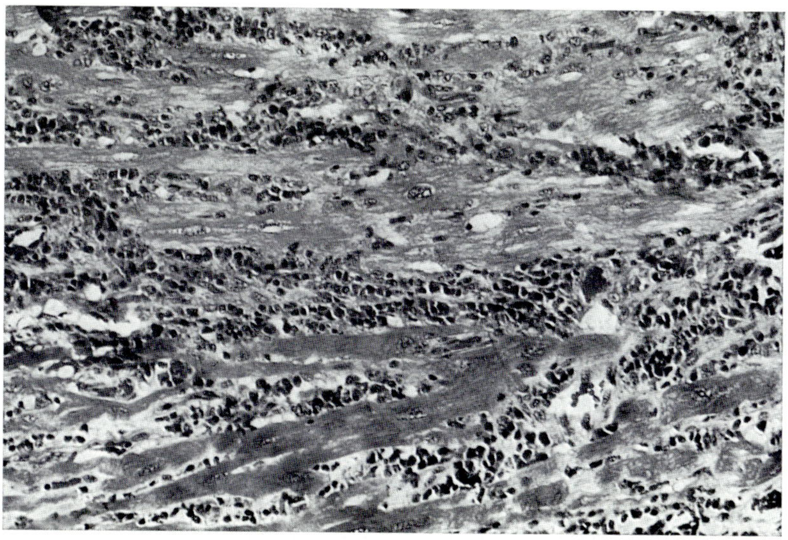

Fig. 3. ALS, imuran-treated monkey. Chronic cardiac rejection at day 76 showing mononuclear infiltration and muscle degeneration. HPS. ×155.

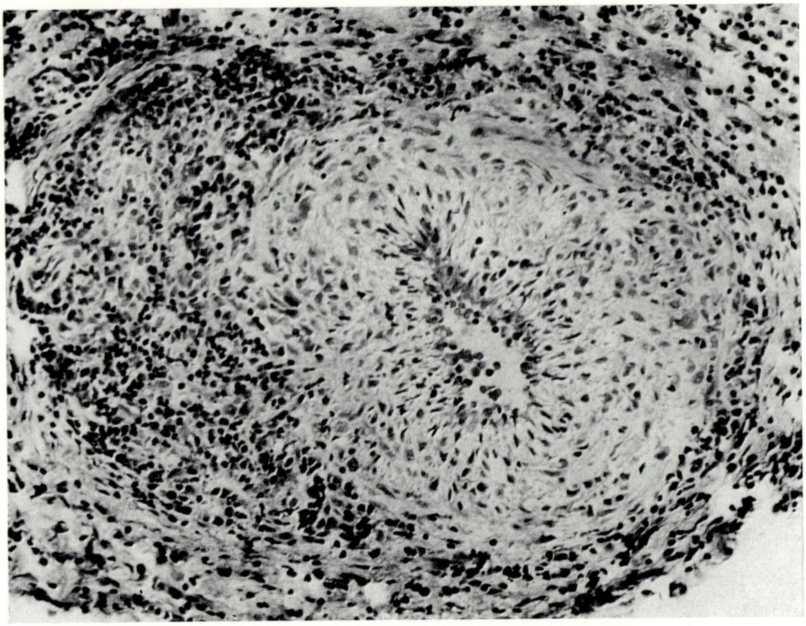

Fig. 4. ALS, imuran-treated monkey. Chronic cardiac rejection at day 76 (same case as figure 3) showing severe focal obliterative arteritis of main branch of coronary artery. HPS. ×170.

myocardium by mononuclear cells (lymphocytes, immunoblasts, plasma cells), myocytolysis and small areas of myocardial necrosis or scar tissue and minute hemorrhages (fig. 3).

Severe focal obliterative arteritis of the main branches of the coronary arteries with disappearance of the muscular wall and infiltration with mononuclear cells was found in all monkeys surviving beyond day 14 (fig. 4). In some cases no infiltrate was observed and the lumen of the vessel was nearly totally occluded. In the longest survivors even small intramural branches of the coronary arteries were affected (fig. 5). No atherosclerotic changes were found in the affected parts of the coronary arteries, nor in the intact parts.

The survival time of ten animals *treated with Imuran, ALS and prednisone* was the same, viz. mean survival 45 days, ranging from 10 to 104 days. Also, the histological picture was identical to that described for animals treated with imuran and ALS alone.

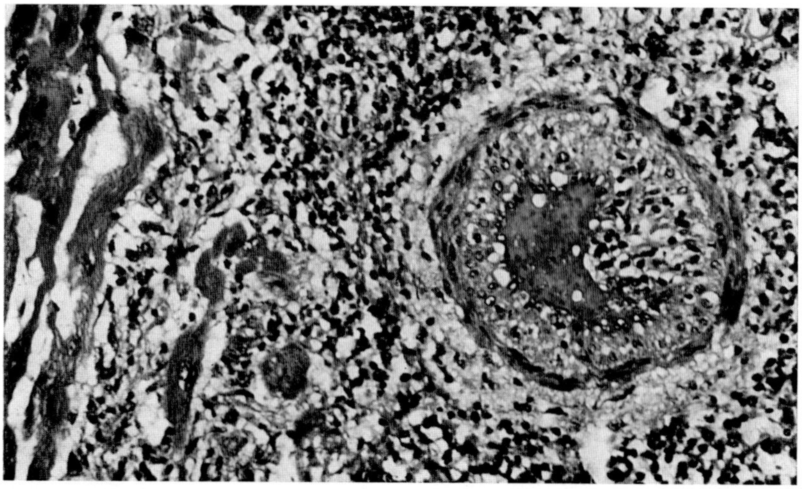

Fig. 5. ALS, imuran-treated monkey. Chronic cardiac rejection at day 128 showing focal lesion of intramural branch of a coronary artery. HPS. ×190.

Renal Allograft Survival and Morphology

The survival time of six *untreated animals* ranged from seven to 19 days (mean survival, 11 days). One additional monkey survived for 47 days. Grossly the kidneys were enlarged with patchy red and yellow areas on the cut surface. Histological examination of the kidneys showed intertubular capillary hemorrhage, interstitial edema and massive or patchy interstitial and perivascular, mononuclear infiltrates (fig. 6a and b). The infiltrate consisted of lymphocytes, immunoblasts and plasma cells, which penetrated the epithelium lining the tubules. Various degrees of tubular epithelium degeneration including frank necrosis were seen.

The endothelium of small vessels was swollen and the lumen narrowed. Interlobular and arcuate arteries showed swelling and vacuolization of the endothelium and increased eosinophilia of the media. In a number of cases, nuclear debris was observed in the media as well as vacuolization of muscle cells. A perivascular mononuclear infiltrate was seen. The interlobar arteries showed no changes. In a few cases the afferent arteriole was obstructed with fibrinoid material which extended into the glomerular tuft. Most glomeruli showed some increase of mesangial cells, swelling of the endothelium of the capillaries and swelling of the visceral and parietal epithelium. Signs of rejection of the ureter, probably due to vascular rejection, were observed in the

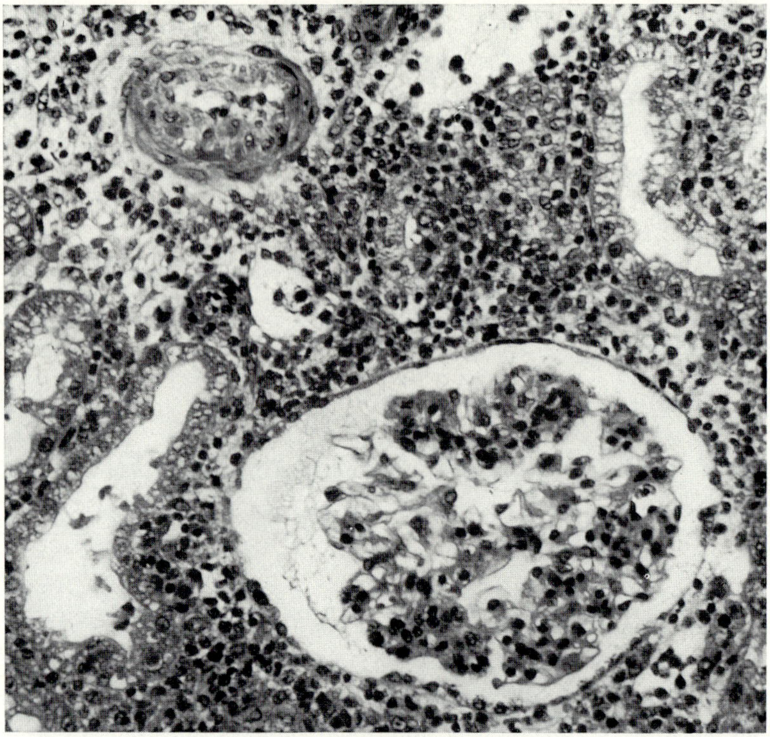

6a

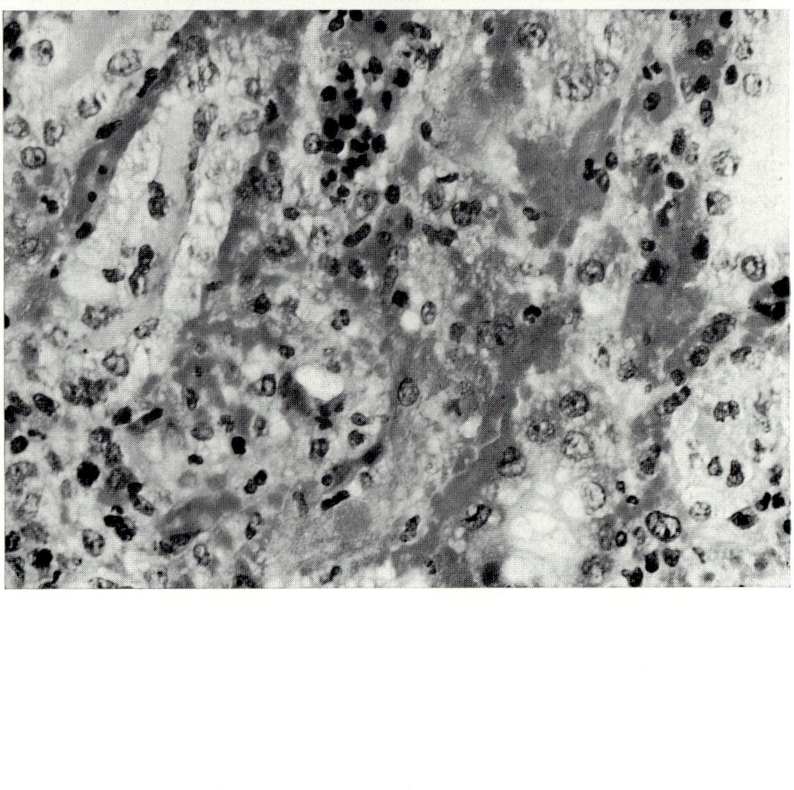

6b

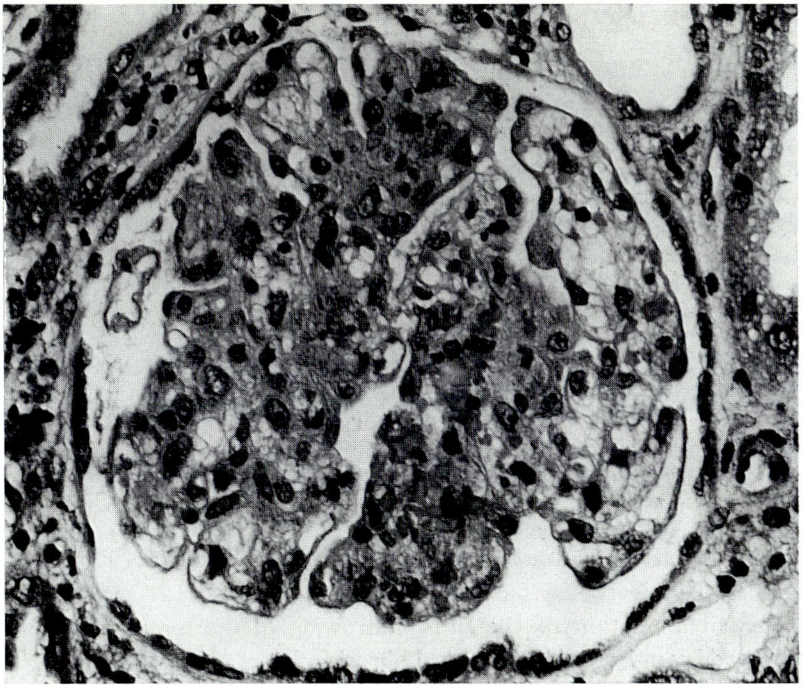

Fig. 7. ALS, imuran-treated monkey. Kidney rejection at day 91 showing chronic rejection glomerulopathy (cf. text). HPS. ×375.

untreated and treated monkeys [DICKE *et al.*, 1971]. The kidney surviving 47 days showed signs of chronic parenchymatous rejection without vascular pathology. In retrospect, this case appeared to be a good match.

The survival time of seven animals *treated with imuran and ALS* ranged from 13 to 91 days (mean survival, 35 days). In addition, two monkeys are still alive, i.e. more than two and a half years after transplantation. At autopsy, the kidneys of animals surviving beyond day 14 were enlarged and pale. Sometimes the surface was granular. Microscopic examination of kidneys of animals dying within 14 days showed a picture of acute rejection. However, the interstitial infiltrate was less than was found in untreated animals and was more localized. In the remaining cases, chronic rejection

---

Fig. 6. Untreated monkey. Acute kidney rejection at day 7 showing mononuclear infiltration, edema and inflammation of a small artery (*a* HPS; × 200) and hemorrhage (*b* HPS; ×350).

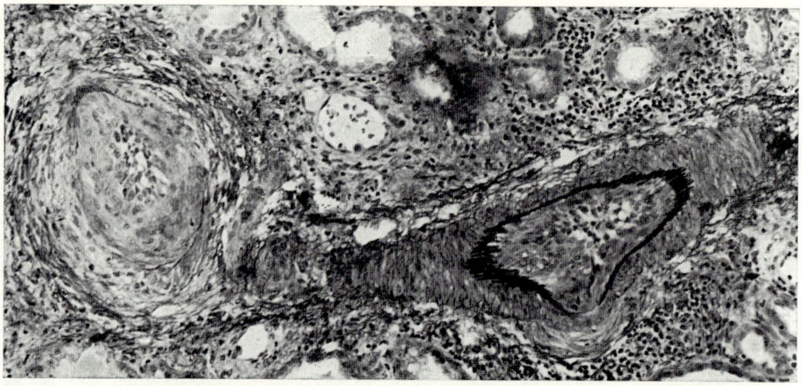

Fig. 8. ALS, imuran-treated monkey. Chronic kidney rejection at day 91 (same case as figure 7) showing interlobular artery with extensive focal obliterative arteritis and local involvement of media. Note also interstitial mononuclear infiltration. Elastic stain. ×130.

was observed with some features of acute rejection. Glomerular pathology consisted of increased mesangial area with some local hypercellularity, increased lobulation of the tufts, narrowing of the lumen of capillaries, thickened capillary walls and local fibrosis of tufts (fig. 7). These lesions are identical to those found in man [PORTER *et al.*, 1967; BUSCH, *et al.*, 1971].

Interlobular, arcuate and interlobar arteries often showed focal intimal proliferation with extensive narrowing of the lumen (fig. 8). Endothelial cells were swollen with polymorphic nuclei. In some instances, mononuclear cells or eosinophilic granulocytes were found lying within the affected areas (fig. 9).

In the interlobar arteries also the media and adventitia were often involved. Destruction of the internal elastic membrane as well as replacement of muscle fibers by fibrous tissue were seen (fig. 8). Some arteries showed only extensive adventitial fibrosis. Foam cells, described as occurring in man [DEODHAR and BENJAMIN, 1971; BUSCH *et al.*, 1971], were not observed. Only slight reduplication of the internal elastic membrane was seen.

Kidney biopsies taken on day 178 and 185, respectively from the animals still alive showed only slight chronic parenchymatous rejection.

Discussion

The lesions observed in acute and chronically-rejected hearts and kidneys in the rhesus monkey are similar to those described for acute and chronically-

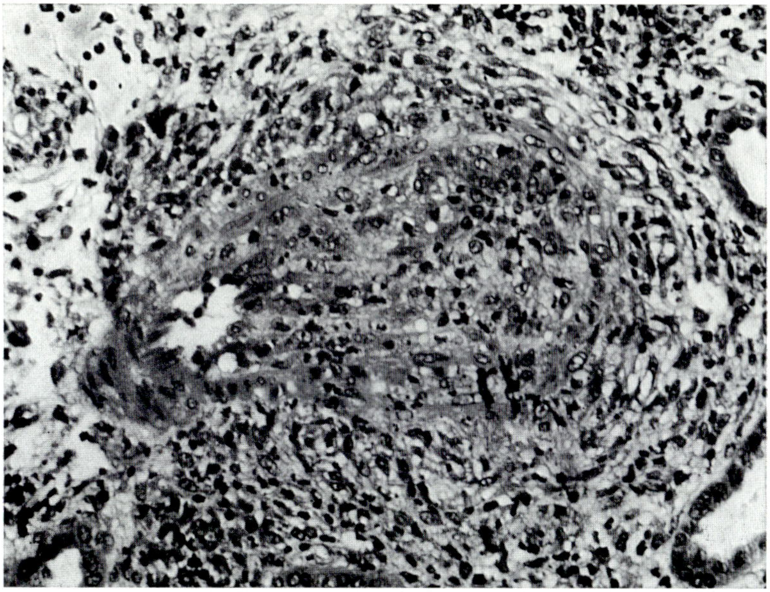

Fig. 9. ALS, imuran-treated monkey. Chronic kidney rejection at day 91 (same case as figures 7 and 8) showing interlobular artery with acute phase of arteritis. HPS. × 240.

rejected hearts and kidneys in man [BIEBER *et al.*, 1970; KNIERIEM *et al.*, 1971; MILAM *et al.*, 1970; STINSON *et al.*, 1969; DEODHAR and BENJAMIN, 1971; PORTER *et al.*, 1966, 1967]. As seems to be the case in man, an obliterative arteritis also appears to emerge as the limiting factor in cardiac allograft survival as well as in renal allograft survival in the rhesus monkey. No influence of the immunosuppressive treatment regimens on the occurrence of this obliterative arteritis has been observed. Because of similarities in the arterial lesions observed in chronically-rejected hearts and kidneys, the renal allograft model might be further explored to devise an immunosuppressive treatment regimen which will effectively prevent obliterative arteritis.

Acknowledgements

These studies were performed in collaboration with H. SNIJDERS, J. VINCENT, H. J. STOL, Z. AYTUG and H. F. L. BROM, Department of Thoracic Surgery, H. W. DICKE, Department of Surgery, University of Leyden, and D. W. VAN BEKKUM, H. BALNER, R. L. MARQUET, and G. A. HEYSTEK, Radiobiological Institute TNO, Rijswijk Z.H.

References

BIEBER, C. P.; STINSON, E. B.; SHUMWAY, N. E.; PAYNE, R., and KOSEK, J.: Cardiac transplantation in man. VII. Cardiac allograft pathology. Circulation *41:* 753–772 (1970).

BUSCH, G. J.; GALVANEK, E. G., and REYNOLDS, E. S., jr.: Human renal allografts. Analysis of lesions in long-term survivors. Hum. Pathol. *2:* 253–298 (1971).

DEODHAR, S. D. and BENJAMIN, S. P.: Pathology of human renal allograft rejection. Surg. Clin. N. Amer. *51:* 1141–1159 (1971).

DICKE, H. W.; MARQUET, R. L.; HEYSTEK, G. A., and BALNER, H.: Effect of immunosuppressive treatment and leukocyte antigen matching on kidney allografts in rhesus monkeys. Transplant. Proc. *3:* 484–487 (1971).

KENNEDY, L. J., jr. and WEISSMAN, I.: Dual origin of intimal cells in cardiac-allograft arteriosclerosis. N. Engl. J. Med. *285:* 884–887 (1971).

KNIERIEM, H. L.; MEESSEN, H. und SCHULTE, H. D.: Morphologische Befunde bei Herztransplantationen. Klin. Wschr. *49:* 837–852 (1971).

MILAM, J. D.; SHIPKEY, F. H.; LIND, C. J., jr.; NORA, J. J.; LEACHMAN, R. D.; ROCHELK, D. G.; BLOODWELL, R. D.; HALLMAN, G. L., and COOLEY, D. A.: Morphologic findings in human cardiac allografts. Circulation *41:* 519–535 (1970).

PORTER, K. A.; RANDALL, J. M.; STOLINSKI, C.; TERASAKI, P. I.; MARCHIORO, T. L., and STARZL, T. E.: Light and electron microscopic study of biopsies from thirty-three human renal allografts and isografts $1^3/_4$–$2^1/_2$ years after transplantation. Ann. N.Y. Acad. Sci. *129:* 615–636 (1966).

PORTER, K. A.; DOSSETOR, J. B.; MARCHIORO, T. L.; PEART, W. S.; RANDALL, J. M.; STARZL, T. E., and TERASAKI, P. I.: Human renal transplants-glomerular changes. Lab. Invest. *16:* 153–181 (1967).

STINSON, E. B.; DONG, E., jr.; BIEBER, C. P.; POPP, R. L., and SHUMWAY, N. E.: Cardiac transplantation in man. II. Immunosuppressive therapy. J. thorac. cardiovasc. Surg. *58:* 326–337 (1969).

Authors' address: C. F. HOLLANDER and C. ZURCHER, Institute for Experimental Gerontology TNO, Lange Kleiweg 151, *Rijswijk Z.H.* (The Netherlands)

Medical Primatology 1972. Proc. 3rd Conf. exp. Med. Surg. Primates, Lyon 1972, part II, pp. 117–124 (Karger, Basel 1972)

The Marmoset – A Tool for the Study of Transplantation Immunity Mechanisms

G. B. MAHOUY

Division of Infant Development, Clinical Research Centre, Harrow, Middlesex and Laboratoire d'Expérimentation animale, Institut de Recherches sur les Maladies du Sang, Paris

Introduction

Mechanisms of immunological 'tolerance' have recently been questioned. Although mice injected in the neonatal period with allogeneic bone marrow cells subsequently fail to reject donor strain skin grafts [3, 4], the original concept of tolerance in this situation appears somewhat doubtful. Similar 'tolerant' mice have been shown to react *in vitro* against target donor fibroblasts in cytotoxicity tests, but more important is the fact that this reaction can be inhibited by serum from these mice [8]. Tolerance can also be induced in adult animals by the injection of bone marrow following either irradiation or drug induced immunosuppression [11, 22]. Again, and in the case of irradiated dogs injected with allogeneic bone marrow, serum factors appear important in maintaining tolerance in these chimaeras [9]. Other artificially-derived chimaeras have been obtained by ovum fusion. TARKOWSKI [21] was the first to take fertilized morulae from two different parental strains and fuse these together, a technique that has been simplified by the use of pronase in the removal of the zona pellucida [12]. Following *in vitro* incubation, transformation into a 'giant' fused blastocyst and transplantation into a pseudopregnant recipient, tetraparental chimaeras were obtained with dual parental strain cell populations. Upon the basis of the persistence of parental strain skin grafts, MINTZ [13, 15] considered these mice to be models of 'permanent and intrinsic tolerance', but this view is now disputed. In such chimaeras a change in the proportion of parental strain cell populations occurs [14] and although conceivably this reflects differential rates of cell growth, this finding might be explained by a breakdown in tolerance. In similar chimaeras, parental strain skin grafts have been rejected [2, 14] and furthermore on occasion, tolerance has been replaced by graft-*versus*-host (GvH) disease [2]. Apart from the obvious lack of permanence, the 'intrinsic' nature of tolerance in this situation has also recently been questioned. Lymphoid cells from these chimaeras have been shown to react against parental strain cell lines in both cytotoxicity [23] and mixed lymphocyte reactions [18], and in each case the reactions are inhibited by serum factors present in the chimaeras. More recently other evidence has suggested that lymphocytes from these chimaeras act in the absence of serum as presensitized cells in a local GvH assay system [1, 2].

Tolerance has also been presumed in natural chimaeras. Cattle [16], sheep [5] and marmoset [10, 24] twin chimaeras have all been described. Marmosets have been studied here in respect to the mechanisms responsible for maintaining dual cell populations. We have examined marmoset chimaeras to learn whether cells from twins are capable of reacting against each other and, furthermore, whether serum factors play any part in inhibiting this reaction.

Materials and Methods

Marmosets and Establishment of Chimaerism

Seven twin pairs of marmosets of the same species *(Callithrix jacchus)* have been investigated. Chimaerism in the heterosexual twins was confirmed upon the basis of chromosomal analysis, whereas in the case of unisexual twins, skin grafting was performed. Macroscopic skin graft survival was determined and in every case compared with an autograft and occasionally with an allograft from another marmoset. The number and sex of the marmosets, together with the index of chimaerism, are shown in table I.

Cell Interactions

Peripheral blood was obtained by femoral puncture. The blood was used for chromosomal analysis, obtaining serum and a lymphocyte-rich cell suspension. For the preparation of the lymphocyte suspension, blood was taken into phenol-free heparin. The red cells were subsequently sedimented by the addition of an equal volume of 'Plasmagel' (Laboratoire Roger Bellon). The lymphocytes were obtained through a 96:40 gradient of 9% Ficoll (Pharmacia Ltd):33% Triosil (Glaxo Ltd.). The cell suspension was spun through the mixture at 3,000 rpm at room temperature for 20 min. The middle layer was carefully removed, washed and a viable cell count was performed using trypan blue. These viable cells were used in both popliteal lymph node weight gain assay and also in mixed lymphocyte reactions.

Table I. Number, sex and index of chimaerism of seven twin pairs of marmosets *(Callithrix jacchus)*

Number of animal	Sex	Age months[1]	Index of chimaerism		
			skin graft, days[1]		chromosomes
33	male	11	+	(223)	
34	female	11	+	(223)	
13	male	dead	+	( )[2]	
15	male	33	+	(204)	
21	male	dead	+	( )[2]	
39	male	22	+	(204)	
144	female	8			+
143	male	8			+
152	male	7			+
153	female	7			+
154	male	7	+	(62)	
155	male	7	+	(62)	
103	male	19	+	(62)	
104	male	19	+	(62)	

1 The age of animals as well as the survival time of the skin grafts are given as the time until the submission of the manuscript of June 21, 1972.
2 The marmosets 13 and 21 died one month after skin grafting.

As shown in figure 1, lymphocytes from one twin were injected into the footpad of the other. Eight days later both popliteal lymph nodes were carefully removed under light anesthesia and their wet weight determined. The weights were subsequently expressed as a ratio of injected/noninjected sides.

In the case of the mixed lymphocyte reactions, various experiments were performed on different occasions dependent upon the number of cells available. The different systems considered are shown in table II. The primary experiment was to investigate and learn of any *in vitro* interreaction of the two cell populations (reflected by their division and tritiated thymidine uptake). Secondly, the reaction was studied in the presence of the chimaeras own serum and compared with the effect of human serum and co-twin serum. The incidence of a foreign marmoset serum in this reaction was also investigated. Furthermore, these results were all compared with the cell-division inhibitory effect of mitomycin. Mitomycin (Glaxo Ltd.) was added to the cells prior to incubation and 0.2 μCi of tritiated thymidine was subsequently added on day 5. Forty-eight hours later the cells were spun and washed three times in Hank's solution. Following the final wash the cells were lysed in distilled water and the lysate applied to a small piece of filter paper. After drying, the papers were fixed and the radioactivity determined in a liquid scintillation counter.

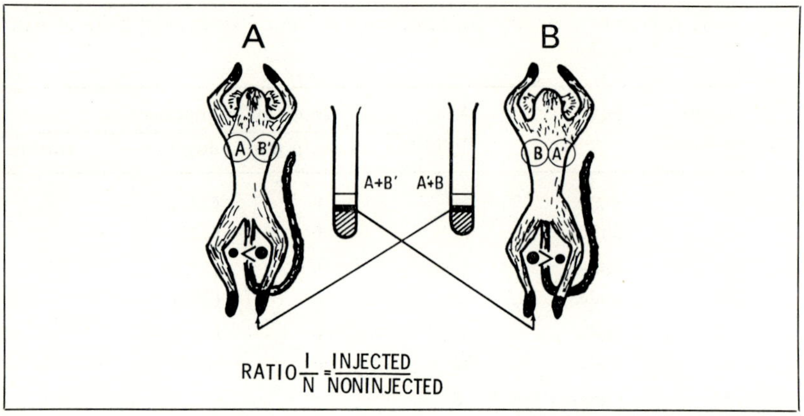

Fig. 1. Popliteal lymph node assay in marmoset chimaera. Weight gain eight days after footpad injection. The chimaeric marmoset A has a dual hemotopoietic cell population: its own cells A and cells B, coming from the co-twin B.

Table II. Various protocols in the mixed lymphocyte reaction

Scheme	Source of cells	Cells Present[1] (M-Mitomycin treated)	Serum[2]
1	A	A + B′	As
2	A	A + B′	Hs
3	A	$(A+B')^{M}$	Hs
4	A	A + B′	Bs
5	A	A + B′	Cs
and vice versa for B			

1 Usually, the total number of cells per culture was 0.5×10^{6} with 0.1 or 0.2 ml serum. But attempts have been successfully made with 0.25×10^{6} cells per culture.
2 As : own marmoset serum; Hs : fresh AB human serum; Bs : co-twin marmoset serum; Cs : foreign marmoset serum.

Results

Establishment of Chimaerism

Skin grafting was successful and unlike allografts, both autografts and twin skin grafts were not rejected. A set of grafts persistent at 46 days

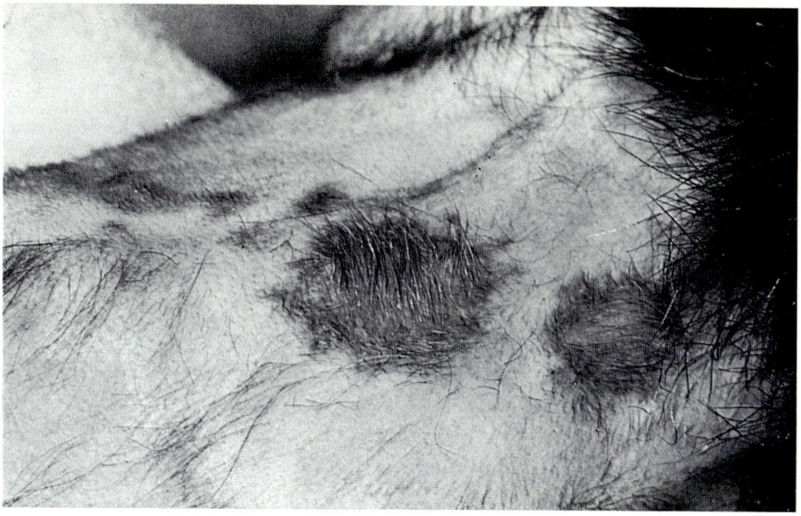

Fig. 2. Forty-six days skin grafts in marmoset. Graft from male into male co-twin, on the left side and autograft on the right side.

is shown in figure 2. In the case of heterosexual co-twin marmosets, the peripheral blood lymphocyte chimaerism was confirmed by sex chromosome analysis.

Popliteal Lymph Node Weight Gain Assay

The limited results in the popliteal lymph node weight assay reflected difficulty in the dissection of the lymph nodes, and for practical reasons this technique was subsequently abandoned. Out of the three attempts we have made, one was negative, and among the rest the best enlargement was noticed in one pair of twins (2.7 and 3.8 enlargement ratio of draining/non-draining popliteal lymph nodes). This enlargement could reflect two different reactions, viz.:

$$\left.\begin{matrix} A' \rightarrow B \\ B' \rightarrow A \end{matrix}\right\} \text{GvH reaction} \quad \text{and also} \quad \left.\begin{matrix} B \rightarrow A' \\ A \rightarrow B' \end{matrix}\right\} \text{homograft reaction}$$

The fact that necessary controls were not available in this previous experiment does not allow a definite conclusion to be drawn. This study needs further investigations with a larger number of animals.

Mixed Lymphocyte Reaction

The results in the mixed lymphocyte reaction, although limited by the numbers of cells available and the various protocols followed, are summarized in table III. These results clearly suggest the activity of a dual cell population *in vitro* and, furthermore, that this reaction can be inhibited by the presence of serum from either twin, but not by human serum. Although the inhibition was variable, this possibly reflected either addition of different amounts of serum or various quantities of serum factors capable of blocking the reactivity of the lymphocytes. However, it was interesting to note that on certain occasions, the inhibition due to the marmoset twin serum was comparable to the effect of mitomycin. Using a serum from a non-co-twin marmoset of the colony, the reaction was partially inhibited. This might suggest nonspecific 'blocking factors'. This problem is currently under investigations.

Table III. Summary of results in mixed lymplocyte reaction

Number of animal	Cells present (M-Mitomycin treated)	Serum added[1]	Radioactivity[2]
143	A+B′	As	6.6
		Hs	172.5
		Bs	35.1
	$(A+B')^M$	Hs	5.8
144	A+B′	As	3.0
		Hs	137.7
		Bs	68.5
	$(A+B')^M$	Hs	5.4
152	A+B′	As	37.5
		Hs	310.0
	$(A+B')^M$	Hs	3.8
153	A+B′	Hs	563.8
		Bs	6.4
		Cs	53.0
	$(A+B')^M$	Hs	6.3

This table shows only some of the results that have been obtained; the data not shown suggest a similar trend.

1 For the meaning, see table II.

2 Tritiated thymidine incorporation is expressed in cpm (count-less background/unit time).

Discussion

Chimaeras have been discribed to occur naturally in many animal species [17]. The marmosets, members of the *Callithricidae (Hapalidae)* family of New World monkeys, are characterized by a high incidence of twinning. Due to common early placental anastomosis (and/or other mechanisms), hemopoietic chimerism between the twins results. As in the case of the tetraparental mouse chimaera, changing cell populations have been noted in marmoset [7], and cattle [20] twins. The slow rejection of twin skin grafts in cattle chimaeras [19] is evidence of the instability of the control-maintaining balance ('tolerance') in the dual population. The preliminary findings suggest an analogous situation in marmosets and in the artificially-derived tetraparental mouse chimaera. In both situations, and in spite of the general persistence of skin grafts, their lymphocytes can react against self (twin and parental strain, respectively) antigens and it seems that this interreaction can be inhibited by an as yet unidentified serum factor. It is an attractive hypothesis to consider that this serum factor is an enhancement antibody and this is being currently investigated.

Acknowledgements

This study was carried out in Dr. R. D. Barnes' Laboratory at the Clinical Research Centre, Northwick Park, Harrow, Middlesex, in collaboration with Miss Maureen Tuffrey and Drs. A. C. Allison and C. Ford.

References

1 Barnes, R. D. and Tuffrey, M.: Tolerance in ovum fusion derived tetraparental mouse chimaeras. Proc. 4th Int. Congr. on Lymphatic Tissue and Germinal Centers in Immune Reactions, B. D. Jankobic, Ed. (Plenum press, New York 1972).

2 Barnes, R. D.; Tuffrey, M.; Kingman, J.; Thornton, C., and Turner, M. W.: The disease of the NZB mouse. Examination of ovum fusion derived tetraparental NZB-CFW chimaeras. Clin. exp. Immunol. *11:* 605–628 (1972).

3 Billingham, R. E.; Brent, L., and Medawar, P. B.: Actively acquired tolerance of foreign cells. Nature, Lond. *172:* 603 (1953).

4 Billingham, R. E.; Brent, L., and Medawar, P. B.: Quantitative studies on tissue transplantation immunity. III. Actively acquired tolerance. Philos. Trans. B, *239:* 357 (1956).

5 Dain, A. R. and Tucker, E. M.: Chromosome and blood group studies of sheep twin chimaeras. J. Physiol., Lond. *200:* 37P (1969).

6 FORD, W. L.; BURR, W., and SIMONSEN, M.: A lymph node weight assay for the graft-*versus*-host activity of rat lymphoid cells. Transplantation *10:* 258–266 (1970).

7 GENGOZIAN, N.; BATSON, J. S.; GREENE, C. T., and GOSSLEE, D. G.: Hemopoietic chimerism in imported and laboratory bred marmosets. Transplantation *8:* 633–652 (1969).

8 HELLSTRÖM, I.; HELLSTRÖM, K. E., and ALLISON, A. C.: Neonatally induced allograft tolerance may be mediated by serum-borne factors. Nature, Lond. *230:* 49–50 (1971).

9 HELLSTRÖM, I.; HELLSTÖM, K. E.; STORB, R., and THOMAS, E. D.: Colony inhibition of fibroblasts from chimeric dogs mediated by the dogs' own lymphocytes and specifically abrogated by their serum. Proc. nat. Acad. Sci., Wash. *66:* 65–71 (1970).

10 HILL, J. P.: Second Croonian lecture. The developmental history of the primates. Philos. Trans. B, *221:* 45–178 (1932).

11 LINDSLEY, D. L.; ODELL, T. T., and TAUSCHE, F. G.: Implantation of functional erythropoietic elements following total-body irradiation. Proc. Soc. exp. Biol. Med. *90:* 512 (1955).

12 MINTZ, B.: Incorporation of nucleic acid and protein precursors by developing mouse eggs. Amer. Zoologist *2:* 432 (1962).

13 MINTZ, B. and PALM, J.: Erythrocyte mosaicism and immunological tolerance in mice from aggregated eggs. J. Cell. Biol. *27:* 66A (1965).

14 MINTZ, B. and PALM, J.: Gene control of hematopoiesis. I. Erythrocyte mosaicism and permanent immunological tolerance in allophenic mice. J. exp. Med. *129:* 1013–1027 (1969).

15 MINTZ, B. and SILVERS, W. K.: 'Intrinsic' immunological tolerance in allophenic mice. Science *158:* 1484–1487 (1967).

16 OWEN, R. D.: Immunogenetic consequences of vascular anastomoses between bovine twins. Science *102:* 400–401 (1945).

17 OWEN, R. D.: Immunological tolerance. Fed. Proc. *16:* 581 (1957).

18 PHILLIPS, S. M.; MARTIN, W. J.; SHAW, A. R., and WEGMANN, T. G.: Serum-mediated immunological non-reactivity between histocompatible cells in tetraparental mice. Nature, Lond. *234:* 146–148 (1971).

19 STONE, W. H.; CRAGLE, R. G.; SWANSON, E. W., and BROWN, D. G.: Skin grafts. Delayed rejection between pairs of cattle twins showing erythrocyte chimerism. Science *148:* 1335–1336 (1965).

20 STONE, W. H.; FRIEDMAN, J., and FREGIN, A.: Possible somatic cell mating in twin cattle with erythrocyte mosaicism. Proc. nat. Acad. Sci., Wash. *51:* 1036 (1964).

21 TARKOWSKI, A. K.: Mouse chimaeras developed from fused eggs. Nature, Lond. *190:* 857–860 (1961).

22 THOMAS, E. D.; STORB, R.; EPSTEIN, R. B., and RUDOLPH, R. H.: Symposium on bone-marrow transplantation. Experimental aspects in canines. Transplant. Proc. *1:* 31 (1969).

23 WEGMANN, T. G.; HELLSTRÖM, I., and HELLSTRÖM, K. E.: Immunological tolerance 'Forbidden clones' allowed in tetraparental mice. Proc. nat. Acad. Sci., Wash. *68:* 1644–1647 (1971).

24 WISLOCKI, G. B.: Observations on twinning in marmosets. Amer. J. Anat. *64:* 445–483 (1939).

Author's address: Dr. G. MAHOUY, Laboratoire d'Expérimentation animale Hôpital Saint-Louis, *F-75010 Paris* (France)

Medical Primatology 1972. Proc. 3rd Conf. exp. Med. Surg. Primates, Lyon 1972, part II, pp. 125–133 (Karger, Basel 1972)

Heterotopic Heart and Kidney Transplantation in Rhesus Monkeys

R. L. MARQUET, G. A. HEYSTEK and D. W. VAN BEKKUM

Radiobiological Insitute TNO, Rijswijk Z. H.

Introduction

In previous investigations from this laboratory it has been shown that among different immunosuppressive regimens thus far studied, the best immunosuppression to prolong kidney allograft survival in rhesus monkeys consisted of the combined treatment with antilymphocyte serum and imuran [8].

The effect of treatment with ALS and imuran on heterotopic heart allograft survival has been studied extensively in rats [4]. It was found that prolongation of heart allograft survival was less pronounced than the survival of kidney grafts. This discrepancy in survival pattern was attributed to specific properties of the respective organs, rather than to differences in surgical techniques. To investigate whether the similar organ specific differences would exist in rhesus monkeys, a comparative study on the survival times of kidney and heterotopic heart allografts in immunosuppressed monkeys was undertaken. Furthermore, the value of the heterotopic heart transplantation model was tested by comparing its results with those obtained in orthotopic heart allografting in rhesus monkeys.

In our institute, unrelated monkeys typed for the RhL-A specificities have been employed in various transplantation experiments [2]. Without immunosuppressive treatment, skin grafts exchanged between incomplete phenotypically identical (IPI) monkeys showed marginal prolongation of survival times, while the severity of the graft *versus* host reaction in bone marrow transplantation experiments between IPI individuals was not modified. Preliminary results about kidney allografting between IPI monkeys have been published by DICKE *et al.* [8]; in a small group of four untreated animals one monkey survived for 35 days. The present series of ten kidney allografted, untreated, unrelated, IPI monkeys is intended as an extension of the above-mentioned group.

Material and Methods

Animals

Rhesus monkeys *(Macaca mulatta)* of both sexes, weighing 2.5–4 kg, were used in the transplantation experiments. After having been imported the animals were held in quarantine for two months during which they were freed from infective diseases [3]. Twenty-four hours before grafting, food was withheld from the monkeys; after transplantation no food limitations were imposed. Apart from immunosuppression and antibiotics, no additional medication was given to the animals.

During kidney transplantation experiments we have observed remarkable differences in survival time between grafted rhesus monkeys, derived from India or Pakistan (fig. 1). Untreated monkeys, imported from India (ten animals) had a mean kidney allograft survival time of 11.5 days, while the mean survival of ten untreated Pakistan derived animals was 22 days. In the Pakistan group, no correlation with the occurrence of compatible RhL-A specificities between donor and host could be determined and thus far no explanation for this finding could be given. All the results to be reported were obtained with rhesus monkeys imported from India, because these animals showed a consistent survival pattern, comparable with kidney graft survival in other animal species.

Transplantation Techniques and Graft Evaluation

Kidney transplantation was performed as has been described previously by DICKE *et al.* [6, 7]. After removal from the donor the kidney grafts were perfused during 10 min at 4 °C with a low molecular weight dextran solution which contained heparin and procaine. Left kidneys were exchanged and contralateral nephrectomy was performed at the time of operation. The total ischemic time could be limited to 20 min. The technical success rate was 70%; failures were mainly due to arterial thrombosis. Functional evaluation of the renal grafts was assessed by regular determinations of blood urea nitrogen (BUN) and plasma potassium levels. Rejection and functional failure of the allografts was reflected by a progressive rise in BUN (up to 450 mg/100 ml) associated with rises in plasma potassium which in terminal cases could increase to 8.0 mEq/l. The hematological response to grafting, kidney function and immunosuppression was assessed by differential blood counts and determinations of hematocrit values. The death of the transplanted animal was taken as the endpoint of graft survival.

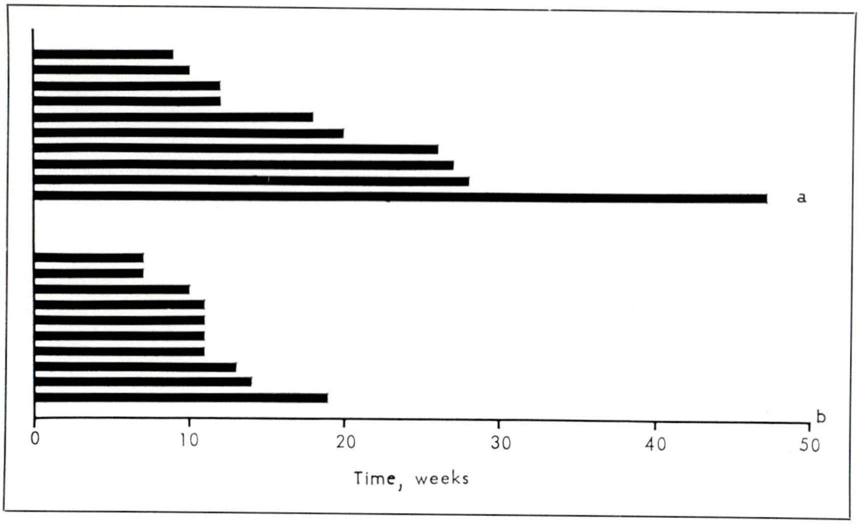

Fig. 1. Renal allograft survival in rhesus monkeys. A comparison between the survival times obtained in animals from *a* Pakistan and *b* India. The animals were not immunosuppressed or selected.

Heterotopic heart allografting was performed intra-abdominally, as a modification of the technique described by MANN *et al.* [10], ressembling the ectopic heart transplantation technique which is routinely employed in rats [4]. The donor animal was heparinized. After careful ligation of caval and pulmonary veins and transsection of the aortic root and pulmonary artery, the donor heart was excised and cooled by perfusion of the coronary arteries with Hanks balanced salt solution at 4 °C. In the recipient monkey, the abdominal aorta and inferior caval vein were dissected and clamped separately. Appropriate elliptical openings in both vessels were made to receive the grafts left subclavian artery and pulmonary artery in end-to-side fashion, employing 6–0 silk sutures. Air was expelled from the coronary system by terminal perfusion with Hanks solution after which the innominate artery was ligated and the aortic arch was closed by continuous suturing.

Rhythmic heart contractions occurred spontaneously as the graft regained circulation following removal of the clamps. The transplant was embedded retroperitoneally. The total ischemic time ranged from 45 to 60 min. Function of the heterotopic transplanted heart was assessed by regular ECG recordings and by palpating through the anterior abdominal wall. Palpation appeared to be a more reliable indicator of the state of the graft, since ECG changes in the course of graft survival were rather inconstant and difficult to interpret. Only when precordial leads were recorded from fixed points of the abdominal wall, measurements of R-wave voltages could sometimes give additional information about impending rejection. Terminal rejection was assumed to have taken place when there was an extremely weakened or absent beat with an associated abnormal or flat ECG.

Immunosuppression

Antilymphocyte serum (ALS) was produced as a horse antirhesus thymocyte serum according to the adjuvant method of MONACO and co-workers as described by VAN BEKKUM *et al.* [5].

The immunosuppressive potency of the ALS was tested in skin allografted rhesus monkeys. Injections of 4 ml/kg on days –2, 0, +2 and +4 (day 0 being the day of transplantation), produced a mean skin graft survival time of 35 days. Treated heart or kidney allografted monkeys received 2 ml/kg ALS s.c., 3 times/week for 60 days, starting on the day of transplantation. Azathioprine (imuran) was given from the day of grafting in daily dosages of 4 mg/kg, i.m. for 36 days.

Donor-Host Selection

IPI rhesus monkeys were selected by typing for leukocyte antigens with a panel of 60 isoantisera, defining 14 RhL-A groups and subsequently with 50 sera of as yet undefined specificity [1]. Pairs of animals were considered to be IPI when compatible distribution patterns with both groups of antisera could be obtained.

Results

Nine kidney allografted untreated rhesus monkeys survived 7–47 days (fig. 2). The median survival time amounted to 11 days. In this group, after transplantation, kidney function never returned to normal but deteriorated increasingly, resulting in death of the animals.

In the series in which untreated IPI combinations were employed (ten animals), kidney allograft survival ranged from eight to 35 days, 50% of the animals survived for 11 days; as compared to the survival pattern in the untreated, unselected group no differences could be observed (fig. 2).

The survival times of ten kidney allografted animals treated with ALS and imuran varied from nine days to two and a half years; 50% of the monkeys survived more than 30 days (fig. 3). In the two extremely long surviving animals BUN levels regained their normal values (44±8 mg %) within three weeks after transplantation, whereas it took seven weeks before the serum potassium contents dropped to normal pretransplant values (3–3.8 mEq/l). The rest of the treated monkeys showed poor kidney function during the whole course of graft survival.

Rhesus monkeys grafted heterotopically with allogeneic hearts survived 5, 7, 9, 9 and 10 days, respectively. From days 3–5 onwards the heartbeat

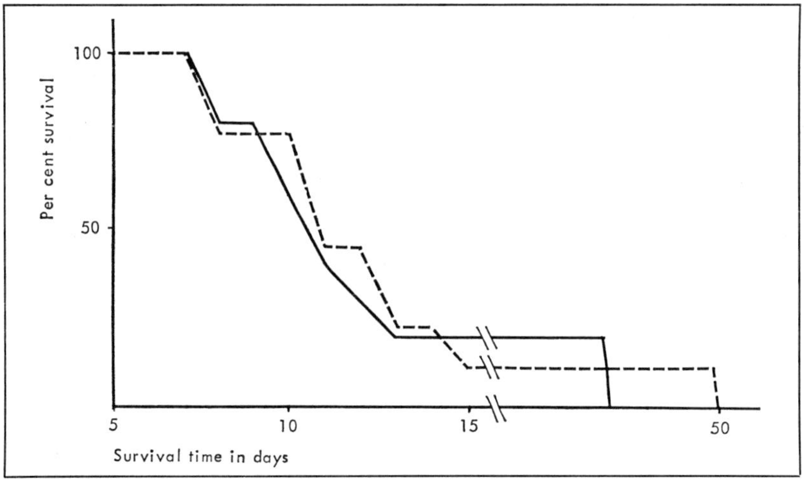

Fig. 2. Comparison between the survival times of untreated, unselected kidney allografted rhesus monkeys and the survival of kidneys exchanged between untreated, unrelated IPI (matched) monkeys. —— Well matched (n = 10), ---- controls (n = 9).

frequency of the grafts began to decrease steadily, followed by deterioration of the ECG pattern as (inconsistently) evidenced by disappearance of P-waves, increasing widening of QRS complexes, ST elevations and diminution of R-wave voltage.

A total number of 14 monkeys was transplanted ectopically with allogeneic hearts and treated with ALS and imuran.The survival times ranged from ten to 64 days. Unfortunately, six transplants had to be excluded from this group owing to the fact that at the time of heartbeat cessation they appeared to be almost totally necrotic. Whether this necrosis was the result of rejection or due to late technical consequences (thrombosis of the arterial anastomosis) could not be distinguished histologically. The survival curve of the remaining eight ectopic grafts is depicted in figure 3. The ECG changes and heart rate alterations, which could be observed in the immunosuppressed group were similar as those described for the untreated allografts, except that they occurred with inconstant fluctuations during the prolonged course of graft function and developed progressively in the last week of heart survival. In some animals a minor difficulty concerning the evaluation of graft function arose, in that after a considerable time of graft survival, palpation revealed only very weak rhythmic beatings, while electrocardio-

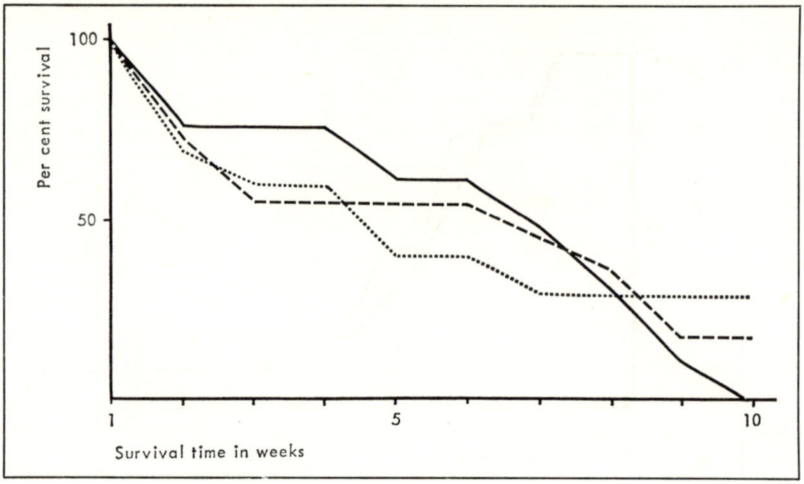

Fig. 3. Comparison between the proportional survival times of kidney and heart (orthotopically and heterotopically-transplanted) allografts in rhesus monkeys, treated with antilymphocyte serum and imuran. In the kidney group the ultimate survival of animals surviving beyond 70 days was: 91 days (one animal) and more than two and a half years (two animals). Two orthotopically-transplanted monkeys surviving beyond 70 days, lived for 76 and 128 days. —— Heterotopic hearts (n = 8); ---- orthotopic hearts (n = 11); .... kidneys (n = 10).

graphic tracings still showed an acceptable picture. When those cases occurred, palpation criteria were considered to be decisive.

Apart from the fact that 30% of the kidney allografts survived longer than the maximal survival time that could be obtained with ectopic hearts, it can be concluded from figure 3 that there exists no marked difference between the general survival pattern of heterotopic heart and kidney allografts in optimally-treated monkeys. The same conclusion holds when the comparison is made with the ALS plus imuran-treated group of 11 orthotopic heart allografted monkeys (fig. 3), being employed in collaboration with the surgical department of Leiden University [9]. Though slightly different during the first weeks, beyond a survival time of 35 days almost identical proportional survival curves of heterotopic and orthotopic hearts could be drawn, which supports the assumption that as far as survival times obtained with immunosuppression are concerned, orthotopic heart allografting does not differ strikingly from heterotopic allografting in rhesus monkeys.

Histological Observations

Since ectopic heart transplantation is associated with a different kind of blood supply and a reduced function, lesions not due to rejection are more liable to develop in heterotopic than in orthotopic grafts. The main aberrant histological features in this respect, which could be observed in the heterotopic hearts, were extensive areas of fresh infarction in the untreated group and marked myocardial fibrosis in the treated group. Morphological features as a result of rejection such as infiltration of mononuclear cells and typical vascular lesions (arteritis) could be perceived in similar degree in comparable heterotopic and orthotopic heart series.

Discussion and Conclusions

Many investigators involved in ectopic heart transplantation in experimental animals have stressed the feasibility of this model to study rejection phenomena in general. Whether this heart transplantation model provides reliable immunological information, which can be transferred to the situation in which the heart is transplanted orthotopically has thus far, to our knowledge, only been investigated in dogs [JONGSMA, personal commun.]. It was found that in randomly-selected dogs immunosuppressed with imuran and prednisolone, orthotopically-transplanted hearts survived much longer than heterotopic hearts. Furthermore, essential morphological discrepancies in rejection pattern between both graft types were noticed. None of these differences occurred in hearts transplanted in rhesus monkeys. Surgical as well as species specific peculiarities might as well be the causes of the observed differences. From our results it can be concluded that heterotopic heart allografting appears to be a suitable transplantation model to investigate the immunological behaviour of allogeneic hearts in rhesus monkeys. Though, after exclusion of the doubtful cases in the heterotopic heart series the ultimate technical success rate in both studied heart groups is roughly the same (60%), it is obvious that from a surgical point of view the ectopic heart transplantation model has many advantages.

The observation that in rhesus monkeys kidney allograft survival showed only minor differences with the survival of heart allografts, when immunosuppression was applied, contrasts with results obtained in rats in which, concerning antigenicity, the kidney seems to be a privileged organ. Since it thus appears that in rhesus monkeys, immunosuppressed with ALS and imuran, the organs thus far investigated are immunologically comparable, further studies on the rejection of vascularized grafts might equally well be performed on the easiest and cheapest transplantation model, namely on kidney allografting. The possibility remains that heart-kidney discrepancies may become apparent also in rhesus monkeys, when immunosuppressive regimens, superior to those employed in this investigation, will be obtained and applied.

The relevance of the RhL-A system for histocompatibility has been well established. Although selection of 'full house' combinations is not possible yet, phenotypic identity between unrelated rhesus monkeys has proved to influence skin graft survival. Our investigations on kidney allografts exchanged between untreated IPI monkeys indicate that there exists no significant correlation between kidney allograft survival and matching according to the current knowledge of the RhL-A system. Whether the influence of donor-host selection on kidney graft survival can be expressed when in addition immunosuppression is applied, is currently under investigation.

Acknowledgements

This study was performed in collaboration with C. F. HOLLANDER, Institute for Experimental Gerontology TNO, Rijswijk Z. H., and H. BALNER, Radiobiological Institute TNO, Rijswijk Z. H.

References

1 BALNER, H.; LEEUWEN, A. VAN; VREESWIJK, W. VAN; DERSJANT, H., and ROOD, J. J. VAN: Leukocyte antigens of rhesus monkeys (RhL-A) and chimpanzees (ChL-A). Similarities with the human HL-A system; in Tissue antigens, vol. 1, p. 229 (Munksgaard, Copenhagen 1971).
2 BALNER, H.; DERSJANT, H.; VREESWIJK, W. VAN, and GABB, B. W.: Tissue typing of rhesus monkeys. Application in transplantation research. Transplant. Proc. *4:* 3 (1972).
3 BEKKUM, D. W. VAN; WAAIJ, D. VAN DER, and PUTTEN, L. M. VAN: Need for specific pathogen free monkeys in certain radiobiological studies. Ann. N.Y. Acad. Sci. *162:* 363 (1969).
4 BEKKUM, D. W. VAN; HEYSTEK, G. A., and MARQUET, R. L.: Effects of immunosuppressive treatment on rejection of heart allografts in rats. Transplantation *8:* 678 (1969).
5 BEKKUM, D. W. VAN; BALNER, H.; DICKE, K. A.; BERG, F. G. VAN DEN; PRINSEN,

G. H., and HOLLANDER, C. F.: The effect of pretreatment of allogeneic bone marrow graft recipients with antilymphocytic serum on the acute graft *versus* host reaction in monkeys. Transplantation *13:* 400 (1972).

6 DICKE, H. W.; HEYSTEK, G. A.; MARQUET, R. L., and VREESWIJK, W. VAN: Allogeneic kidney transplantations in rhesus monkeys. Annual Report of the Radiobiological Institute TNO (1968).

7 DICKE, H. W.; HEYSTEK, G. A.; MARQUET, R. L., and VEN, M. G. VAN DER: Allogeneic kidney transplantations in rhesus monkeys. Effect of immunosuppression. Annual Report of the Radiobiological Institute TNO (1969).

8 DICKE, H. W.; MARQUET, R. L.; HEYSTEK, G. A., and BALNER, H.: Effect of immunosuppressive treatment and leukocyte antigen matching on kidney allografts in rhesus monkeys. Transplantation Proc. *3:* 484 (1971).

9 HOLLANDER, C. F.; SNIJDERS, H.; VINCENT, J.; STOL, H. J., and BROM, H. L. F.: Cardiac allograft pathology in rhesus monkeys. Annual Report of the Radiobiological Institute TNO, Experimental Gerontology Unit and Primate Centre (1970).

10 MANN, F. C.; PRIESTLY, J. T.; MARKOWITZ, I., and YATER, W. M.: Transplantation of the intact mammalian heart. Arch. Surg. *26:* 219 (1933).

Author's address: Dr. R. L. MARQUET, Radiobiological Institute TNO, 151 Lange Kleiweg, *Rijswijk Z.H.* (The Netherlands)

Medical Primatology 1972. Proc. 3rd Conf. exp. Med. Surg. Primates, Lyon 1972, part II, pp. 134–141 (Karger, Basel 1972)

Extracorporeal Immunosuppression with L-Asparaginase in the Baboon

D. Sampson, T. Han, L. S. Hersh and G. P. Murphy

Department of Experimental Surgery and Department of Medicine, Roswell Park Memorial Institute, New York State Department of Health, Buffalo, N.Y., State University of New York at Buffalo, Buffalo, N.Y., and Research and Development Laboratories, Corning Glass Works, Corning, N.Y.

Introduction

The use of immunosuppressive agents in the treatment of patients with organ allografts is invariably accompanied by undesirable toxic side effects and current dosage schedules aim at minimizing these effects, while providing the maximum immunosuppression consistent with the safety of the patient. Such regimes are at best a compromise, and infection still remains the major cause of mortality in the allograft recipient. A similar dilemma occurs in the chemotherapy of the cancer patient.

The problem could be partially resolved by the use of agents whose immunosuppressive or chemotherapeutic effects depended on the modification of the chemical environment of the immunologically competent or malignant cell by enzymic means. Such an agent could be immobilized in an extracorporeal circulation, and the desired chemical changes could be brought about in the patient or experimental animal without the agent escaping into the circulation.

The enzyme L-asparaginase fulfills these criteria well, and the present report documents the immunosuppressive effects of an extracorporeal chemotherapy unit containing plates coated with this drug. The baboon was chosen as the experimental animal most readily available, whose pattern of immunological response closely resembles that of man.

The immunosuppressive properties of L-asparaginase have been well documented. Prolonged renal allograft survival has been achieved in the dog [Rapaport *et al.*, 1971], an increased survival of guinea pig to rabbit skin xenografts has been observed [Etheredge *et al.*, 1971], and an increased skin allograft survival has been reported in mice [NELSON *et al.*, 1970]. *In vitro* immunosuppression has been demonstrated by the inhibition of lymphocyte blastogenesis in phytohemagglutinin stimulated cultures [ASTALDI *et al.*, 1969] and other *in vitro* assays of the immunological effects of L-asparaginase have been recently reviewed [PRAGER, 1971].

The enzyme L-asparaginase has been of benefit in the treatment of human leukemia [HILL *et al.*, 1967; OETTGEN *et al.*, 1967], but direct toxic effects of the drug as well as sensitization to the protein were observed [OETTGEN *et al.*, 1970].

A new technology has been evolved whereby enzymes and other proteins can be covalently bonded to insoluble matrices such as glass or methacrylate plates and yet still maintain their chemical activity [WEETALL, 1971] and recently an extracorporeal unit containing ten plates coated with L-asparaginase has been described [SAMPSON *et al.*, 1972]. Using *E. coli* L-asparaginase, the enzyme was coupled with glutaraldehyde and a silane, γ-amino triethoxysilane. (The triethoxysilanes have the general formula $R\,CH_2\,CH_2\,CH_2 - Si\,[OCH_2\,CH_3]$.) Each plate carries approximately two international units of L-asparaginase and thus the device as a whole contains approximately 20 units.

Material and Methods

Adult baboons *(Papio anubis)* were used for the study. Access to the circulation for the extracorporeal perfusion was gained by cannulation of the femoral vessels with tapered polyvinyl chloride catheters under phencyclidine (4 mg/kg) anesthesia. The schema of the perfusion circuit is shown in figure 1. Prior to the perfusion the circuit was primed with lactated Ringer's solution, and following connection of the device to the circulation, the animals own blood pressure provided a blood flow rate between 150 and 200 ml/min. Blood flow was monitored electromagnetically and pressures on both the arterial and venous sides of the machine were measured with aneroid manometers. A bubble trap was included for safety. At the commencement of the perfusion the animals were heparinized with 15,000 units of heparin.

The perfusions were carried out for one to three hours during which time serial blood sampling was performed via the blood lines. Assays of serum asparagine and L-asparaginase was carried out [COONEY *et al.*, 1970] and lymphocyte reactivity was assayed on whole blood samples [HAN and PAULEY, in press]. This assay consists of a five-day culture of the whole blood sample, incubating with the vegetable mitogens phytohemagglutinin, pokeweed mitogen, Concanavalin A and Staphylococcal filtrate. At the end of the incubation period, stimulation by these mitogens was measured by tritiated thymidine uptake. The uptake of stimulated cells was compared with that of unstimulated cells to give a 'blastogenic index'.

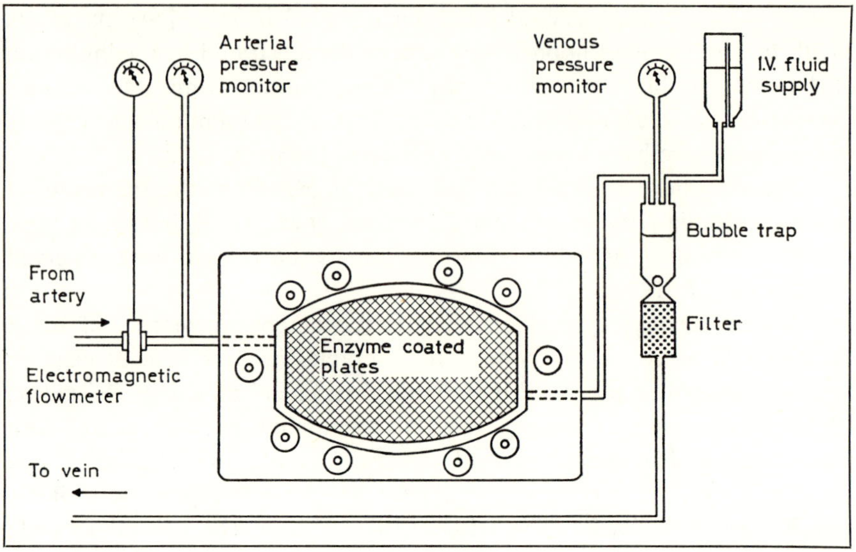

Fig. 1. The perfusion circuit of the extracorporeal chemotherapy device. The coated plates are held in a plexiglass container, small spacers allowing blood to pass between them. The device measures 20 × 10 cm.

In order to be sure that any immunosuppressive effects observed were indeed due to the L-asparaginase, control perfusions were carried out using methacrylate plates. Further control perfusions were carried out using plates coated with silane agent only or with both the silane coupling agent and glutaraldehyde, to assess the contribution if any, of these compounds to the immunosuppressive effect.

Following the perfusion the animals were allowed to recover; some of them were subjected to repeated perfusions to evaluate toxicity or possible sensitization. Toxicity was monitored by the following determinations: serum SGOT, SGPT, blood urea nitrogen creatinine, electrolytes, hematocrit, white blood cell and platelet counts for up to one month after perfusion. A total of 20 perfusions was carried out, including the various controls.

Results

All the animals survived the perfusions well. The only complication seen during perfusion was that of high output cardiac failure if flow rates exceeded 200 ml/min. This was readily reversed by slowing down blood flow with a screw clamp.

2
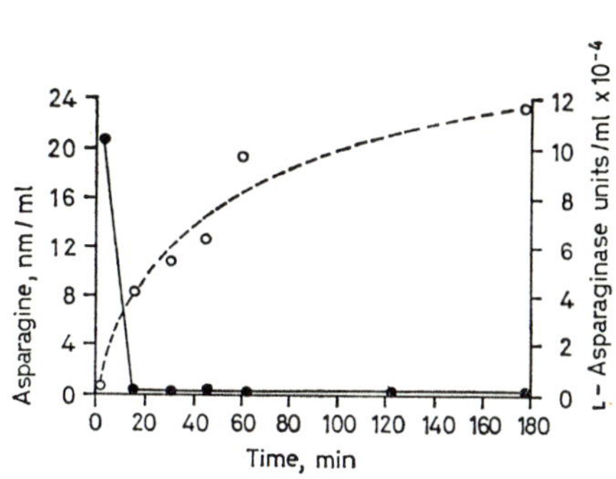

3
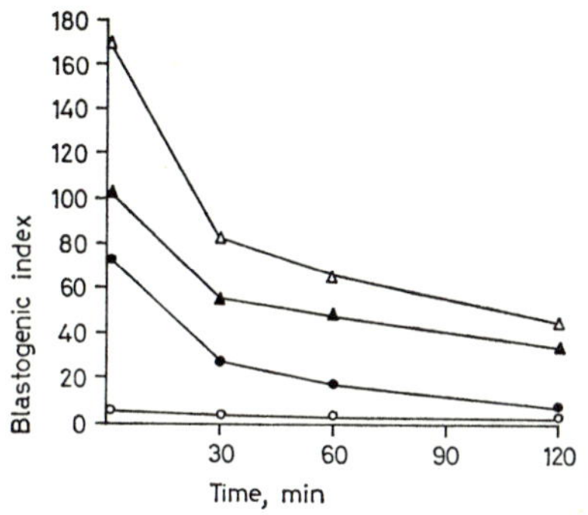

Fig. 2. A rapid depletion of asparagine is shown in this experiment. During the three-hour perfusion some L-asparaginase accumulates in the circulation, but these levels are not sufficient to produce immunosuppression. ● Serum asparagine; ○ serum L-asparaginase.

Fig. 3. A progressive inhibition of lymphocyte reactivity to all mitogens is seen, but residual activity persists even after three hours perfusion. Mitogens: ● CCA; ○ PHA; ▲PWM; △ SF.

The results of some typical experiments are shown in figures 2–5. A profound fall in serum asparagine levels occurred in all but the control experiments, and in figure 2 it can be seen that there was no detectable asparagine in the circulation within 15 min of commencement of the perfusion. This figure also illustrates the leakage of L-asparaginase from the plates. Although this is minute, it can be seen that serum levels of L-asparaginase steadily rise to a value of 12×10^{-4} units/ml after three hours.

Figure 3 demonstrates that a progressive immunosuppressive effect was seen but that even after three hours some lymphocyte reactivity remains as judged by the 'blastogenic index'.

Figure 4 shows a slower clearance of asparagine than that of the previous experiment, but the leakage of enzyme from the plates was very similar. However, figure 5 shows an intense immunosuppression such that the lymphocytes were virtually unresponsive to all mitogens within an hour of the commencement of the perfusion.

Studies carried out serially after the termination of the perfusions showed that lymphocyte reactivity returned to normal values within about three hours, indicating rapid reversibility.

In the control experiments, no immunosuppressive effect was seen following perfusion over plain methacrylate plates, however the silane

4

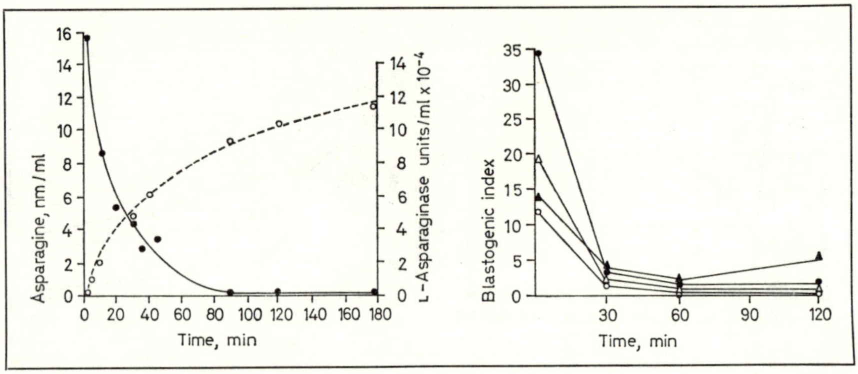

5

Fig. 4. Asparagine levels were not cleared as rapidly as those in the previously-illustrated experiment, but did eventually fall to zero. Leakage of L-asparaginase into the circulation was similar to that shown previously. ● Serum asparagine; ○ serum asparaginase.

Fig. 5. The intense immunosuppression observed in this experiment rendered the lymphocytes virtually unreactive to all stimulants within an hour of the commencement of the perfusion. Mitogens: ● CCA; ○ PHA; ▲ PWM; △ SF.

coupling agent and the glutaraldehyde did produce some immunosuppression, but not to the extent seen when L-asparaginase was used.

During the perfusions, following an initial fall in total white blood count, a rise was seen to about one and a half times starting value. All the toxicity studies were negative, and serial perfusions showed no fall off of either biochemical or immunological effect.

Discussion

The results of the extracorporeal perfusions through a device containing plates carrying immobilized L-asparaginase show that the *in vitro* immunosuppressive activity of this enzyme [Astaldi *et al.*, 1969; Ohno and Hersh, 1970; Simberkoff and Thomas, 1970] is not impaired in this situation. However, the mechanism of the immunosuppression is not clear. It has been postulated that asparagine is essential for the protein metabolism of the lymphocyte which is deficient in the enzyme asparagine synthetase, and that the absence of this amino acid would impair lymphocyte function [Prager, 1971]. The results presented do not support this view, in that the most profound immunosuppression was seen in those experiments in which serum asparagine was most slowly depleted. Also, this is not the most likely explanation, since the assay procedure used in our experiments involved a five-day culture of cells in a medium containing asparagine, and it is unlikely that a few hours' exposure to low asparagine levels could exert an immunosuppressive effect measured five days later. It is more plausible that direct contact of the cell with L-asparaginase may modify its metabolism in a way which depresses reactivity to antigenic stimulation. Recently, it has been suggested that the immunosuppressive activity of L-asparaginase may not be due to that enzyme at all, but rather to a contaminant glutaminase which results in a depletion of glutamine [BENEZRA *et al.*, 1972]. It is noteworthy that some of the immunosuppression we have observed is due to the action of one of the coupling agents, glutaraldehyde. However, cell viability as judged by the trypan blue dye exclusion test is not impaired in any way.

The leakage of enzyme from the plates is detectable but minute, and even after several hours' perfusion, the serum levels are far below those previously reported to be necessary to produce immunosuppression [HERSH, 1971; BENEZRA *et al.*, 1972]. It seems also, that this leakage of enzyme has not resulted in sensitization of the animals, since serial perfusions show no loss of efficiency as would be expected if anti-L-asparaginase antibody were present [PRAGER, 1971]. The depletion of asparagine is very satisfactory and if the leukemic cell really is dependent on this amino acid for survival the extracorporeal unit tested may be of value in the treatment of human leukemia.

While an *in vitro* immunosuppression has been demonstrated, it remains to be seen whether the extracorporeal device is capable of prolonging allograft survival in a recipient perfused on the unit. The rapid reversibility of the immunosuppression would require long periods of perfusion, but this would be practical using a smaller unit inserted into a Scribner shunt. Such a small unit has been described [HYDEN, 1971] and successfully used in the treatment of Hodgkin's disease. In the clinical situation, the reversibility of the immunosuppression could be advantageous in that in the event of life-threatening sepsis occurring, the immunosuppressed state could be quickly reversed.

The efficiency of the device can be increased in several ways. Thus, the surface area of the insoluble matrix could be increased, or even be made in the form of a methacrylate or glass wool. Secondly, the enzyme activity could be increased by using a different form of L-asparaginase. The present work was carried out using *E. coli* L-asparaginase, but the activity of the plates could be increased by a hundredfold if *Erwinia* L-asparaginase was used. With such an increase in efficiency, long-term treatment with a small device becomes a practical possibility.

Acknowledgements

The authors are grateful for the assistance of Dr. D. A. Cooney of the Laboratory of Toxicology, NCI, NIH, Bethesda, Md., for the performance of the serum asparagine and L-asparaginase assays.

References

Astaldi, G.; Burgio, G. R.; Biscatti, G.; Astaldi, A., jr., and Ferfoglia, L.: L-asparaginase and blastogenesis. Lancet *ii:* 643–644 (1969).

Benezra, D.; Pitaro, R.; Birkenfeld, V., and Hochman, A.: Reversal of the immunosuppressive effect of L-asparaginase by L-glutamine. Nature *236:* 80–82 (1972).

Cooney, D. A.; Capizzi, R. L., and Handschumacher, R. E.: Evaluation of L-asparagine metabolism in animals and man. Cancer Res. *30:* 929–935 (1970).

Etheredge, E. E.; Shons, A.; Harris, N., and Najarian, J. S.: Prolongation of skin xenograft survival by L-asparaginase. Transplantation *11:* 353–354 (1971).

Han, T. and Pauly, J.: Simplified whole blood methods for evaluating *in vitro* lymphocyte reactivity of laboratory animals. Clin. exp. Immunol. (in press).

Hersh, E. M.: Immunosuppression by L-asparaginase and related enzymes. Transplantation *12:* 368–376 (1971).

Hill, J. M.; Roberts, J.; Loeb, E.; Khan, A.; Mac Lellan, A., and Hill, R.W.: L-asparaginase therapy for leukemia and other malignant neoplasms. J. amer. med. Ass. *202:* 882–888 (1967).

Hyden, H.: An extracorporeal shunt apparatus for blood detoxification. Arzneimittelforsch. *21:* 1671–1675 (1971).

Nelson, S. D.; Lee, M. B., and Bridges, J. M.: Immunosuppressive activity of L-asparaginase in mice. Transplantation *9:* 566–570 (1970).

Oettgen, H. F.; Old, L. J.; Boyse, E. A.; Campbell, H. A.; Phillips, F. S.; Clarkson, B. D.; Tallal, L.; Leeper, R. D.; Schwartz, M. K., and Kim, J. H.: Inhibition of leukemias in man by L-asparaginase. Cancer Res. *27:* 2619–2631 (1967).

Oettgen, H. F.; Stephenson, P. A.; Schwartz, M. K.; Leeper, R. D.; Tallal, L.; Tann, C. C.; Clarkson, B. D.; Golbey, R. B.; Krakoff, I. H.; Karnofsky, D. A.; Murphy, M. L., and Burchenal, J. H.: Toxicity of *E. coli* L-asparaginase in man. Cancer *25:* 253–277 (1970).

Ohno, R. and Hersh, E. M.: The inhibition of lymphocyte blastogenesis by L-asparaginase. Blood *35:* 250–262 (1970).

Prager, M. D.: Effect of asparaginase on immunologic phenomena. Colloques int. Cent. nat. Rech. scient. *197:* 179–188 (1971).

RAPAPORT, F. T.; SHIMADA, T., and WATANABE, K.: Prolongation of canine renal allograft survival by L-asparaginase. Transplantation *12:* 217–222 (1971).

SAMPSON, D.; HERSH, L. S.; COONEY, D., and MURPHY, G. P.: A new technique employing extracorporeal chemotherapy. Amer. Soc. Artif. Intern. Org. *18:* 54–58 (1972).

SIMBERKOFF, M. S. and THOMAS, L.: Reversal by L-glutamine of the inhibition of lymphocyte mitosis caused by *E. coli* asparaginase. Proc. Soc. exp. Biol. Med. *133:* 642–644 (1970).

WEETALL, H. H.: Enzymes immobilized on organic carriers. Research Develop. *22:* 18–22 (1971).

Author's address: Dr. D. SAMPSON, Department of Surgery, University of Cambridge, *Cambridge* (England)

Medical Primatology 1972. Proc. 3rd Conf. exp. Med. Surg. Primates, Lyon 1972, part II, pp. 142–148 (Karger, Basel 1972)

Bone Marrow Transplantation in Rhesus Monkeys

Methods to Mitigate the Acute Graft versus Host Disease[1]

U. W. SCHAEFER[2], K. A. DICKE and D. W. VAN BEKKUM

Radiobiological Institute TNO, Rijswijk Z.H.

Rodents and primates differ considerably in their response to transplantation of allogeneic bone marrow cells. Monkeys and human beings develop an acute, mice and rats a delayed form of the graft *versus* host (GvH) disease. The severe immunological reaction in primates – due to a high amount of immunocompetent lymphocytes in the marrow – is nearly always fatal. In rodents a certain percentage of chimaeras will become tolerant against the graft and survive the late GvH which is caused by lymphocytes developed by the differentiation of transplanted stem cells. Attempts to prevent or to alleviate the acute form of GvH disease have included: selection of donor and recipient according to histocompatibility, immunosuppressive treatment of recipient or donor and conditioning of the graft *in vitro* by incubation or cell separation methods.

Theoretically, the animal of choice to investigate the effectiveness of the mentioned approaches is the monkey. In practice, however, explorative work has to be carried out with mice, a species in which the primate type of GvH can be reproduced by transplantation of spleen cells [1]. Methods which have been found to be effective against GvH reaction in this mouse model were tested in monkeys. The results of our investigations in both species have been regularly published [2, 3, 6, 9].

In our primate model which consists of lethally-irradiated rhesus monkeys *(Macaca mulatta)* transplanted with allogeneic bone marrow cells from random donors [4], the following measures mitigate significantly the GvH

1 This research has been sponsored in part by the Commission of European Communities (formerly Euratom), contract No. 079–69–1 BIAC.

2 Fellow of the International Agency for Research on Cancer, WHO, Lyon, in 1970/1971.

Table I. Monkey survival following bone marrow transplantation. Effect of various methods used to mitigate acute GvH

Treatment	Cell dose per kg body weight	Number of animals	30-day survival %	Survival in days
Control group	4×10^8	25	0	7–25
Cyclophosphamide (post transplantation phase)	4×10^8	26	40	11–413
Horse ALS (pre transplantation phase)	4×10^8	13	23	7–51
Stem cell separation	$0.4–0.5 \times 10^8$	10	80	21–51

disease: treatment of the recipient in the early post transplantation phase with cyclophosphamide, methotrexate [9] or antilymphocyte serum (ALS) [3]; short-term treatment of the recipient with horse ALS just prior to transplantation [2]; and the use of purified stem cell concentrates obtained by gradient centrifugation [6].

Table I demonstrates the effect of cyclophosphamide and horse ALS, respectively on the survival time of irradiated and bone marrow-transplanted monkeys. Both approaches decrease the severity and delay the onset of the GvH reaction. Horse ALS did not interfere with the take and proliferation of the graft. ALS produced in rabbits was found to have inhibitory effects on the take, presumably because of cytotoxic effects on the hemopoietic stem cells. Studies are in progress to investigate if preparations from different rabbit strains and/or by different immunisation schedules may prove to be less damaging to stem cells. There are two important disadvantages of ALS. Short-term treatment prior to transplantation can induce thrombocytopenia, long-term administration in the post transplantation phase increases the risk of infections.

Hemopoietic stem cells and immunocompetent lymphocytes can be separated according to density by centrifugation on a gradient which consists of layers of different bovine albumin concentrations [5]. The bone marrow fractions obtained by gradient centrifugation can be tested *in vitro* for proliferation capacity by a cell culture technique, the CFU-C assay [7], and for immunocompetent lymphocytes by the phytohemagglutinin technique. Extensive studies in mice have proved that these *in vitro* assays are reliable parameters for hemopoietic stem cells and GvH active cells [7]. In monkeys

Table II. Storage protocol

Step	Procedure
1	Suspend cells in HBSS (305 mosmol)
2	Add cryoprotective agents (final concentration: 10% DMSO + 20% CAS or 10% PVP/glycerol + 20% CAS)
3	Transfer suspension to glass ampoule
4	Freeze sample at a controlled rate of 1 °C/min to –196 °C
5	Store ampoule in liquid nitrogen
6	Thaw rapidly in a 40 °C waterbath
7	Start slow stepwise dilution by adding a very small amount of HBSS which equals $^{1}/_{50}$ of the thawed cell suspension
8	Pipette to the suspension increasing amounts of HBSS – roughly doubling quantities added each time – so that a ten fold dilution is reached by *one hour;* shake carefully between each step for 3–5 min
9	Treat cell suspension with DNase to avoid gel formation
10	Centrifuge 10 min at 500 *g*
11	Pour off supernatant and resuspend cells in HBSS
HBSS	= Hank's balanced salt solution
CAS	= calf serum
DMSO	= dimethyl sulfoxide
PVP	= polyvinylpyrrolidone
DNase	= desoxyribonuclease

these *in vitro* assays have to be compared with the *in vivo* transplantation response. So far we have found a complete correlation. When a tenfold enrichment of stem cells was measured in a bone marrow fraction *in vitro*, the minimum cell dosage to be transplanted could be ten times reduced. The comparative effect of fractionated and unfractionated marrow can be seen in table I. Small numbers of a stem cell rich fraction (0.4–0.5 × 10^8 cells/kg body weight) were able to restore the recipient without causing acute GvH disease. The control group treated with 4 × 10^8 cells/kg of unfractionated marrow – a minimum number, which protects 95% of the recipients – on the average died from acute GvH reaction within two weeks after grafting.

A disadvantage of the gradient technique is that only 20–30% of the original stem cell population can be collected and that it takes four to seven days until the *in vitro* results can be evaluated. Therefore, we were interested in a reliable storage technique for hemopoietic cells. Such a method would permit to pool cells obtained by repeated punctures of the

Table III. Recovery of hemopoiesis in lethally-irradiated rhesus monkeys by frozen allogeneic bone marrow grafts

Experiment No.	Cell dose per kg body weight	Storage time, days	Cryoprotection[1]	Take	GvH	Survival days
1	8×10^8	1	DMSO	+	+	13
2	4×10^8	1	DMSO	+	+	13
3	4×10^8	1	DMSO	+	+	12
4	4×10^8	1	DMSO	+	+	12
5	4×10^8	14	DMSO	+	+	14
6	4×10^8	40	DMSO	+	+	13
7	4×10^8	2	PVP/glyc.	+	+	8
8	4×10^8	2	PVP/glyc.	+	+	13
9	4×10^8	2	PVP/glyc.	+	+	13

1 Cryoprotective media: 10% DMSO+20% calf serum or 10% PVP+10% glycerol+20% calf serum.

donor and to graft marrow fractions which have been evaluated *in vitro* before transplantation.

It is known that bone marrow cells of rodents can be stored at low temperature under carefully-controlled conditions without marked loss of viability [8, 10]. In primates the results have been disappointing. Only a few quantitative studies have been undertaken. VAN PUTTEN [10] transplanted into lethally-irradiated rhesus monkeys marrow frozen in polyvinylpyrrolidone (PVP) – glycerol or in dimethyl sulfoxide (DMSO). In autologous combinations he found a storage efficiency of only 50%. In allogeneic combinations reproducible takes could not be obtained.

We investigated marrow preservation in mice, monkeys and man. We found that not only the choice of cryoprotective agent, the speed of freezing and thawing, but also the technical procedure after thawing are very important to maintenance of stem cell viability. We introduced a very slow stepwise dilution procedure after thawing to remove the cryoprotective agents gradually [11]. The storage and dilution procedure is outlined in table II.

Gradual removal of PVP /glycerol or of DMSO by decreasing the concentration within one hour from 10 to 1% apparently diminishes the osmotic stress imposed upon the cell after thawing.

Applying the stepwise dilution technique we could achieve in mice, monkeys and man, an 80- to 100-percent stem cell recovery as measured

Table IV. Transplantation of stem cell rich allogeneic or autologous marrow fractions frozen in 10% DMSO + 20% calf serum in irradiated monkeys

Experiment No.	Donor	Storage time, days	CFU-C concentration	Cell dose per kg body weight $\times 10^7$	Red. F.	Take	Survival days	Remarks
1	allogeneic	7	13	6.0	3–7	+[1]	>100	reversal
2	allogeneic	1	5	5.0	4–8	+	26	pneumonia, chronic GvH
3+	allogeneic	55	10	5.0	4–8	+	14	sepsis, no GvH
4	allogeneic	14	28	1.4	14–29	+	>32	alive
5	allogeneic	14	28	1.4	14–29	+[1]	13	sepsis, aplasia
6	autologous	8	28	0.28	14–29	+	>15	alive
7	autologous	7	35	0.14	28–57	+	>32	alive
8	autologous	10	52	0.12	33–67	+	22	ulcus ventriculi

The factor (Red. F.), by which the minimum cell number to be transplanted ($2–4 \times 10^8$/kg body weight allogeneic, $4–8 \times 10^7$/kg autologous cells in case of unfractionated marrow) could be decreased, has been calculated according to the CFU-C concentration measured *in vitro.*

+ Fractionation of preserved marrow.

1 Temporary take (experiment 1 finally host type marrow; experiment 5 terminating sepsis and marrow aplasia).

by the spleen colony assay [12], the CFU-C assay [7] and transplantation experiments.

Lethally-irradiated rhesus monkeys have been injected with allogeneic marrow frozen in 10% PVP/glycerol + 20% calf serum or in 10% DMSO + 20% calf serum. The results are listed in table III.

In all cases a take of the graft could be obtained although limited cell numbers have been injected. As mentioned earlier, 4×10^8 cells/kg body weight is a minimum number which yields a take in 95% of the animals when fresh allogeneic cells have been administered. The recipients showed a repopulation of the hemopoietic organs and the peripheral blood with donor cells. They died within eight to 14 days because of a severe acute GvH disease comparable to that observed after grafting of fresh allogeneic cells.

These transplantation experiments in monkeys as well as our *in vitro* CFU-C data in man gave evidence that the stem cell population of primate bone marrow can be preserved almost without loss of viability. Our group transplanted already in a human patient with combined immune deficiency (CID) stem cell rich marrow frozen in DMSO and could demonstrate a take.

The advantage of a storage method which allows a nearly complete recovery of hemopoietic stem cells is obvious. Total marrow or stem cell rich marrow fractions of known antigen pattern from living donors or cadavers as well as autologous marrow could be stored in bone marrow banks for transplantation purposes probably for a long time. We needed a relieble preservation method to simplify and to improve the use of stem cell concentrates obtained by gradient centrifugation. Using the described storage technique and the stepwise dilution after thawing, systematic investigations were started to increase the efficiency of the gradient technique.

Introducing refractionation procedures we were able to obtain stem cell fractions concentrated up to 70 times, as compared with unfractionated marrow. The preservation of the marrow fractions permitted us to calculate the minimal cell dose to be transplanted according to the stem cell concentration measured *in vitro*. Table IV shows the results of a few preliminary experiments. In autologous as well as in allogeneic situations, definitive takes could be achieved grafting very low cell numbers of highly-purified frozen stem cell fractions. Transplantation of preserved purified stem cell concentrates are of therapeutical advantage especially in cases of CID. The risk of an acute GvH reaction would be diminished and the chance for tolerance might by higher. In addition, repeated grafting would be possible by employing gradually-increasing numbers of the preserved stem cells.

References

1 Bekkum, D. W. van: The selective elimination of immunologically competent cells from bone marrow and lymphatic cell mixtures. I. Effect of storage at 4°C. Transplantation *2:* 393–404 (1964).

2 Bekkum, D. W. van; Balner, H.; Dicke, K. A.; Berg, F. G. van den; Prinsen, G. H., and Hollander, C. F.: The effect of pretreatment of allogeneic bone marrow graft recipients with antilymphocytic serum on the acute GvH reaction in monkeys. Transplantation *13:* 400–407 (1972).

3 Bekkum, D. W. van; Putten, L. M. van; Balner, H., and Dicke, K. A.: Mitigation of secondary disease following allogeneic bone marrow transplantation by ALS and other immunosuppressive substances (studies in mice and subhuman primates). Coll. Int. CNRS *190:* 377–395 (1969).

4 BEKKUM, D. W. VAN and VRIES, M. J. DE: Radiation chimaeras (Logos/Academic Press, London 1967).

5 DICKE, K. A.: Bone marrow transplantation after separation by discontinuous albumin density gradient centrifugation; thesis Leiden (1970).

6 DICKE, K. A. and BEKKUM, D. W. VAN: Allogeneic bone marrow transplantation after elimination of immunocompetent cells by means of density gradient centrifugation. Transplant. Proc. *3:* 666–668 (1971).

7 DICKE, K. A.; PLATENBURG, M. G. C., and BEKKUM, D. W. VAN: Colony formation in agar. *In vitro* assay for haemopoietic stem cells. Cell Tissue Kin. *4:* 463–477 (1971).

8 LEWIS, J. P.; PASSOVOY, M., and TROBAUGH, F. E. jr.: The transplantation efficiency of marrow cooled to –100 °C at 2 °C per minute. Cryobiology *3:* 47–52 (1966).

9 MULLER-BÉRAT, C. N.; PUTTEN, L. M. VAN, and BEKKUM, D. W. VAN: Cytostatic drugs in the treatment of secondary disease following homologous bone marrow transplantation. Extrapolation from the mouse to the primate. Ann. N.Y. Acad. Sci. *129:* 340–354 (1966).

10 PUTTEN, L. M. VAN: Quantitative aspects of the storage of bone marrow cells for transplantation. Europ. J. Cancer *1:* 15–22 (1965).

11 SCHAEFER, U. W.; DICKE, K. A., and BEKKUM, D. W. VAN: Recovery of haemopoiesis in lethally irradiated monkeys by frozen allogeneic bone marrow grafts. Rev. Europ. Étud. clin. biol. *17:* 483–488 (1972).

12 TILL, J. E. and MCCULLOCH, E. A.: A direct measurement of the radiation sensitivity of normal mouse bone marrow cells. Radiat. Res. *14:* 213–222 (1961).

Authors' address: Dr. U. W. SCHAEFER, Dr. K. A. DICKE and Prof. D. W. VAN BEKKUM, Radiobiological Institute TNO, Lange Kleiweg 151, *Rijswijk Z. H.* (The Netherlands)

Medical Primatology 1972. Proc. 3rd Conf. exp. Med. Surg. Primates, Lyon 1972, part II, pp. 149–155 (Karger, Basel 1972)

Orthotopic Heart Transplantation in Rhesus Monkeys

J. Vincent, H. J. Stol, H. Snijders, Z. Aytug, H. L. F. Brom, C. F. Hollander, R. L. Marquet and G. A. Heystek

Department of Thoracic Surgery, University of Leiden and REP-TNO Institutes, Rijswijk

Orthotopic homologous heart transplantations were performed in rhesus monkeys *(Macaca mulatta)*, weighing from two to five kg, in order to develop the surgical technique and to study the survival rate, in relation to immunosuppressive treatment in subhuman primates.

The well-known superficial immersion cooling technique without extracorporeal circulation was adapted for this experiment and appeared to be the technique of choice [2, 4, 7–9].

The unrelated, random animals were matched only by their weight. The heart of a donor weighing 10–15% more than the recipient was considered to be preferable for grafting. The monkeys were premedicated with 0.1–0.2 mg atropine with ketamine (ketalar), 2 mg/kg i.m. This dosage permitted us to prepare and canulate the radial or posterior tibial artery and femoral vein to keep the arterial pressure and fluid balance controlled. Thereafter, alloferine, 0.5 mg/kg body weight was given and the monkey intubated and ventilated with the infant Amsterdam (Keuskamp) respirator with oxygen and 2.5% CO_2.

The monkey was fixed by four points on a simple frame in a children's bath, so that about 6 × 20 cm of the sternal area remained exposed. The bath was filled with ice pieces and cold water and covered with a sterile adhesive foil. The operating field was kept in this way, both dry and sterile (fig. 1, 2).

Median sternotomy was performed and the large vessels and caval veins exposed. At this time the donor sternotomy was done at normothermia and 3 mg heparine/kg body weight was administered to both monkeys. Then, the donor heart was excised and cooled in 4 °C Ringer solution.

When the recipients temperature reached 22 °C, the caval veins were closed, a small Pilling clamp was placed on the aorta and the pulmonary artery and both ventricles were removed by cutting through the A-V junction and at the base of the big vessels. The ventilation was then stopped.

The interatrial septum, the right atrium, the left atrium, the pulmonary artery and the aorta were joined by continuous suture using 4–0 double-armed atraumatic silk suture (fig. 3). The air was removed from

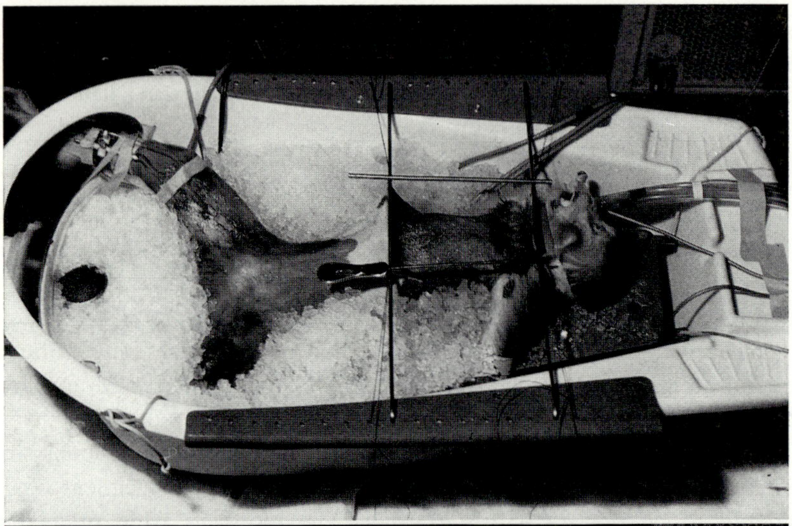

1

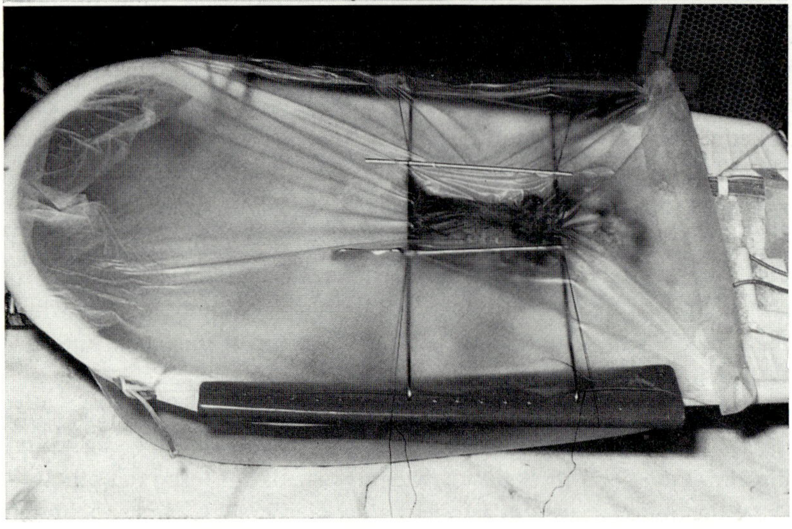

2

Fig. 1. The monkey fixed on a frame in the ice bath.
Fig. 2. The bath covered with a sterile adhesive foil ready for the operation.

the right heart by temporarily opening the cava veins just before the pulmonary artery was closed. At this moment the water in the bath was reset at 40 °C [1, 5, 10]. The air in the left heart was evacuated by filling with saline by means of a catheter introduced into the left ventricle through the aortic anastomosis, just before it was finished. In the meantime the lungs were ventilated to remove the air from the pulmonary veins and left atrium. Then the apex of the left and of the right ventricle were punctured. The clamp was removed and the circulation was restored with a gentle manual 'finger' massage. The blood, sodium bicarbonate 8% (3–4 cm^3), protamine (1.5 heparine dose) and isoprenaline were given i.v. The thorax was washed out with 37 °C sterile saline. The heart started to beat after a few minutes, adrenaline and calcium were given intracardially, when necessary, to insure a good arterial pressure and cardiac output.

Time schedule:	cooling 37–22 °C	±25 min	circulatory arrest	±30 min
	ischemia	±50 min	massage	3–10 min
			rewarming 22–37 °C	±30 min

The mediastinum was drained, and the sternotomy closed. With the temperature at 37–38 °C the monkey was removed from the bath and dried.

Each monkey received 0.2 mg atropine and 0.1–0.2 mg neostigmine. Spontaneous ventilation and consciousness returned between 10 min and

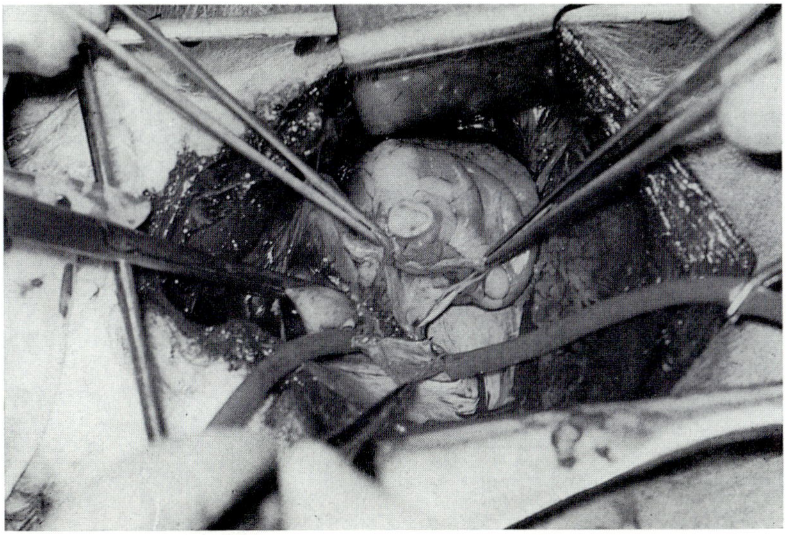

Fig. 3. The first point of the interatrial anastomosis.

2 h after this injection. Once awake, the monkey would not tolerate any drains, tubes, canulas, ECG or EEG electrodes and fought to free himself.

We are able to report about our five series:

(1) A series of animals previously used for other experiments with immunosuppressive treatment, skin transplantation, bowel sterilisation, etc. We operated on 28 animals. During surgery and during the first 4 h we lost 15 monkeys. The remaining 13 animals had an average survival time of 4.1 days (1–20 days survival).

Note that none of these unselected animals were treated with immunosuppressives. Postoperative care included cardiotonics, diuretics and antibiotics. The causes of death based on the clinical follow-up and histopathological study were:

Table I. Survival time of orthotopic heart transplantation on rhesus monkeys with different immunosuppressive treatment.

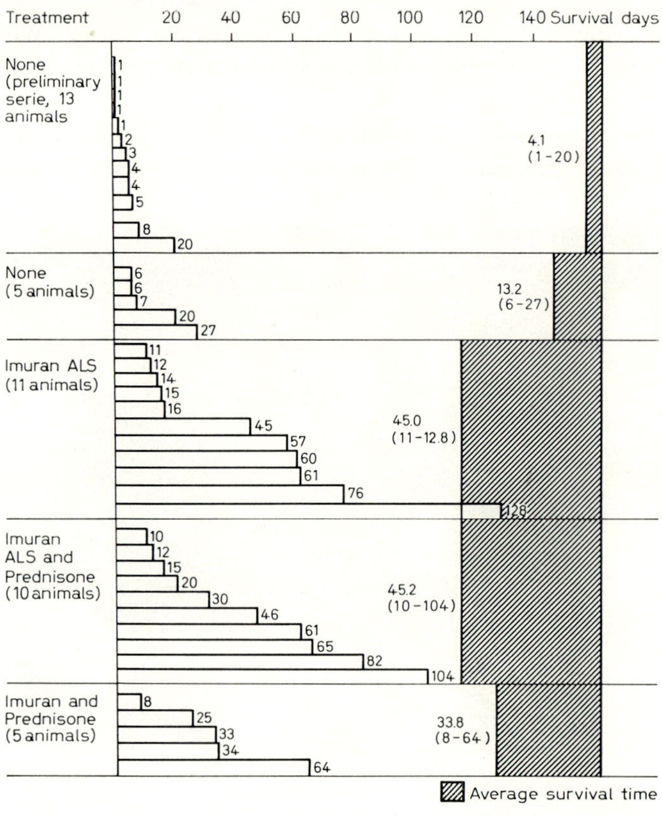

7 acute rejection; 1 heart decompensation; 1 pneumothorax, 1 atelectasis; 1 atrial thrombosis; 1 surgical hemorrhage; 1 unknown.

(2) In the second series 'clean' unused monkeys were transplanted. These five monkeys served as control animals and received no immunosuppressives. Their average survival time was 13.2 days (6–27). The cause of death in every case was acute rejection.

(3) This series comprised 11 'unused' monkeys treated for 36 days with 4 mg Imuran/kg body weight and for 60 days with antimonkey lymphocyte serum (ALS) 3 times/week. Their average survival time was 45.0 days (11–128). The cause of death in these monkeys was chronic rejection and arteritis, the severity of which corresponded with the survival time. Severe ALS glomerulonephritis and sepsis was seen in one case (12 days).

(4) This series comprised ten 'unused' monkeys treated with Imuran (i.m. 4 mg/kg body weight/day for 20 days and 1 mg from the 21st day), ALS (s.c. 2 ml/kg body weight 3 times/week) and Prednisone (2 mg/kg body weight/day from the 29th day). Their average survival time was 45.2 days (10–104). The cause of death in nearly every case was chronic rejection and arteritis.

(5) This series comprised five 'unused' monkeys treated with Imuran (i.m. 4 mg/kg body weight/day for 20 days and 1 mg from the 21st day) and Prednisone (2 mg/kg/day from the 29th day). Their average survival time was 33.8 days (8–64). The cause of death in nearly every case was chronic rejection and arteritis.

Comment

The surgical technique is patterned after that of SHUMWAY *et al.* [10]. Because of the low weight of the monkeys (2–5 kg) we eliminated the extracorporal circulation and choose the superficial immersion cooling technique. The atrial septum was suture only once, because of the total size of the hearts (3 × 6 cm and less). Technically the anastomosis of the pulmonary artery remains the most difficult part due to the extreme fragility of the tissues. It is preferable in the postoperative period to keep the pleural cavities closed.

We have found that in order to accurately monitor this surgical operation we have to control the following parameters: temperature of the monkey and of the bath, ECG, EEG, arterial pressure, blood gas level and electrolytes.

The body temperature was measured by placing a thermometer in the esophagus. Routine EEG recording was not possible during immersion. Therefore, we developed a method utilizing water insulated electrodes, placed percutaneously. The EEG recordings obtained were of similar quality to those obtained by routine methods (fig. 4).

These recordings gave us a detailed view of brain activity during the anesthesia, hypothermia, circulatory arrest and rewarming.

Arterial blood pressure measurements were obtained by placing a canula in the femoral artery; however, occasional monkeys developed gangrene and, therefore, we currently place a canula either in the radial or the posterior popliteal artery. Continuous arterial blood pressure is recorded and blood samples are frequently analyzed for blood gas values and electrolytes during the critical times of the experiment.

The postoperative care and follow up remain a critical part of the experiment, since a conscious monkey fights against any attempt to help him. Intramuscular medication an a partially-closed cage filled with oxygen are the only possible postoperative care. Fortunately, the animals are usually in an amazingly good condition a few hours after the operation, and do not require any acute or continuous treatment. Depending on their clinical condition they daily receive antibiotics, cardiotonics and diuretics. A blood count is done weekly, as well as ECG and chest X-rays. The

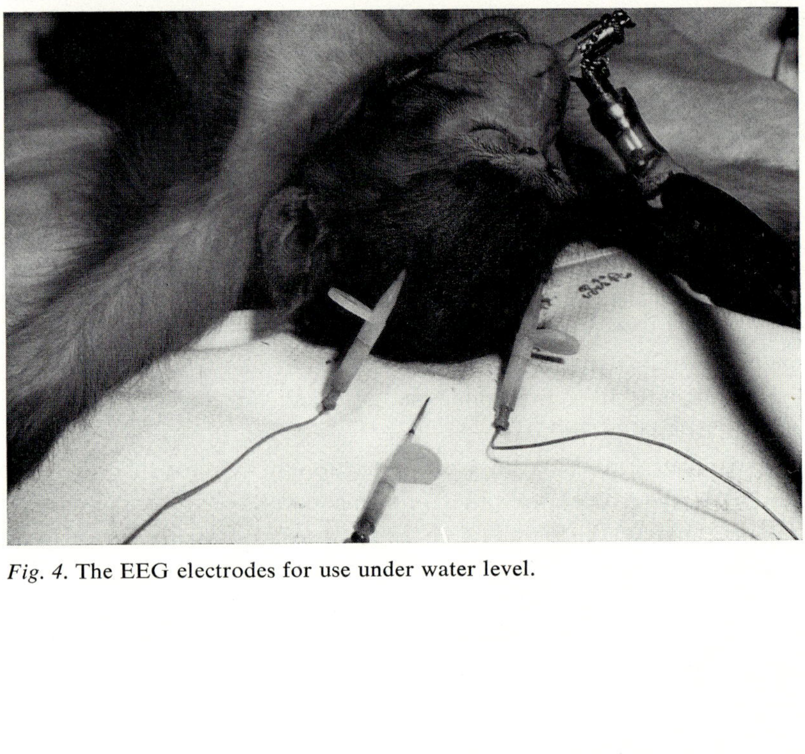

Fig. 4. The EEG electrodes for use under water level.

treated animals received the standard immunosuppressive treatment regardless of their clinical condition. The diagnosis of the rejection can be seen on some of our ECG.

The results of our experiment may be summarized as follows:

(1) Heart transplantations in monkeys from 2 kg body weight are successful without circulatory by-pass under hypothermia.

(2) The danger of acute rejection of the graft can be avoided with proved therapy.

The unsolved problem remains the obliterative arteritis due to chronic rejection – this seems to be the major factor of the survival of the graft [3].

References

1 Barnard, C. N.: A human cardiac transplant. Sth. afr. med. J. *41:* 1271 (1967).

2 Barrat-Boyes, B. G.; Simpson, M., and Neutzer, J. M.: Intracardiac surgery in neonates and infants using deep hypothermia and limited cardiopulmonary bypass. Circulation *43/44:* suppl 1, pp. 1–25/1–30 (1971).

3 Bieber, Ch. P.; Stinson, E. B.; Shumway, N. E.; Payne, R., and Kosek, J.: Cardiac transplantation in man. VII. Cardiac allograft pathology. Circulation *41:* 753 (1970).

4 Callaghan, J. C.; McOwens, D. A.; Scott, J. W., and Bigelow, W. G.: Cerebral effects of experimental hypothermia. Arch. Surg. *68:* 208 (1954).

5 Cooley, D. A.; Bloodwell, R. B.; Hallman, G. L., and Nora, J. J.: Transplantation of the human heart. J. amer. med. Ass. *205:* 479 (1968).

6 Dicke, H.; Marquet, R.; Heystek, G., and Balner, H.: Effect of immunosuppressive treatment and leucocytes antigen matching on kidney allografts in rhesus monkeys. Transplant. Proc. *3:* 484 (1971).

7 McMurray, J. D.; Bernhard, W. F.; Taren, J. A., and Bering, E. A., jr.: The effect of hypothermia on the prolongation of permissible time of total occlusion of the afferent circulation of the brain. Surg. Gynec. Obstet. *202:* 75 (1956).

8 Motri, H.; Dillard, D. H.; Crawford, E. W.; Martin, W. E., and Merendino, K. A.: Method of surface-induced deep hypothermia for open heart surgery in infants. J. thorac. cardiovasc. Surg. *58:* 262 (1969).

9 Niijima, S.: Experimental study on hemodynamics during hypothermic anesthesia. J. jap. surg. Soc. *62:* 102 (1961).

10 Shumway, N. W.; Angell, W. W., and Wuerflein, R. D.: Progress in transplantation of the heart. Transplantation *5:* 900 (1967).

Authors' addresses: J. Vincent, Department of Thoracic and Cardiovascular Surgery, Catholic University, *Nijmegen*; H. Snijders, Department of Surgery, St. Elisabeth Hospital, *Amersfoort*; H. Stol, Z. Aytug, and H. L. F. Brom, Department of Thoracic Surgery, University of Leiden, *Leiden;* C. F. Hollander, R. L. Marquet, and G. A. Heystek, REP-TNO Institute, *Rijswijk Z.H.* (The Netherlands)

Oral Medicine

Medical Primatology 1972. Proc. 3rd Conf. exp. Med. Surg. Primates, Lyon 1972, part II, pp. 159–168 (Karger, Basel 1972)

Histological Changes in the Skin of Baboons Deprived of Riboflavin

H. Foy, T. Gillman[1] and A. Kondi

Wellcome Trust Research Laboratories, Nairobi and Institute of Animal Physiology, Babraham, Cambridge

Introduction

Attention has recently been focused on the vulnerability of the esophagus [13–15], and particularly its association with the epidemiology and etiology of esophageal cancers [1, 2]. The present report concerns our experimental findings in riboflavin-deficient baboons which we think could be relevant to these issues.

The integrity of stratified squamous epithelium is dependent on a variety of dietary components. Riboflavin deficiency, in particular, has been suspected as predisposing man to severe cutaneous and oral lesions. The skin lesions have been briefly reported in riboflavin-deprived baboons and it was shown that, clinically, these resolved with riboflavin therapy [5, 7].

We describe here the profound macroscopic and microscopic architectural disorganizations of keratinization and the associated lesions that occur in the skin, mouth and esophagus of riboflavin-deprived baboons. The esophageal lesions are particularly important as they seem to resemble similar pathological ones in man. The relevance of these findings in baboons to the puzzling etiology of the world distribution of primary esophageal cancers in man is considered.

Materials and Methods

Details of the diet and experimental procedures for producing severe lesions in baboons fed a balanced synthetic diet lacking only riboflavin have been reported elsewhere [6, 8].

1 T. Gillman regrettably died towards the end of 1971.

Observations on 19 young male riboflavin-deprived baboons form the basis of this report. Four pair-fed baboons on the same diet, but with 1 mg riboflavin added daily, served as controls and remained healthy. After eight to 24 weeks the riboflavin-deprived animals lost weight, became anorexic and developed severe cutaneous lesions. Eleven of the 19 deprived baboons died or were sacrificed and autopsied. Hematological changes in these animals have been described previously [5, 8]. Skin biopsies were done on eight animals before and after treatment with 10–15 mg of riboflavin daily, which produced rapid resolution of the skin lesions, appetite gain and recovery.

Histological studies were made using a variety of methods on biopsy or autopsy material. *In vitro* labelling with tritiated thymidine was also done on multiple skin biopsies, from different regions before and after treatment, using the technique of Lachapelle and Gillman [12].

Results

The onset and progress of the skin lesions are shown in figures 1–3. Macroscopically the changes commenced unobtrusively as small focal scaling and depigmentation areas (fig. 1), progressing to nodular lesions, most marked on the muzzle, chin and face. Finally, extensive scaling, 'mud-pack'-like reactions were observed on face (fig. 2 and 3), arms, hands, legs and feet; scrotal and perianal skin was later affected as was the periocular skin leading to severe blepharitis (fig. 2). Microscopically, the cutaneous lesions showed hyperkeratosis, gross derangement of the normally highly organized process of keratinization (fig. 4) with acanthosis, extensive and impressive 'pseudocarcinomatous' hyperplasia (fig. 5), thickened, deep intradermal extensions of widened, branching rete pegs (fig. 4 and 5) and aberrant individual cell keratinization and intraepidermal 'epithelial pearl' formation (fig. 4–6). Nuclear anaplasia was absent, despite strikingly increased epidermal nuclear labelling with tritiated thymidine (fig. 7). Acantholysis was often seen and gross changes in sebaceous glands and ducts (fig. 7).

In the 11 baboons that died or were sacrificed the esophageal mucosa was thin and pale, and in five of them there were large (3–5 cm diameter), chronic, penetrating ulcers with raised epithelial edges in the lower third of the esophagus or at the cardioesophageal junction. One of these animals had been given an antagonist of riboflavin (galactoflavin) as well as the deprived diet for 173 days; the other four had no galactoflavin, and had been riboflavin deprived for 160–241 days. In all five baboons the esophageal ulcers penetrated the muscularis mucosa to submucosa (fig. 12). Secondarily-thrombosed vessels were occasionally seen in the ulcer bases while the raised edges showed gross epithelial hyperplasia, numerous mitotic figures, basally

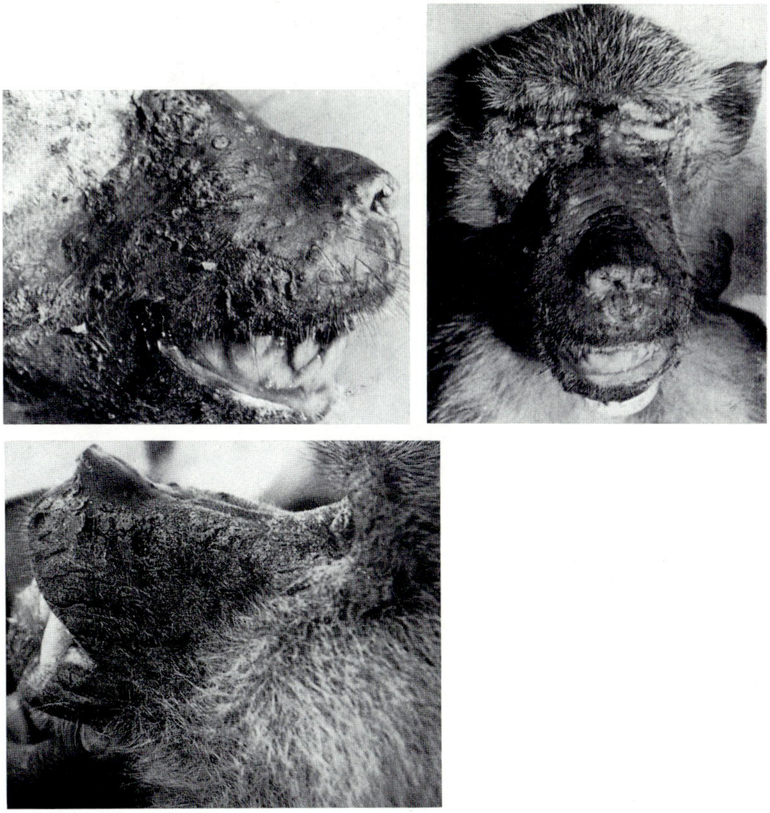

1 2 3

Fig. 1–3. Progression of facial skin lesions from small scales to nodules, and extensive mud-pack-like lesions.

and suprabasally (fig. 13), and were distinguishable from carcinomas only by the absence of muscular invasion by the disorganised, highly active epithelial growth (fig. 12 and 13). Histologically severe changes were seen in oral (cf. fig. 8 and 9) and esophageal epithelia, which showed irregular, patchy hyperplasia (fig. 9) thinning, dysplasia (fig. 11), foci of intraepithelial dyskeratinization and 'pearl' formation (fig. 9 and 13).

Despite the widespread and severe changes in squamous epithelia described above, ocular and respiratory epithelial dysplasia, found in vitamin-A deficiency [10] were not seen in riboflavin-deprived baboons.

All the cutaneous lesions were shown, by serial biopsies, to resolve promptly on giving 5 mg riboflavin for five days with return of epidermal

4

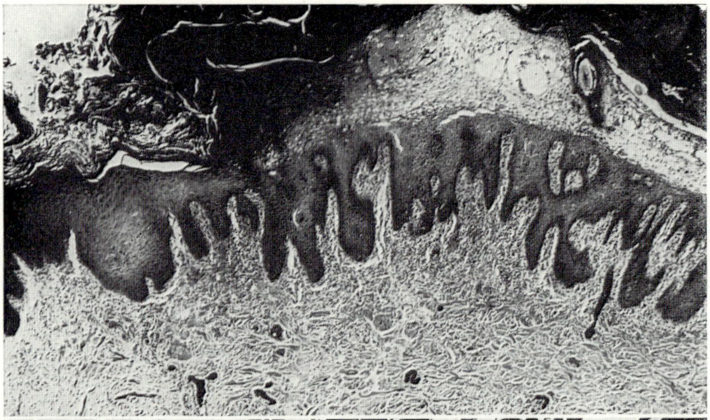

5

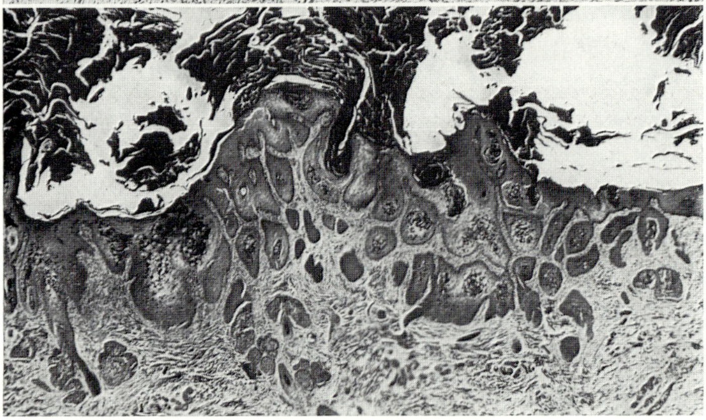

Fig. 4. Tibial skin with extensive epidermal hyperplasia, dysplasia. Increased width and depth of rete pegs with overlying hyperkeratosis; acantholysis and intraepidermal 'epithelial pearl'. × 30.

Fig. 5. Cheek skin gross epithelial hyperplasia dysplasia; pseudocarcinomatous downgrowths of deformed, elongated rete pegs; extensive cell keratinization and 'epithelial pearl' formation. × 20.

thymidine labelling to basal levels such as seen in controls – skin field matched for skin field.

Discussion

The cutaneous reactions in riboflavin-deprived baboons resembled those described in riboflavin-deprived mice painted with carcinogens [18]. The mu-

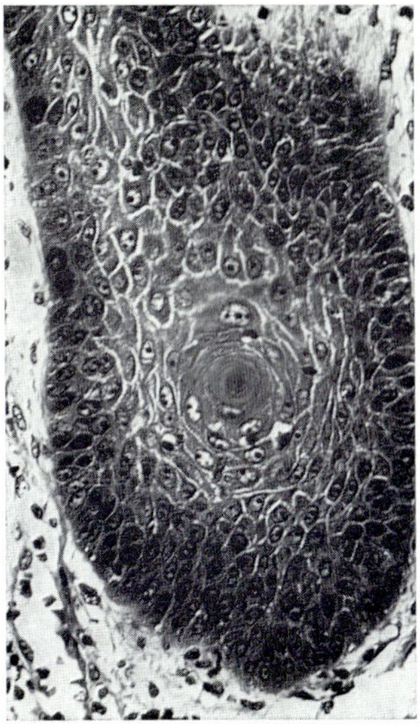

Fig. 6. One 'epithelial pearl' within thickened rete peg. × 300.

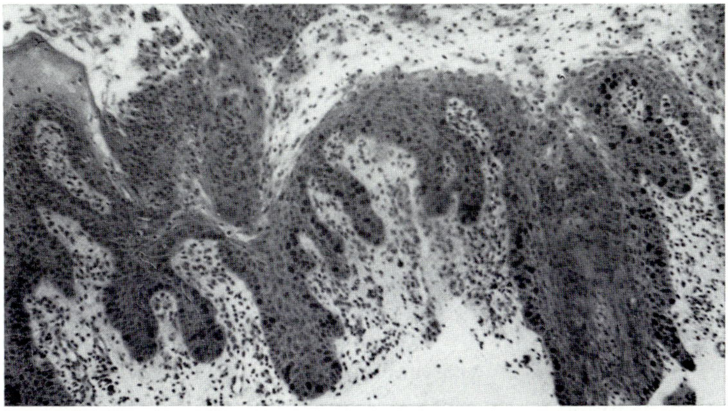

Fig. 7. Skin midforearm with tritiated thymidine showing increased numbers basally and suprabasally, premitotic epidermal cells with intranuclear label. × 75.

rine lesions were, however, not as severe as in baboons but they did show cellular abnormalities similar to those described above (fig. 4–6). According to WYNDER and KLEIN [18], the earliest microscopical changes in riboflavin-deprived mice were in the squamous epithelial-lined surfaces of the esophagus

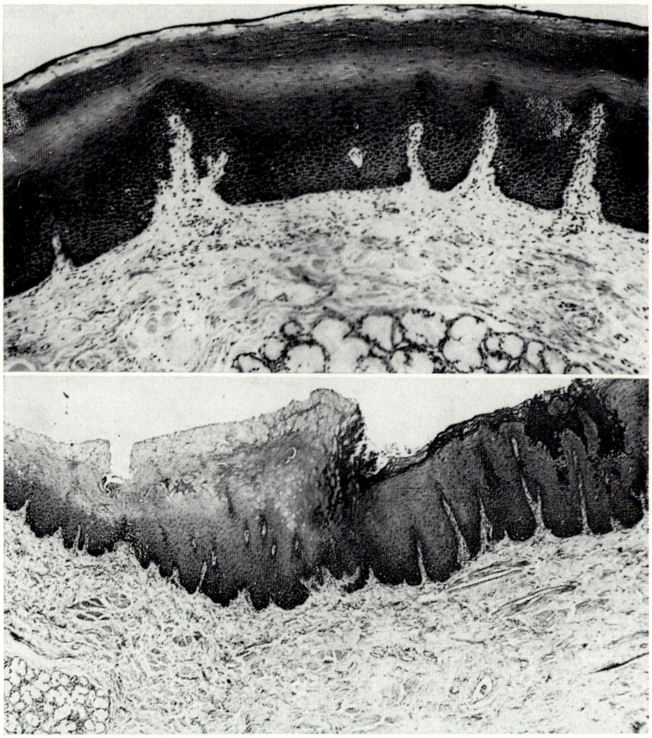

8

9

Fig. 8 and *9*. Buccal mucosa healthy baboon showing ordered keratinization (× 60) to compare with figure 9, which shows patchy thinning of buccal epithelium with aberrant foci of intraepithelial keratinization (right). × 25.

and fore-stomach, the thickness of which was, as in our baboons' esophagi, reduced (fig. 10 and 11). Esophageal ulceration or associated epithelial dysplasias, such as found in our baboons (fig. 12 and 13), did not occur in mice. Wynder and Klein [18] reported that riboflavin deprivation made mouse skin far more susceptible to tumour production by the applications of benzpyrene or croton oil, and Mulay and O'Gara [16] also noted the enhancing of azo-dye carcinogenesis by riboflavin deprivation in rats.

Wynder *et al.* [19] and Warburg [17] remarked on the similarities between the esophageal lesions in riboflavin-deprived mice and those precancerous, oral and esophageal lesions in the human Plummer-Vinson syndrome.

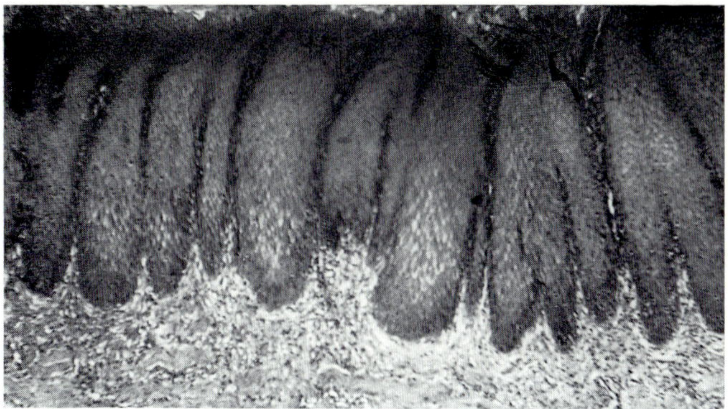

10

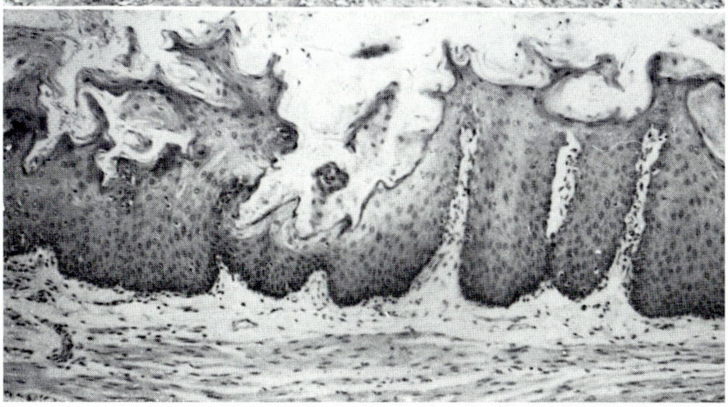

11

Fig. 10 and *11*. Healthy baboon; thickened lower esophageal epithelium to compare with figure 11, showing gross irregularities in thickness and keratinization of riboflavin deprivation.

The etiology of esophageal cancer has attracted considerable attention during the last decade as the global incidence has been shown to vary so widely. It is stated, for example, that the incidence of esophageal cancer in Kazakhstan is almost 200 times higher than in Canada [11] and great variations in incidence are said to exist in various parts of Africa.

Doll [3] considered that the marked differences between high and low incidence areas of esophageal cancer were associated with some unknown locally operative environmental or dietary factor. Whether ingestion of nitrosamines, or other known or unknown natural or synthetic dietary carcinogens, or the presence of riboflavin antagonists, as in ackee nuts [9], are indeed causative agents, clearly awaits further epidemiological and experimental studies.

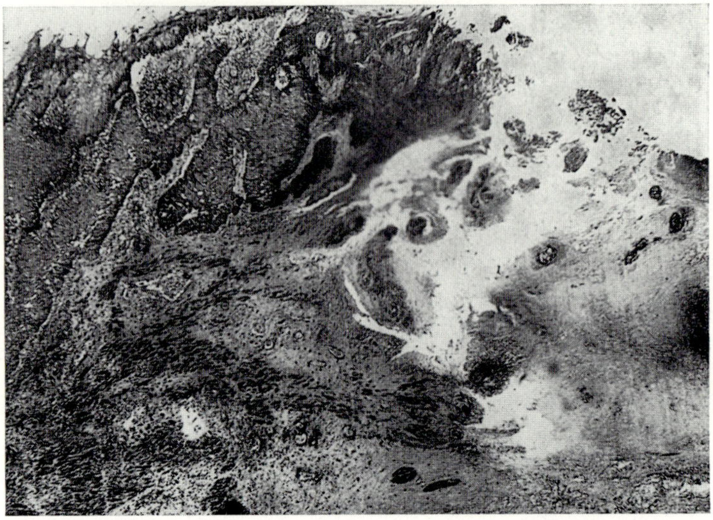

Fig. 12. Chronic deeply-penetrating ulcer of lower esophagus with deep necrosis (through muscularis mucosae) and grossly abnormal hyperplastic epithelium at left of ulcer's edge. × 35.

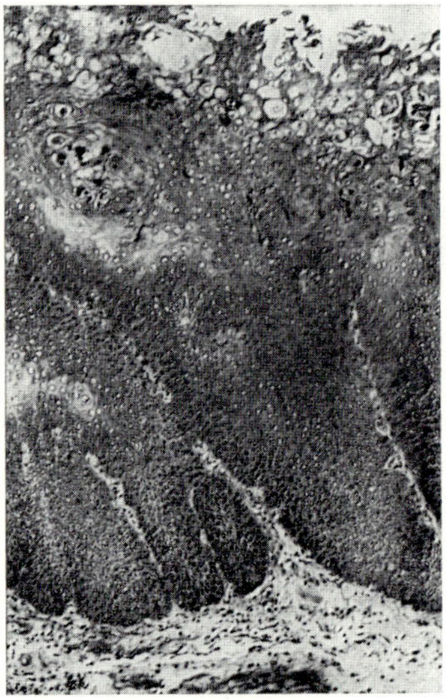

Fig. 13. Higher power of figure 12, showing very thickened hyperplastic and dysplastic esophageal epithelium at edge of ulcer with numerous widely-scattered nuclei and intraepithelial 'pearl' formation.

Our findings in baboons seem to underline the need to consider severe riboflavin deprivation as playing some role in the etiology of esophageal dysplasia and cancers. The buccal and esophageal lesions in our baboons would clearly be regarded by some as precancerous, or at least likely to increase the susceptibility of a nutritionally-deranged buccal or esophageal epithelia to the effects of carcinogens.

The high incidence in our riboflavin-deprived baboons of epithelial dysplasia, associated with chronic penetrating ulcers in the lower esophagus, and the findings of WYNDER and KLEIN [18] and MULAY and O'GARA [16], suggests that further work is necessary to elucidate the epidemiological, clinical, histological and biochemical part that riboflavin deficiency plays in this syndrome. Riboflavin deficiency is considered to be widespread in Africa, where, in addition, severe nutritional deficiency also has a very high incidence in certain areas [4].

The riboflavin-deprived baboon is a valuable experimental animal for the production of nutritionally-induced buccal and esophageal damage and one in which the synergistic effects of carcinogens, presently only suspect, could be more critically examined.

Acknowledgements

One of us (T. G.) wishes to thank the Wellcome Trust for travel grants to Nairobi which permitted him to participate in these studies. Our thanks are also due to Messrs. P. G. W. SMITH for the histological preparations, L. J. WRIGHT and A. L. GALLUP for assistance with preparation of photomicrographs, Mrs. D. HERBERT, Miss Z. A. M. MACDOUGALL, Miss J. M. JOHNSTON and Mrs. V. R. B. DAVIDSON for their patient secretarial assistance and to Mr. D. W. BUTCHER, the Librarian at the ARC Institute, for his help.

References

1 British Medical Journal: Leading article. Tumours of the oesophagus. Brit. med. J. *ii:* 779 (1968).
2 British Medical Journal: Leading article. Gastric and oesophageal cancer in the Welsh. Brit. med. J. *ii:* 590 (1969).
3 DOLL, R.: The geographical distribution of cancer. Brit. J. Cancer *23:* 1 (1969).
4 DU PLESSIS, L. S.; NUNN, J. R., and ROACH, W. A.: Carcinogen in a Transkeian Bantu food additive. Nature, Lond. *222:* 1198 (1969).
5 FOY, H. and KONDI, A.: Comparison between erythroid aplasia in marasmus and Kwashiorkor and the experimentally induced erythroid aplasia in baboons by riboflavin deficiency. Vitam. Horm., N. Y. *26:* 653 (1968).

6 Foy, H.; Kondi, A., and Mbaya, V.: Effect of riboflavine deficiency on bone marrow function and protein metabolism in baboons. Preliminary report. Brit. J. Nutr. *18:* 307 (1964).

7 Foy, H.; Gillman, T.; Kondi, A., and Preston, J. K.: Hepatic injuries in riboflavin and pyridoxine deficient baboons. Possible relations to aflatoxin induced hepatic cirrhosis and carcinoma in Africans. Nature, Lond. *212:* 150 (1966).

8 Foy, H.; Kondi, A., and Vergee, Z.: Relation of riboflavin deficiency to corticosteroid metabolism and red cell hypoplasia in baboons. J. Nutr. *102:* 571 (1972).

9 Gaskin, R. St. C. and Persaud, T. V. N.: The anti-tumour effect of the active principle from *Blighia sapida* (ackee). West Indian med. J. *16:* 198 (1967).

10 Hayes, K. C.; McCombs, H. L., and Faherty, T. P.: The fine structure of vitamin A deficiency. 1. Parotid duct metaplasia. Lab. Invest. *22:* 81 (1970).

11 Kramchaninov, N. F.: Materialy k izucheniiu smertnosti ot raka pishchevoda sredi naseleniia Gur'evskoi oblasti. Vop. Onkol. *14:* 38–42 (1968).

12 Lachapelle, J. M. and Gillman, T.: Tritiated thymidine labelling of normal human epidermal cell nuclei. A comparison, in the same subjects, of *in vivo* and *in vitro* techniques. Brit. J. Derm. *81:* 603 (1969).

13 Lancet: Leading article. Cancer in Africa. Lancet *i:* 371 (1967).

14 Lancet: Leading article. The vulnerable oesophagus. Lancet *ii:* 267 (1968).

15 Lancet: Leading article. Oesophageal cancer in Africa. Lancet *ii:* 1178 (1969).

16 Mulay, A. S. and O'Gara, R. W.: Enhancing effect of a riboflavin analog on azo-dye carcinogenesis in rats. J. nat. Cancer Inst. *40:* 731 (1968).

17 Warburg, O.: Nobel Laureates Lecture, Lindau, Germany.

18 Wynder, E. L. and Klein, U. E.: The possible role of riboflavin deficiency in epithelial neoplasia. 1. Epithelial changes of mice in simple deficiency. Cancer *18:* 167 (1965).

19 Wynder, E. L.; Hultberg, S.; Jacobson, F., and Bross, I. J.: Environmental factors in cancer of the upper alimentary tract. A Swedish study with special reference to Plummer-Vinson (Patterson-Kelly) syndrome. Cancer *10:* 470 (1957).

Author's address: Dr. Henry Foy, Wellcome Trust Research Laboratories, *Nairobi* (Kenya)

Medical Primatology 1972. Proc. 3rd Conf. exp. Med. Surg. Primates, Lyon 1972, part II, pp. 169–176 (Karger, Basel 1972)

Betel Quid Inducement of Carcinoma *in situ* in the Buccal Mucosa of Baboons

J. E. Hamner, III and O. M. Reed

Section of Oncology, National Institute of Dental Reserach, National Institutes of Health, Bethesda, Md. and NIDR Unit, Southwest Foundation for Research and Education, San Antonio, Tex.

Introduction

Numerous epidemiological surveys, cancer registries, and individual investigator's reports have drawn attention to the relationship between the high prevalence of oral carcinoma in Southeast Asia and the commonly-practiced habit of chewing betel [1, 5, 6, 8, 10–13, 16–21, 23–26, 30].

The standard betel quid, known in India as 'pan', is composed of betel leaf, areca palm nut, lime, and catechu [17]. Tobacco is frequently added to this chew. Each of the five ingredients may play some role in cancer induction, either directly, as a carcinogenic substance, or indirectly, as a cocarcinogen. Buccal carcinoma in India has been found to occur far more frequently among the betel chewers who incorporate tobacco in their basic quid [1, 10, 18, 23, 30].

The present investigation was undertaken as a long-range study to test the carcinogenic capacity of betel quid ingredients by chronic exposure of the baboon buccal mucosa to these substances, closely duplicating its customary usage by humans, since either the occurrence of spontaneous tumors or the induction of experimental tumors are rare in nonhuman primates [2, 7, 14, 15, 22, 27–29, 31].

Materials and Methods

One young adult male and 11 female baboons *(Papio cynocephalus)* were conditioned to a deficient dietary protein level of 7.2% in an effort to emulate Indian dietary standards. The protein deficient dietary patty was composed of 64% protein deficient primate chow made by Nutritional Biochemicals Corporation, Cleveland, Ohio, 27% cottonseed hulls, and 9% baboon biscuits made by Ralston Purina Co., St. Louis, M. The diet contained 7.2% protein on a dry weight basis. Each baboon was fed one pound of this diet per day in three portions. The animals were housed individually in single, elevated standard baboon cages, with water always available.

The baboons' weight dropped slightly in the initial phase, but was maintained thereafter on a fairly even scale. In contrast, their average serum protein level was lowered from a normal of 7.4 to 6.4% and has been kept at the reduced level for over three years.

To insure freshness of the betel leaves in order to closely duplicate the '*pan*' habit, selected leaves were routinely flown air-mail special delivery from Bombay, India to San Antonio, Tex. and kept under moist refrigeration. Using a mortar and pestle, the betel quid was prepared as a ground mixture of fresh betel leaves, areca palm nut, and USP calcium hydroxide. Three times per week, five of the baboons had approximately 3.5 g of the basic betel quid administered via a plastic open-end 3-ml syringe, while seven other animals had the basic betel quid plus Maharashtran sun-dried tobacco placed against their right buccal mucosa, which had been modified originally by two semilunar incisions and sutures to create a retention pouch. The opposite left buccal mucosa of each animal served as a control.

After 48 months of this regimen, half of the baboons in each group were placed on normal diet, and the betel quid treatment was ceased. Clinical observation and biopsy wia

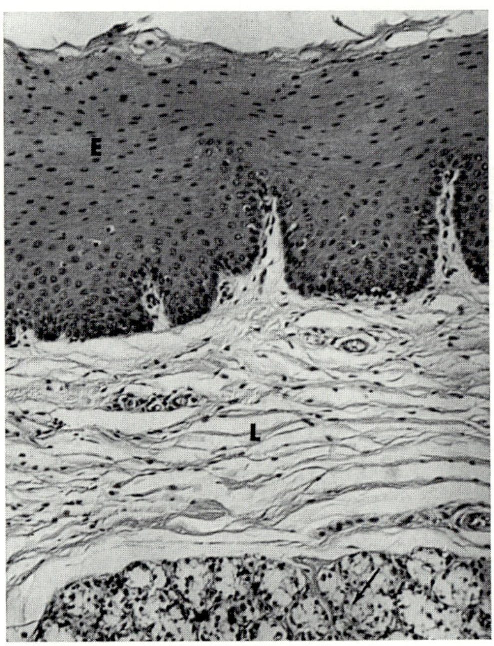

Fig. 1. Normal baboon buccal mucosa. E = epithelium; L = lamina propria; arrow = mucous gland. HE. × 130.

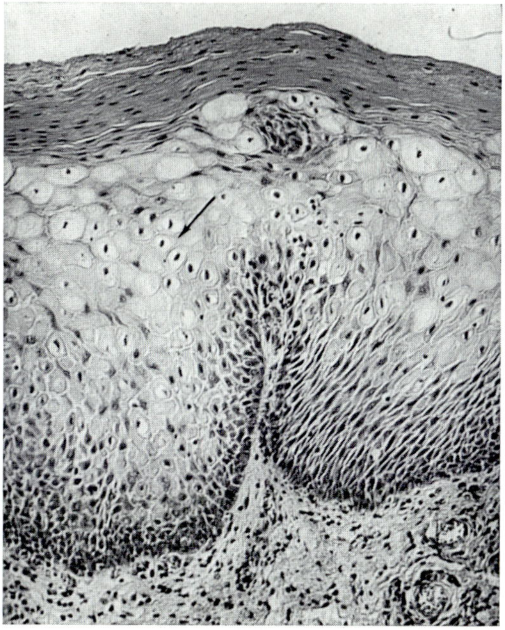

Fig. 2. Betel quid + tobacco pouch treatment for 12 months. Hyperparakeratosis, spindling, hyperchromatic-nucleated atypical cells in the lower strata, and marked ballooning of the intervening cells (arrows). HE. × 180.

be used to follow these animals to detect any change of their previously-treated buccal mucosa in a return toward normal.

Sernylan (phencyclidine hydroxhloride) was used to tranquilize the animals prior to each treatment (dose: 1 ml/kg body weight). Pentobarbital sodium (dose: 30 mg/kg body weight) was utilized to anesthetize the baboons sufficiently to permit the initial surgical modification of the buccal mucosa.

Clinical observations were made thrice weekly prior to quid insertion into the buccal pouch. Biopsies were taken after 1, 6, 9, 12, 16, 23, 29, 34, 42, 48, and 50 months. The tissue specimens were fixed in 10% buffered formalin, dehydrated in ascending alcohols, embedded in paraffin, sectioned at 6 μm, and stained with hematoxylin and eosin, the Masson trichrome, and the PAS methods.

Results

The preliminary results of the first 34 months of this study have been previously reported [9]. After 12 months of treatment with betel quid + to-

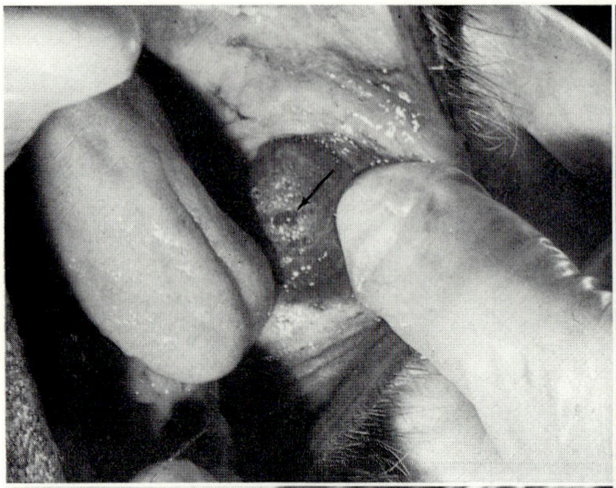

3

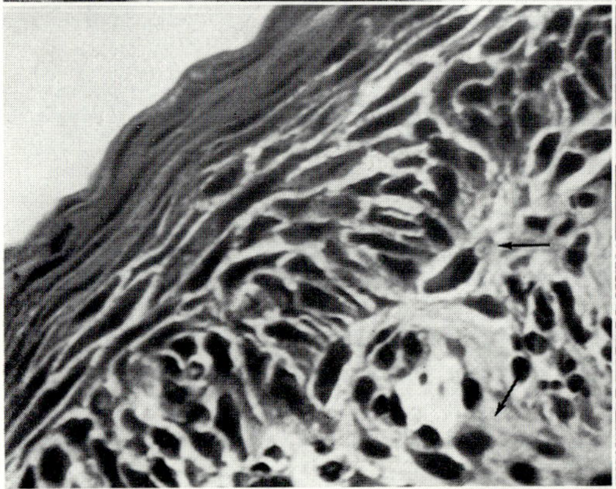

5

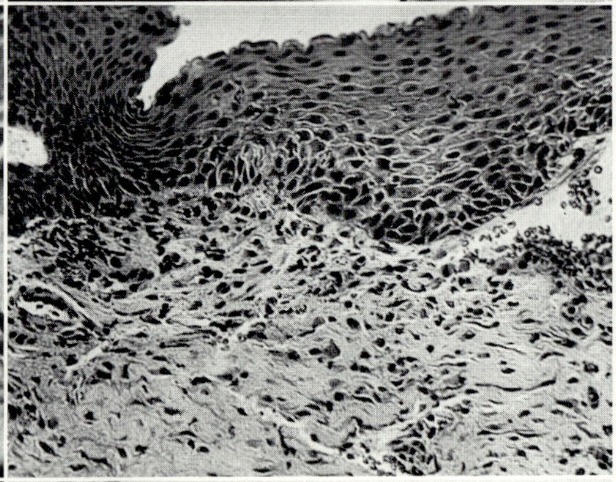

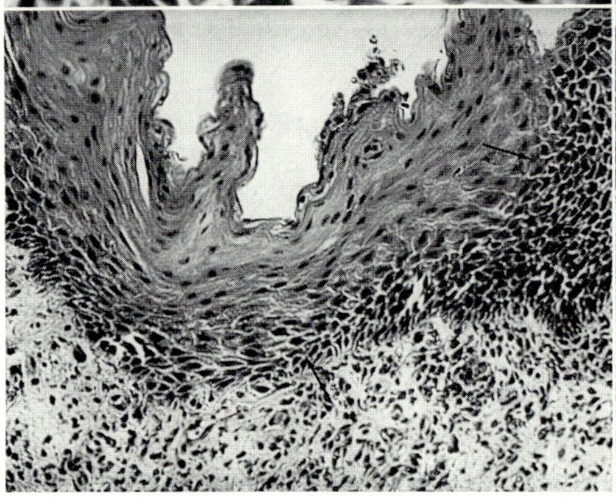

7

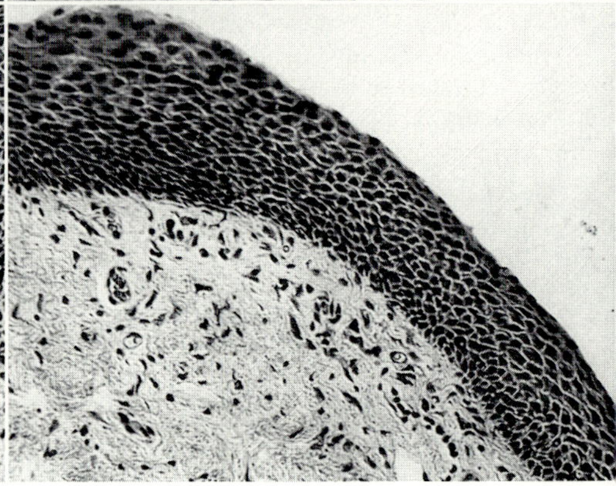

bacco three times per week, the buccal mucosa demonstrated a striking histologic change from its normal appearance (fig. 1 and 2). The principle changes of epithelial atypia included nuclear hyperchromatism, shrivelling of the nuclei in an elongated fashion, variance in cellular size and shape, ballooning of cells, and cellular proliferation. The group which received betel quid without tobacco exhibited these changes to a lesser degree. No histologic changes were noted in the opposite control buccal mucosal pouches.

Clinical buccal mucosal lesions were readily apparent in all animals of both groups after 30 months. These chronic ulcerative hemorrhagic lesions were detected where the betel quid was customarily placed in the buccal pouch (fig. 3).

After 34 months $^1/_7$ baboons in the betel + tobacco group revealed marked epithelial atypia in the biopsy specimen, diagnosed as carcinoma *in situ*. Beneath a narrow band of surface parakeratin, the lower strata epithelial cells assumed many bizarre forms, which had lost cohesion in many areas. In several places of this particular biopsy specimen the basement membrane was not demonstrable after PAS staining, and the atypical epithelial cells were proceeding downward (fig. 4).

A 42-month biopsy from the buccal mucosal pouch of another baboon receiving betel quid + tobacco demonstrated similar histological changes characteristic for carcinoma *in situ*.

Figure 6 depicts the clinical appearance of one baboon's buccal mucosa after 48 months when it was removed from the betel quid + tobacco treatment. Epithelial atypical changes were readily evident in the histopathology at this same time (fig. 7).

Having the betel quid + tobacco treatment removed for two months did not cause the buccal mucosa of the same animal to return to a normal state (fig. 8).

Fig. 3. Ulcerative, hemorrhagic appearance of buccal pouch mucosa treated for 30 months with betel quid + tobacco (arrow).

Fig. 4. Carcinoma *in situ* with loss of a demonstrable basement membrane and superficial downward thrust of markedly atypical epithelial cells (arrows); betel quid + tobacco, 34 months. HE. × 450.

Fig. 5. Note the loss of intercellular cohesion (arrow) and the abrupt change into carcinoma *in situ* (arrow) within the epithelium; betel quid + tobacco, 42 months. HE. × 200.

Fig. 6. Hemorrhagic, ulcerative buccal mucosal lesions (arrow) after 48 months treatment with betel quid + tobacco.

Fig. 7. Carcinoma *in situ*, 48 months betel + tobacco treatment HE. × 190.

Fig. 8. Histologic appearance of the same baboon's buccal mucosa two months later exhibiting no appreciable reversal toward normality.

Discussion

Results from the present study indicate that the severe caustic action of lime damages both the epithelium and underlying connective tissue, allowing tobacco apparently to exert a weak carcinogenic effect on the epithelial tissue. The cycle of chemical trauma and tissue breakdown with inflammation is followed in sequence by epithelial regeneration, which is eventually histologically characterized by bizarre cells with extensive atypical features.

The histopathological findings of epithelial atypia in this reported study are similar to those reported by COHEN [3, 4], who sutured 'Mainpuri' tobacco into surgically-created buccal pouch areas in monkeys.

Our investigation will be continued along the lines described to determine if the lesions appearing as carcinoma *in situ* after 48 months will progress on to extensive, invasive carcinoma. Likewise, it will be interesting to note whether continued abstinence from the betel quid + tobacco treatment in the other group of baboons will produce a reversal from the epithelial atypical state to normal in their buccal mucosa.

Acknowledgements

The author wishes to express deep appreciation to Mr. M. A. PARELKER and the Tata Institute of Fundamental Research in Bombay, India for faithfully supplying the fresh betel leaves, areca nuts, and tobacco utilized in this investigation as well as to Mrs. MARY ALICE LARSEN at the National Institute of Dental Research and Mrs. ESTHER RODRIGUEZ at the Southwest Foundation for Research and Education for the preparation and staining of the histologic materials, to Mr. I. GOMEZ and Mr. J. MARIN for the animal care, and to Mrs. BARBARA WILLIAMSON for the manuscript preparation.

References

1 ATKINSON, L.; CHESTER, I. C.; SMYTH, F. G., and TEN SELDAM, R. E. J.: Oral cancer in New Guinea. Cancer *17:* 1289–1298 (1964).

2 BONNE, C.; LODDER, J. und STTEEF, G. M.: Das Teercarcinom beim Affen *(Macacus cynomolgus)*. Z. Krebsforsch. *32:* 310–326 (1930).

3 COHEN, B. and SMITH, C. J.: Aetiological factors in oral cancer. Experimental investigation of early epithelial changes. Helv. odont. Acta *2:* 112–124 (1967).

4 COHEN, B.; POSWILLO, D. E., and WOODS, D. A.: The effects of exposure to chewing

tobacco on the oral mucosa of monkey and man. Ann. roy. Coll. Surg. Engl. *48:* 255–273 (1971).

5 COOPER, M. A.: Cancer Registry for the Kanyakumari District, Madras, India (1965–1967).

6 DAVIS, G. D.: Buyo cheek cancer, J. amer. med. Ass. *44:* 711–718 (1915).

7 ENGLE, E. T.; KRAKOWER, C., and HAAGENSEN, C. D.: Estrogen administration to aged female monkeys with no resultant tumors. Cancer Res. *3:* 858–866 (1943).

8 FRIEDELL, H. L. and ROSENTHAL, L. M.: The etiologic role of chewing tobacco in cancer of the mouth. J. amer. med. Ass. *116:* 2130–2135 (1941).

9 HAMNER, J. E., III and REED, O. M.: Betel quid carcinogenesis in the baboon. J. med. Prim. *1:* 75–85 (1972).

10 HIRAYAMA, T.: An epidemiological study of oral and pharyngeal cancer in central and Southeast Asia. Bull. Wld Hlth Org. *34:* 41–69 (1966).

11 HOBBS, L.: India India. A personal report, 1st ed. (McGraw-Hill, New York 1967).

12 HORTWICH, C.: Die menschlichen Genussmittel, pp. 524–553 and 569–591 (Tauchnitz, Leipzig 1911).

13 KHANOLKAR, V. R. and SURYABAI, B.: Cancer in relation to usages. Arch. Path. *40:* 351–361 (1945).

14 KLUVER, H. and BRUNSCHWIG, A.: Oral carcinoma in a monkey colony. A report of two additional cases. Cancer Res. *7:* 627–633 (1947).

15 KROTKINA, N. A.: A case of malignant bone tumor in a monkey induced with methylcholanthrene. Arkh. Pat. *10:* 38–42 (1948).

16 MALHOTRA, S. L.: Geographical distribution of gastrointestinal cancers in India with special reference to causation. Gut *8:* 361–372 (1967).

17 MEHTA, F. S.; PINDBORG, J. J., and HAMNER, J. E., III: Oral cancer and precancerous conditions in india, 1st ed. (Tata Press, Bombay 1971).

18 MEHTA, F. S.; PINDBORG, J. J.; GUPTA, P. C., and DAFTARY, D. K.: Epidemiologic and histologic study of oral cancer and leukoplakia among 50,915 villagers in India. Cancer *24:* 844–861 (1969).

19 MODY, J. K. and RANADIVE, K. J.: Biological study of tobacco in relation to oral cancer. Indian J. med. Sci. *13:* 1–23–1037 (1959).

20 ORR, I. M.: Oral cancer in betel nut chewers in Tranvancore. Lancet *225:* 575–580 (1933).

21 PAYMASTER, J. C.: Cancer and its distribution in India. Cancer *17:* 1026–1034 (1964).

22 PFEIFFER, C. A. and ALLEN, E.: Attempts to produce cancer in rhesus monkeys with carcinogenic hydrocarbons and estrogens. Cancer Res. *8:* 97–127 (1948).

23 PINDBORG, J. J.: Frequency of oral carcinoma, leukoplakia, leukokeratosis, leukoedema, submuccous fibrosis, and lichen planus in 10,000 Indians in Lucknow, Uttar Pradesh, India. J. dent. Res. *44:* 615 (1965).

24 SANGHVI, L. K.; RAO, K. C. M., and KHANOLKAR, V. R.: Smoking and chewing of tobacco in relation to cancer in the upper alimentary tract. Brit. med. J. *i:* 1111–1114 (1955).

25 SHANTA, V. and KRISHNAMURTHI, S.: A study of aetiological factors in oral squamous cell carcinoma. Brit. J. Cancer *13:* 381–388 (1959).

26 SINGH, A. D. and ESSEN, C. F. VON: Buccal mucosal cancer in South India. Amer. J. Roentgenol. *96:* 6–14 (1966).

27 SHIMKIN, M. B. and SHOPE, R. E.: Some observations on cancer research in the Soviet Union. Cancer Res. *16:* 915–917 (1956).

28 STEINER, P. E.; KLUVER, H., and BRUNSCHWIG, A.: Three carcinomas of the tongue in two monkeys. Cancer Res. *2:* 704–709 (1942).

29 SUGIURA, K.; SMITH, W. E., and SUNDERLAND, D. A.: Experimental production of carcinoma in rhesus monkeys. Cancer Res. *16:* 951–955 (1956).

30 WAHI, P. N.; LAHIRI, B., and KEHAR, H.: Epidemiology of oral and oropharyngeal cancer. A study of regional factors in Uttar Pradesh. J. amer. med. Ass. *46:* 175–181 (1966).

31 ZUCKERMAN, S.: A rhesus macaque *(Macaca mulatta)* with carcinoma of the mouth. Proc. zool. Soc., Lond. 59–61 (1930).

Author's address: Dr. JAMES E. HAMNER, III, Chief, Section of Oncology, National Institute of Dental Research, NIH, *Bethesda, MD 20014* (USA)

Medical Primatology 1972. Proc. 3rd Conf. exp. Med. Surg. Primates, Lyon 1972, part II, pp. 177–181 (Karger, Basel 1972)

Pyrolite Carbon Dental Implants in Baboons

A Preliminary Report

J. E. HAMNER, III and O. M. REED

Section of Oncology, National Institute of Dental Research, National Institutes of Health, Bethesda, M. and NIDR Unit, Southwest Foundation for Research and Education, San Antonio, Tex.

Introduction

Replacement of missing teeth by artificial means has been a long search, dating from man's earliest recorded history.

Recently, scientific interest and efforts for replacement of natural teeth have been concentrated on utilizing modern artificial materials which have proven successful for other original purposes.

Plastic in the form of polymethylmethacrylate has been employed as a rapid dental implant replacement material with varying reports of success [5, 8, 10, 11, 13, 14].

Calcium aluminate, as a high-fired ceramic with an adequate pore size of 100–150 μm to allow for tissue ingrowth, has been shown to be highly successful as an implantable dental root material in dogs [15], rhesus monkeys [7], and baboons [9].

Based upon its excellent properties of biocompatibility, strength, and ease of construction in various forms, it was decided to utilize pyrolite carbon as a dental implant material [1–3].

Materials and Methods

Pyrolite carbons have strengths that range from 50,000–100,000 p.s.i. and moduli of elasticity in the range of $3–6 \times 10^6$ p.s.i., i.e., in the same range as that of the bone [2]. Pyrolite carbons in thickness in excess of about 0.002 in are impermeable. The as-deposited surface is very rough and porous. This porosity extends to a depth of approximately 0.0005 in (12 μm), and under the scanning electron microscope the surface exhibits a morphology quite similar to corral [4].

1
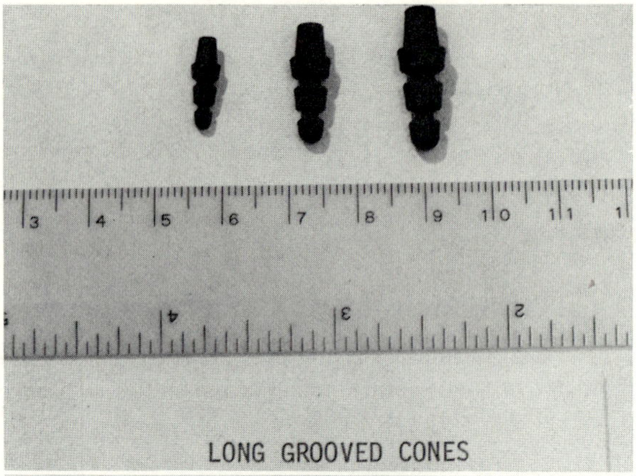

2
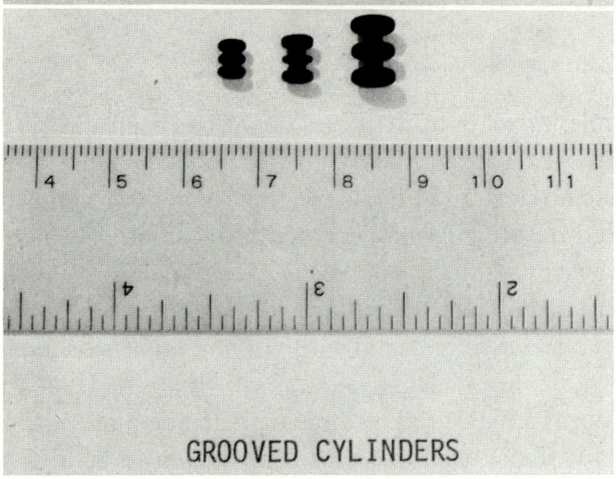

Fig. 1. Pyrolite carbon dental root implants in the form of long-grooved cones.
Fig. 2. Pyrolite carbon dental root implants in the form of grooved cylinders.

Following extraction of the necessary natural teeth, pyrolite carbon dental implants were immediately inserted into the fresh socket area of the jaws in the baboons *(Papio cynocephalus)*. These implants consisted of various sized and designed grooved cones, grooved cylinders, and blade-vents (fig. 1–3). In certain animals a two-step procedure was performed, using a two-piece grooved cone carbon implant (fig. 4). First, the carbon root, containing a stainless steel sleeve, was inserted, and the overlying gingiva was sutured to place. After healing was accomplished, the carbon crown portion was connected by the steel screw to the implanted root after reflecting a periosteal flap.

A total of 36 pyrolite carbon implants have been placed to date with the longest retained implant being eight months.

Histopathologic specimens have not yet been prepared, so this preliminary report will be concerned with the clinical and radiographic appearance of the implants.

3

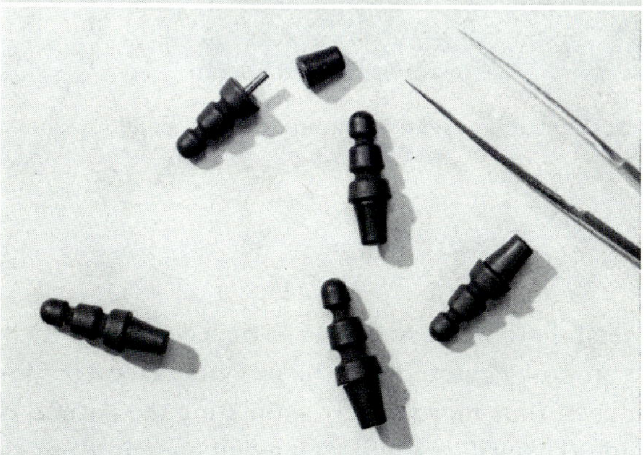

4

Fig. 3. Pyrolite carbon dental implants in the form of blade-vents.

Fig. 4. A two-piece (crown and root portions) grooved cone pyrolite carbon dental implant.

Preliminary Results

Some of the inverted carbon, grooved cones and cylinders were lost due to their improper conformity to the tooth socket's size and shape.

The carbon blade-vents, especially the thin ones, performed remarkedly well. They became quite stable within several months, and the gingival and periodontal tissue acceptance of the pyrolite carbon implant was excellent (fig. 5). Inflammation was not a problem. Several of the crowns of the blade-vents were too high, producing traumatic occlusion, which was corrected by slicing off the excess occlusal portion. Tissue compatibility was also quite satisfactory for the two-stage grooved cone carbon implants. Radiographs were of little use, since this material is not radiopaque.

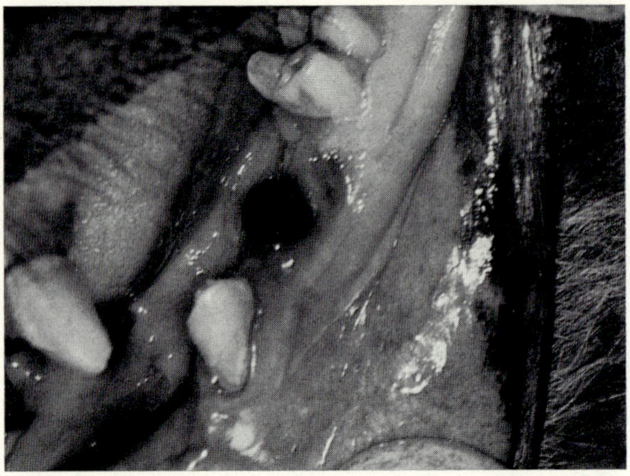

Fig. 5. Note the excellent gingival appearance around a three-month pyrolite carbon blade-vent.

Discussion

Based on the successful preliminary results of time compatibility, retension, and stability, we propose next to insert the pyrolite carbon blade-vents in edentulous baboon jaws, thus more closely simulating the natural condition in man. These blade vents will resemble the metal ones now used in patients [6, 12]. Both the blade-vent carbon implants and the two-stage grooved cone implants will be placed in edentulous jaw areas and subjected to occlusal force, as full gold crowns will be constructed. Simultaneously-implanted commercial metal blade-vent and anchor implants in the same baboon will serve as the corresponding control.

References

1 Bokros, J. C. and La Grange, L. D.: The compatibility of carbon and blood. Annual Report Contract PH 43-67-1411, National Heart and Lung Institute (1969).

2 Bokros, J. C. and Akins, R. J.: Applications of pyrolitic carbon in artificial heart valves. A status report. Presented at the Buhl International Conference, Pittsburgh, Penn., 1971.

3 Bokros, J. C.; La Grange, L. D., and Schoen, F. J.: Control of structure of carbon for use in bioengineering; in Engineering in medicine-bioceramics (Wiley, New York 1972).

4 Bokros, J. C. (personal commun.).

5 Brown, B. H.; Neff, P. A., and Tylenda, A.: Acrylic resin tooth implants in rhesus monkeys. A preliminary report. J. prosth. Dent. *22:* 367–370 (1969).

6 Cranin, A. N.: Oral implantology (Thomas, Springfield 1970).

7 Driskell, T. D.; O'Hara, M. J., and Sheets, H. D., jr.: Development and application of ceramics and ceramic composites for implant dentistry. Clemson University Symposium on Implant Materials for Dentistry, Clemson, S.C. 1970.

8 Hamner, J. E., III; Reed, O. M., and Hand, A. R.: Clinical, radiographic, histologic, and electron microscopic observations of plastic tooth implantations in baboons. Oral Surg. *30:* 555–570 (1970).

9 Hamner, J. E., III; Reed, O. M., and Greulich, R. C.: Ceramic root implantation in baboons. J. biomed. Mater. Res. Symposium: Biomaterials for skeletal and cardiovascular applications (Wiley-Interscience, Part 3, pp. 1–13, New-York 1972).

10 Hodosh, M.; Povar, M., and Shklar, G.: The dental polymer implant concept. J. prosth. Dent. *22:* 371–380 (1969).

11 Iwaschenko, G. M.: On the possibility of implanting acrylic teeth and roots. Dent. Abstr. *3:* 391 (1958).

12 Linkow, L. I.: The blade vent. The most promising tooth abutment. Oral Implant. *1:* 175–198 (1971).

13 Tobin-White, A.: Implantation of acrylic teeth in the jaws. Int. dent. J. *8:* 15 (1958).

14 Waerhaug, J. and Zander, H.: Implantation of acrylic roots in tooth sockets. Oral Surg. *9:* 46 (1956).

15 Young, F. A.; Hulbert, S. F., and Lawson, B. F.: Ceramic tooth implants. IADR Progr. Abstr., p. 216 (1970).

Author's address: Dr. James E. Hamner, III, Chief, Section of Oncology, National Institute of Dental Research, NIH, *Bethesda, MD 20014* (USA)

Medical Primatology 1972. Proc. 3rd Conf. exp. Med. Surg. Primates, Lyon 1972, part II, pp. 182–190 (Karger, Basel 1972)

Oral Sensory and Proprioceptive Innervation in the Squirrel Monkey

K. KUBOTA and T. MASEGI

Section of Anatomy, Institute of Stomatognathic Science, School of Dentistry, Tokyo Medical and Dental University, Yushima, Tokyo

Introduction

A need exists for description of the sensory innervation of the oral region from the viewpoint of the neuromuscular mechanism involved in masticatory movement. This neurohistological investigation of the oral sensory innervation in the squirrel monkey was designed to gain more morphological information of the neuromuscular mechanism involved in jaw movement of primates, including man. Such knowledge may have phylogenetic significance important for our understanding of the innervation of the human oral region. In this presentation, the neurohistological features will be described for the tongue, masticatory muscles, dental tissues, lip, palate, and temporomandibular joint.

Materials and Methods

Three fresh tongues and five heads of squirrel monkeys, fixed in 10% formalin, were used for this investigation. After freezing, two tongues were sectioned at 30 μm. The other tongue was embedded in celloidin and serial sections were cut at 30 μm.

Following hemisection in the medial line and decalcification in a solution of 5% HCl, the heads were embedded in celloidin. Serial sections were prepared at 30 μm from one half sagittally, and frontal sections were cut from the other half. Every other section was stained with two different staining techniques: hematoxylin-eosin (HE) and silver impregnation [19]. Counting of the muscle spindles was performed by tracing the position of the spindles on paper from the serial cross-sections stained with HE under a microscope. The total number measured on one side of each of the masticatory muscles was calculated from the number of sections in which the spindle could be observed.

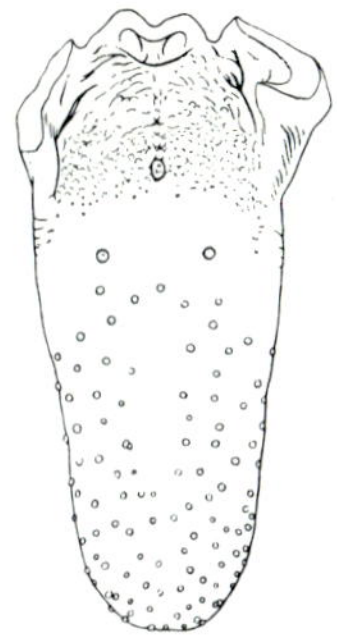

Fig. 1. Dorsal aspect of the squirrel monkey tongue.

Results

Tongue

Three circumvallate papillae arranged in an inverted V-shape, as found in squirrel monkey (fig. 1), represent the primitive pattern of the papillary arrangement on the dorsum of the mammalian tongue. There is a relationship between the number of vallate papillae and phylogenetic evolution of primates. As shown in figure 2, tongues of the lower monkeys from Tupaioids to Ceboids have three circumvallate papillae arranged in an inverted V-shape. Tongues of Simioids have three to six circumvallate papillae, arranged also in an inverted V-shape. Those of Anthropoids, including man, increase to 13 in number. Prosimiae and Platyrrhini reveal the basic mammalian arrangement of vallate papillae.

The dorsum of the tongue is covered with filiform papillae, consisting of a connective tissue core innervated by free nerve endings (fig. 3). The nerves evolve from the main nerve bundles which have abundant nerve cells in ganglion and isolation form, coursing forward to the apical region of the tongue through the space between the transverse and inferolongitudinal muscles.

Fungiform papillae are sparsely distributed on the dorsum (fig. 1) and are innervated by numerous nerve fibers. The papillae are furnished with few taste buds having some embryonic characteristics. No end formation of the glomerular type is seen in the secondary papillae. These findings are quite similar to those of the marmosets [15] and slow loris [12], but are different from the higher primates' tongues possessing the glomerular type [14, 17].

Subdivision of eight major groups of primates	Number of vallate papillae											Life			Eating habits				
	3	4	5	6	7	8	9	10	11	12	13	A	AT	T	OM	OMi	Fr	H	S
Antropoids																			
man							●	—	—	—	●			●	●				
orang							●	●				●					●	●	
chimpanzee					●	—	●						●		▲	▲	●	●	▲
gorilla					●									●			?	●	
gibbon	●	—	—	—	—	●						●			●	●	●	●	●
Simioids																			
papio	●													●	▲	●	(●)	●	●
pig-tailed				●									●		▲	●	●	●	●
Japanese m.		●	●										●		▲	●	●	●	●
patas			●											●		●	(●)	●	●
crab-eating	●	●										●	▲		●	●	●	●	●
colobus	●											●					▲	●	
Ceboids																			
wooly m.	●											●			?	?	●	●	
squirrel m.	●											●			▲	●	●	●	?
Hapaloids																			
Pinche's	●											●			●	●	●	●	?
marmoset	●											●			●	●	●	●	?
Tarsioids																			
tarsier	●											●			●	●	●	●	?
Lemuroids																			
aye-aye	●											●			●	●	●	●	
Lorisoids																			
slow loris	●											●			●	●	●	●	?
Tupaioids																			
tree-shrew	●												●		●	●	●	●	?

Fig. 2. Diagram showing relationship between the number of circumvallate papillae and other characteristics of primates. A = arboreal; AT = arboreal-terrestrial; T = terrestrial; OM = omnivorous; OMi = insectivorous plus vegetarian; Fr = frugivorous; H = herbivorous; S = seed-eater

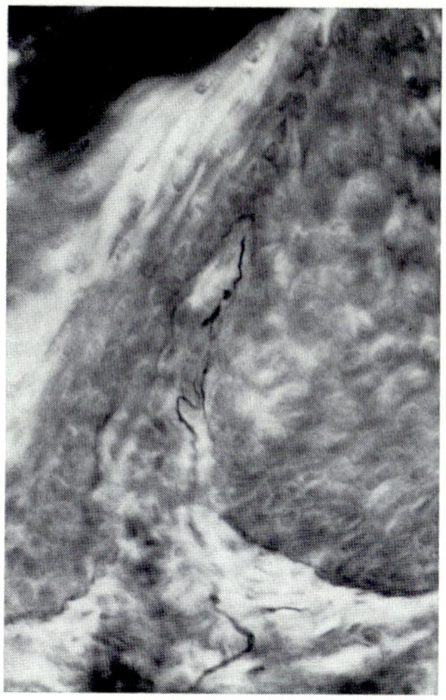

Fig. 3. Photomicrograph demonstrating sensory innervation in the filiform papillae. Silver-impregnation stain. × 500, reproduced at 63%.

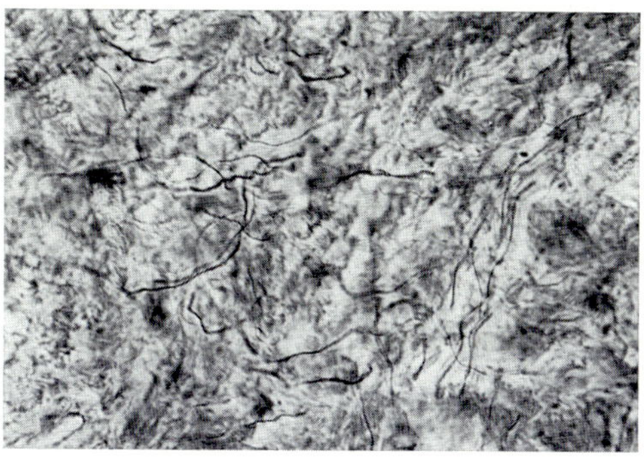

Fig. 4. Photomicrograph illustrating rich innervation in the circumvallate papillae. Silver-impregnation stain. × 380, reproduced at 63%.

The circumvallate and foliate papillae are supplied with abundant plexiform nerves and glandular tissues similar to those of the higher primates (fig. 4). The abundant thin and a few thick nerve fibers concentrated in the lateral wall area of the papilla form a complicated plexus. The lateral wall lined by this nerve plexus protrudes into the trench and is furnished with a large number of taste buds. This tendency of protrusion of the lateral wall is more conspicuous in the squirrel monkey than in the marmoset [15]. Such protrusion is caused by a bulk of the plexiform nerves being concentrated in the lateral area of the papilla and is much more prominent in the higher primates [14] than in the squirrel monkey [16].

No muscle spindles were found in the tongue of the squirrel monkey, even though muscle spindle-like structures have been demonstrated in the slow loris tongue [12]. The problem of whether muscle spindles are present in the tongue musculature has been investigated by many workers. CORBIN and HARRISON [4] obtained degeneration of from 2 to 5% of the myelinated fibers in the most peripheral portion of the hypoglossal nerve following removal of the dorsal root ganglion of C_2 in the rhesus monkey and of C_1 in the rabbit. They concluded that the degenerating fibers are sensory and proprioceptive to the tongue musculature. COOPER [3] has reported the presence of muscle spindles in human and monkey tongues, but he found an absence of such spindles in the cat and lamb. He postulated that the hypoglossal nerve served as the pathway of the afferent impulses from these organs. Muscle spindles have been reported as occurring symmetrically in the genioglossus muscles of the Japanese shrew-mole [13].

Masticatory Muscles

Of the four masticatory muscles, innervated by the mandibular division of the trigeminal nerve, in the squirrel monkey only three muscles (masseter, medial pterygoid, and temporalis) are innervated by the muscle spindles. The lateral pterygoid muscle is devoid of spindles.

The masseter muscle, consisting of the zygomatico-mandibular, deep, and superficial layers (fig. 5), contains 53 muscle spindles. The muscle fibers of the former and the upper part of the middle extend horizontally from the zygomatic arch to the lateral surface of the mandibular ramus, being separated from each other incompletely by a nerve bundle. They appear to act in close relation to side movement of the mandible. Six muscle spindles were counted along the lower surface in the zygomatico-mandibular muscle and 24 spindles in the upper part of the deep layer. The superficial layer is separated from the deep layer by a thin collagenous septum arising from the lower surface of the arch. Twenty-three spindles are concentrated in the restricted area close to the lower portion of the septum.

The medial pterygoid muscle contains 57 spindles. They are concentrated in the restricted area of the inferomedial part of the muscle (fig. 5).

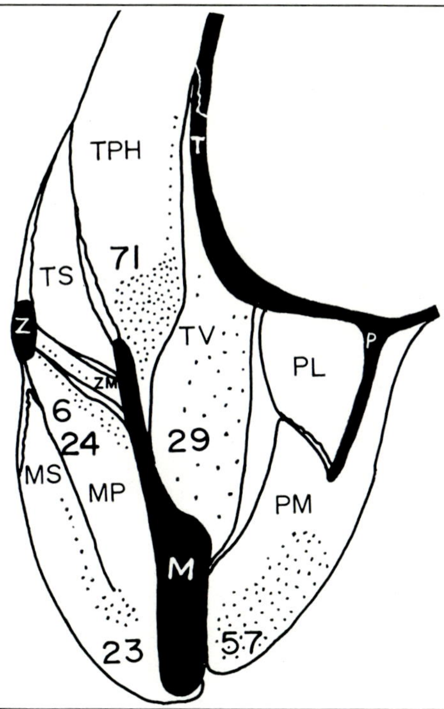

Fig. 5. Diagram of the number and functional localization of muscle spindles in the masticatory muscles of the squirrel monkey. M = mandible; MP = deep layer of masseter; MS = superficial layer of masseter; P = pterygoid process; PL = lateral pterygoid; PM = medial pterygoid; T = temporal bone; TPH = horizontal portion of the deep layer of temporal; TS = superficial layer of temporal; TV = vertical layer of temporal; Z = zygomatic arch; ZM = zygomaticomandibular muscle.

The temporal muscle, composed of three layers (vertical, deep, and superficial), is endowed more abundantly with spindles than are the other masticatory muscles (fig. 5). It contains 107 spindles: 78 in the lower part of the horizontal portion of the deep layer, lying on the roof of the joint (fig. 6) and 29 in the vertical layer.

In the literature pertaining to muscle spindles, it is found that they are present in the mammalian masseter, medial pterygoid, and temporal muscles [6, 9, 13, 25]. As shown in figure 5, it is highly probable that the spindles might have a functional localization in the muscles related to the jaw movements, e.g. in the horizontal portion of the deep layer of the temporal muscle (forward movement of jaw), in the superficial layer of masseter, medial ptery-

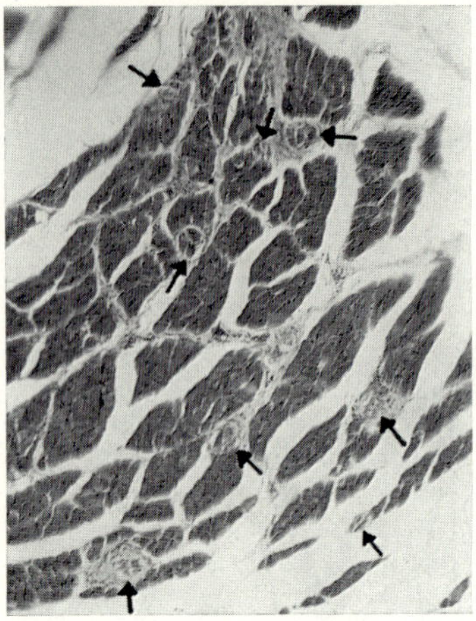

Fig. 6. Photomicrograph revealing muscle spindle distribution in the horizontal portion of the deep layer of the temporal muscle (arrows). HE. × 120, reproduced at 63%.

goid and vertical layer of temporal muscle (opening movement of jaw), and in the zygomatico-mandibular and deep layers of masseter muscle (side movement of jaw). There are, however, two different opinions as to whether the lateral pterygoid muscle is devoid of spindles [5, 6, 9, 13, 25] or has spindles [2, 7, 10, 22, 23]. The presence of muscle spindles has been reported in the lateral pterygoid muscle of *Macaca fascicularis* [10].

It is of a noteworthy phylogenetic significance that the neuromuscular findings in squirrel monkey are quite similar to findings in the Japanese shrew-mole [13]. The characteristic localization of muscle spindles in the masticatory muscles suggests that even with minimal movement of the jaw, the spindles convey sensory information about the jaw movement to the central nervous. The conspicuous concentration of a large number of muscle spindles in the horizontal portion of the deep layer of the temporal muscle is very significant in regard to the proprioceptive function in compensating for the absence of muscle spindles in its antagonist (lateral pterygoid muscle) for retrusion and protrusion of the jaw.

Dental Tissues

Tooth pulp of the squirrel monkey is supplied with abundant nerve fibers, distributed freely in the region of the coronal pulp. KUBOTA and KUBO-

TA [18] have already demonstrated the same finding in the *Macaca fascicularis*. The periodontal membrane consists of strong bundles of collagenous fibers. Nerve supply to the periodontal tissue is characterized by a free ramifying ending of the thin and thick nerve fibers. No specialized receptors as supposed by KIZIOR *et al.* [11] were found in this area. The *gingiva* has well-developed (tall) papillae, which are innervated by a free ending of thin nerve fibers branching from the large nerve bundle in the mucosa. The lip, especially the red area, is richly innervated by a free ending of thin nerve fibers. The *Palate* shows several elevated transverse ridges of mucosa. The papilla has a comparatively rich supply of free terminations of thin nerve fibers. The elevated ridges seem to be especially furnished with much more nerve elements than the other papillae.

Temporomandibular Joint

Careful observation of the joint preparation stained by silver impregnation produced no evidence of that the myelinated and nonmyelinated nerve fibers distributed in the joint tissue of the squirrel monkey. The joint innervation, however, has been demonstrated in the rhesus monkey [21] and various other mammals [1, 5, 8, 20, 24].

References

1 BERNICK, S.: Blood supply and innervation of the teeth, periodontium and temporomandibular joint of the guinea pigs. J. dent. Res. *43:* 812 (1964).

2 CHRISTENSEN, L. V.: Muscle spindles in the lateral pterygoid muscle of miniature swine. Arch. oral Biol. *12:* 1203–1204 (1967).

3 COOPER, S.: Muscle spindle in the intrinsic muscles of the human tongue. J. Physiol. *122:* 193–202 (1953).

4 CORBIN, K. B. and HARRISON, F.: The sensory innervation of the spinal accessory and tongue musculature in the rhesus monkey. Brain *62:* 191–197 (1939).

5 FRANKS, A. S. T.: Studies on the innervation of the temporomandibular joint and lateral pterygoid muscle in mammals. J. dent. Res. *43:* 947–948 (1964).

6 FREIMANN, R.: Untersuchungen über Zahl und Anordnung der Muskelspindeln in den Kaumuskeln des Menschen. Anat. Anz. *100:* 158–264 (1953–1954).

7 GILL, H. I.: Neuromuscular spindles in human lateral pterygoid muscles. J. Anat. *109:* 157–167 (1971).

8 HORMADA, J.: Die Innervation des Kiefergelenks und einige anatomisch-klinische Bemerkungen. Dtsch. Zahn-, Mund-Kieferheilk. *34:* 19–28 (1960).

9 KARLSEN, K.: The location of motor endplates and the distribution and histological

structure of muscle spindle in jaw muscles of the rat. Acta odont. scand. *23:* 521–547 (1965).

10 KARLSEN, K.: Muscle spindles in the lateral pterygoid muscle of a monkey *(Macacus irus)*. Arch. oral Biol. *14:* 1111–1112 (1969).

11 KIZIOR, J. E.; CUOZZO, J. W., and BOWMAN, P. C.: Functional and histological assessment of the sensory innervation of the periodontal ligament of the cat. J. dent. Res. *47:* 59–64 (1968).

12 KUBOTA, K.: Comparative anatomical and neurohistological observations on the tongue of slow loris *(Nycticebus coucang)*. Anat. Rec. *158:* 163–176 (1967).

13 KUBOTA, K.: Particular proprioceptive innervation of snout and masticatory muscles in Japanese shrew-mole. Symp. Morphology on Maxillomandibular Apparatus at the 9th Int. Congr. Anatomists, Leningrad 1970.

14 KUBOTA, K. and HAYAMA, S.: Comparative anatomical and neurohistological observations on the tongues of primates. Primates *3:* 74–75 (1961–1962).

15 KUBOTA, K. and HAYAMA, S.: Comparative anatomical and neurohistological observations on the tongues of pigmy and common marmosets. Anat. Rec. *150:* 473–486 (1964).

16 KUBOTA, K.; HAYAMA, S., and IWAMOTO, M.: Comparative anatomical and neuro histological observations on the tongue of the squirrel monkey *(Saimiri sciurea)*. Primates *7:* 381–389 (1966).

17 KUBOTA, K.; ITO, K.; AMEMIYA, A., and KOMORI, A.: Chronological changes of the nerve supply, especially of the pattern of the sensory nerve endings in the mucous membrane of the apex linguage of the human tongue. Bull. Tokyo med. dent. Univ. *9:* 457–469 (1962).

18 KUBOTA, K. and KUBOTA, J.: On the formation of the so-called cell-rich zone in the human dental pulp. (Innervation of Tooth, No. 5.) Okajimas Folia anat. jap. *37:* 29–47 (1961).

19 KUBOTA, K.; SATO, Y., and KUBOTA, J.: A silver impregnation method of peripheral nerves with frozen sections. (Japanese with English abstract.) J. Jap. stomat. Soc. (Kokubyo Zassi) *23:* 167–274 (1956).

20 MASSON, G.: Structure et innervation du ménisque temporo-mandibulaire. Schweiz. Mschr. Zahnheilk. *63:* 739–763 (1953).

21 MOFFETT, B. C., jr.: The distribution of blood vessels and nerves in the tempormandibular joint of the rhesus monkey. Anat. Rec. *118:* 431 (1954).

22 RAKHWAY, M. T.; SHEHATA, S. H., and BADAWY, Z. H.: The proprioceptive innervation of the lateral pterygoid muscle in man and some other mammals. Acta Anat. *79:* 581–598 (1971).

23 SMITH, R. D. and MARCARIAN, H. Q.: The neuromuscular spindles of lateral pterygoid muscle. Anat. Anz. *120:* 47–53 (1967).

24 THILANDER, B.: Innervation of the temporomandibular disc in man. Acta odont. scand. *22:* 151–156 (1961).

25 VOSS, H.: Ein besonders reichliches Vorkommen von Muskelspindeln und der Anthropoiden. Anat. Anz. *81:* 290–292 (1935–1936).

Author's address: Dr. KINZIRO KUBOTA, Section of Anatomy, Institute of Stomatognathic Science, Tokyo Medical and Dental University, *Tokyo 113* (Japan)

Medical Primatology 1972. Proc. 3rd Conf. exp. Med. Surg. Primates, Lyon 1972, part II, pp. 191–197 (Karger, Basel 1972)

Delayed Hypersensitivity Reaction in the Periodontium of the Marmoset

A Preliminary Report[1]

B. M. Levy, D. C. Nelms, S. Dreizen and S. Bernick

University of Texas at Houston,
Dental Science Institute and Graduate School of Biomedical Sciences.
Houston, Tex., and Department of Anatomy, School of Medicine,
The University of Southern California, Los Angeles, Calif.

Introduction

The gingival reaction to Freund's complete adjuvant (CFA) has received surprisingly little attention, although delayed hypersensitivity reactions in the gingiva of guinea pigs have been compared with those in the skin [Adams *et al.*, 1969]. Baer and Lorenz [1970] showed that the marmoset, *Saguinus mystax*, was capable of mounting a delayed hypersensitivity skin reaction to tuberculin, but the lesions were not identical to those elicited in sensitized humans or guinea pigs. No descriptions of the histological reaction to CFA in marmosets have been reported.

The histopathology of periodontal disease in the marmoset has been under intensive investigation in our laboratories. Because the destructive phase of chronic periodontitis in these animals might have a delayed hypersensitivity reaction component, a study was designed to determine whether delayed hypersensitivity reactions would, in fact, destroy the periodontal tissues. The experiments reported here deal with the periodontal tissue responses to combined CFA and purified protein derivative of the tubercle bacillus (PPD) and to PPD alone in tuberculin-sensitized animals.

1 Supported in part by USPHS, NIDR Grant DE 02232.

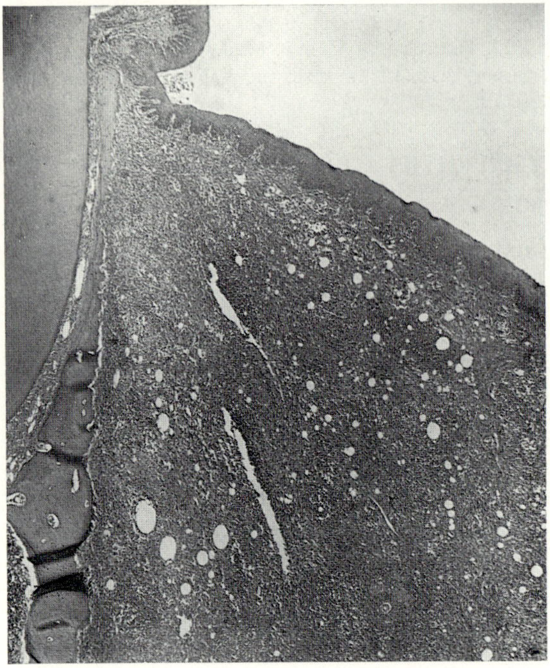

Fig. 1. Low power photomicrograph illustrating the adjuvant granuloma limited to the attached gingiva. HE stain. × 10.

Materials and Methods

Ten marmosets, eight *Callithrix jacchus* and two *Saguinus oedipus*, were utilized in this study. All animals were injected with 0.1 ml CFA in the subcutaneous tissues of the inguinal region and in the lower left attached buccal gingiva adjacent to the first and second premolar teeth. Eighteen days later, the animals were challenged by two injections of PPD. Each animal received 125 test units of PPD in the buccal gingiva adjacent to the lower left mandibular second or third molar, distal to the clinical CFA granuloma. A second injection was made into the marrow of the right mandible beneath the molar and premolar teeth. A small hole was drilled through the buccal cortical plate of bone just distal to the lower right canine tooth with a No. ½ round bur. A ¾-inch 25 gauge needle was then inserted in a distal direction as far as possible, and approximately 15 test units of PPD in 0.005 ml saline were deposited. One nonsensitized *C. jacchus* was treated with PPD in similar fashion and served as a control. Forty-eight hours later the animals were killed by the i.v. injection of pentobarbital, isopropyl alcohol and propylene glycol (Sleep-away, Fort Dodge Company). The mandibles were removed. The left half of the lower jaw was fixed in alcohol-formalin-acetic acid (AFA) and processed for celloidin sections in a mesiodistal plane. The right fragment of the mandible was fixed in 10% neutral formalin and processed

2
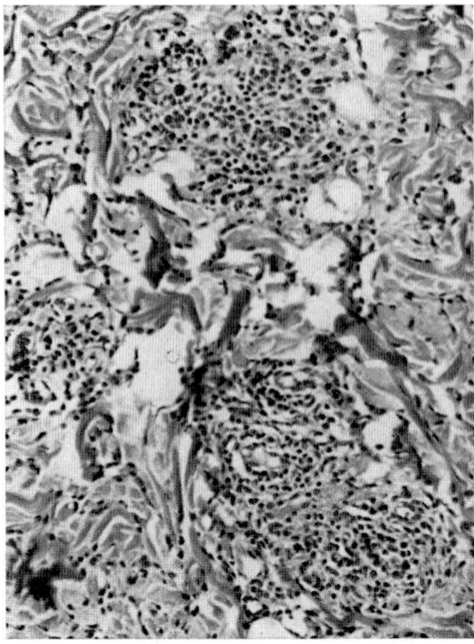
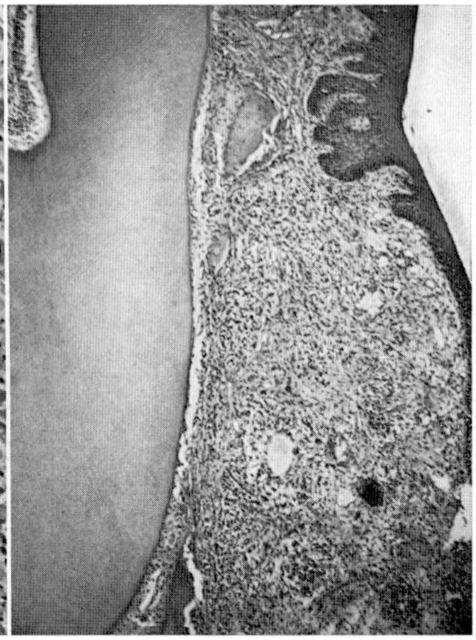
3

Fig. 2. Photomicrograph illustrating perivenular accumulation of immunocytes in the buccal gutter connective tissue. HE stain. × 50.

Fig. 3. Low power photomicrograph of buccal aspect of lower left first molar tooth. The buccal cortical plate of bone has been destroyed by the aggressive granulation tissue, which has also invaded the periodontal ligament. HE stain. × 20.

for paraffin sectioning in a buccal lingual plane. All sections were stained with hematoxylin and erythrosin, hematoxylin and PAS and Giemsa. Other special stains were used where indicated.

Results

The buccal attached gingiva of the left mandible, which was injected with CFA and PPD, was replaced from the surface epithelium to the periosteum by granulation tissue (fig. 1).

The granulation tissue was composed of histiocytes, macrophages, fibroblasts and endothelial cells densely infiltrated by large and small mononuclear cells. Methyl green-pyronin (MGP) stains indicated a large population of pyroninophilic immunocytes. Polymorphonuclear leukocytes were scattered throughout the tissue, forming small focal collections in some areas. Multinucleated giant cells were frequently noted deep in the tissue, usually near the periosteum of the cortical bone. The smaller blood vessels were thickened by proliferating endothelial cells. Areas of myxomatous or edematous-ap-

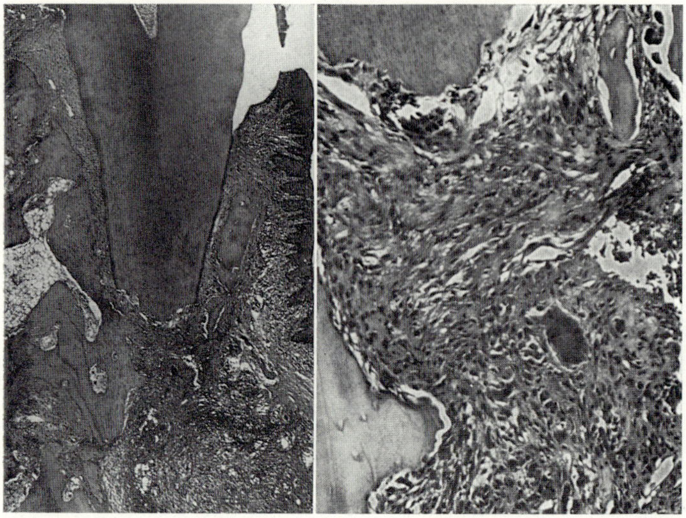

4 5

Fig. 4. Low power photomicrograph illustrating the marked destruction of attached gingiva, cortical plate of alveolar bone and root of lower right second molar. HE stain. × 10.

Fig. 5. High power photomicrograph of area near resorbed root shown in figure 4. The bone, tooth and bone marrow is replaced by fibrous connective tissue infiltrated by mononuclear cells and multinucleated giant cells. HE stain. × 50.

pearing fibrous tissue were frequently observed. Perivenular accumulations of mononuclear cells were noted, especially in the connective tissue of the buccal gutter adjacent to the gingiva and in the edematous areas (fig. 2).

In some animals there was little destruction of bone. In others, there was a scalloping of the buccal cortical bone. Multinucleated osteoclasts were occasionally noted within the resorption lacunae. Most of the lacunae contained immature fibroblasts and large mononuclear cells. In several animals, however (four *C. jacchus* and one *S. oedipus*) there was marked destruction of the buccal plate of bone which exposed the periodontal membrane to the granulation tissue (fig. 3). In these cases, the periodontal ligament was also destroyed. In two *C. jacchus* the root was eroded by the granulation tissue (fig. 4). The areas formerly occupied by bone and/or teeth were replaced by fibroblasts, macrophages, histiocytes, lymphocytes and occasional multinucleated giant cells (fig. 5). The areal bone marrow was essentially normal.

The reaction in the right mandible, in which PPD was injected into the marrow space beneath the premolar teeth in the sensitized animals, was dif-

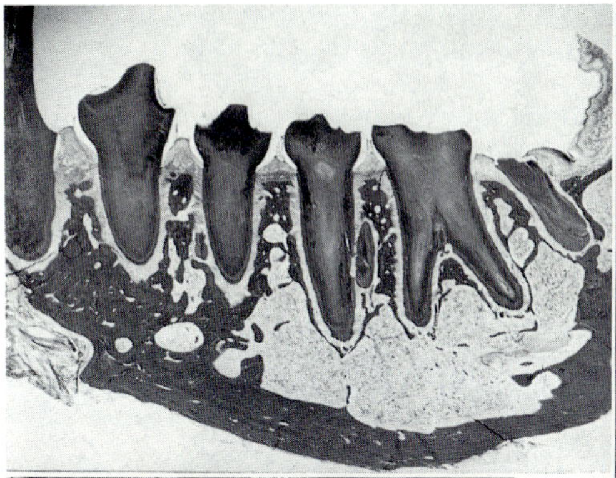

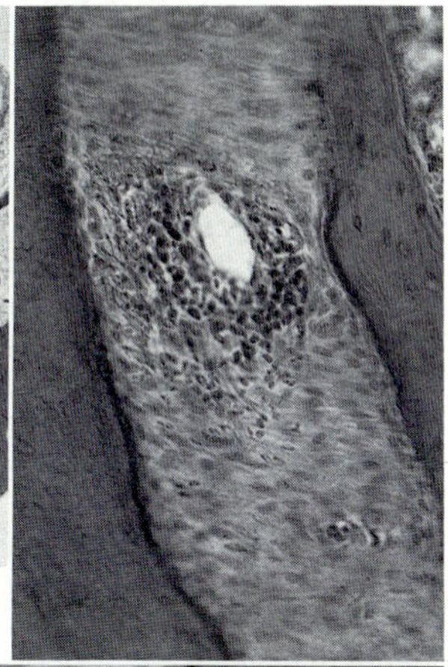

7

8

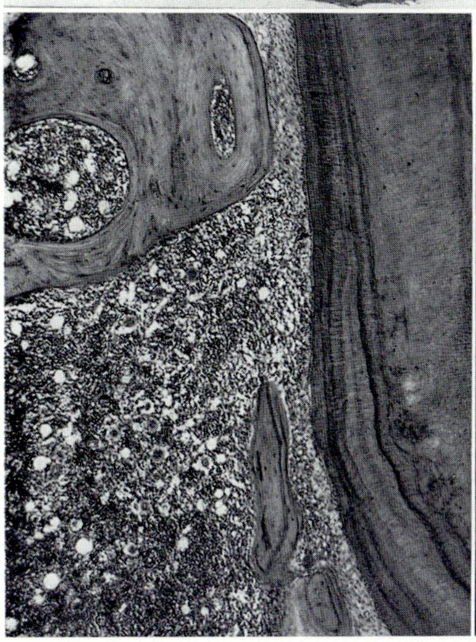

6

9

Fig. 6. Low power photomicrograph of right mandible illustrating the well-demarcated area of destructive reaction beneath the 3rd premolar and 1st molar tooth. Masson trichrome stain. × 10.

Fig. 7. Section of alveolar bone and lower right second molar tooth. The hyperplastic marrow has invaded into the periodontal ligament space, obliterating the attachment fibers. HE stain. × 125.

Fig. 8. Section of periodontal ligament of lower right 2nd molar. Note the perivascular infiltration of mononuclear cells. HE stain. × 200.

Fig. 9. Section of periodontal ligament illustrating lymphoid cells as well as osteoclasts in the resorption lacunae of the periodontal bone. HE stain. × 200.

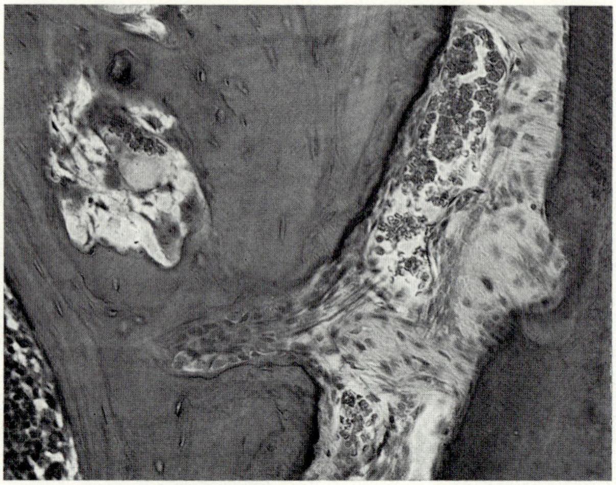

Fig. 10. Section of periodontal ligament in the apical region of a premolar tooth. The periodontal blood vessel appears congested and there is undermining resorption in the alveolar bone. HE stain. × 200.

ferent from that seen on the left side. The trabecular bone was destroyed in a well-circumscribed area (fig. 6). There was a marked proliferation of the hematopoietic elements, which filled the marrow spaces, excluded most of the fatty marrow and resorbed many of the thin bony trabeculae. The proliferation appeared to be due to a stimulation of the myeloid elements of the marrow. There were many macrophages with a foamy cytoplasm scattered among the myeloid elements. In some areas, the proliferating marrow cells invaded the periodontal ligament through the interalveolar spaces (fig. 7). In those areas the attachment fibers were obliterated. In other parts of the periodontal ligament, there were perivenular accumulations of mononuclear cells (fig. 8). Occasionally, hemorrhagic foci were seen. In a few loci, bone was resorbed on both the periodontal ligament and the endosteal surfaces of the alveolus. In these cases the resorption lacunae frequently contained multinucleated osteoclasts (fig. 9 and 10).

Discussion

WAKSMAN [1961] pointed out that massive necrosis of tissue was a characteristic of the intense delayed hypersensitivity reaction in skin. The gingiva apparently reacts differently from skin with regard to this phase of the lesion. No necrosis was noted in the gingiva of marmosets, although large areas of necrosis do occur in the skin of marmosets similarly treated [LEVY *et al.*, unpubl. data]. ADAMS *et al.* [1969] also reported this difference between the gingiva and skin in the delayed hypersensitivity reaction of guinea pigs to 2:4-dinitrochlorobenzene.

The destructive aspects of the adjuvant-associated lesions appear to be related to the dense accumulation of immunocytes (mononuclear cells) and histiocytes. In the skin, the invasive-destructive lesion occurs where the antigen is fixed in tissues outside the blood vessels. The infiltrating histiocytes appear to have a direct destructive effect upon the parenchymal elements [WAKSMAN, 1961]. Whether the mechanism for the destruction of bone and gingival connective tissue is similar to that for the skin remains to be proven.

Not only was the adjuvant-associated lesion destructive of bone, but the delayed hypersensitivity reaction to PPD alone in sensitized animals was also bone destroying. The destruction in this case must have occurred in the short period of 48 h, following the injection of the PPD into sensitized animals. No reaction to PPD occurred in the nonsensitized animal.

The finding of a massive myeloproliferation of the bone marrow as a delayed hypersensitivity response has not been previously reported. This interesting reaction is also a destructive one and warrants further exploration. Since the destructive character of the lesion in both the adjuvant associated and the 'Mantoux' reactions are so striking, experiments on the time related development of these lesions are in progress.

References

ADAMS, D.; WILLIAMSON, J. J., and DOLBY, A. E.: Delayed hypersensitivity response in guinea-pig oral mucosa. J. Path. *97:* 495–501 (1969).

BAER, H. and LORENZ, D.: The induction of immediate and delayed sensitivity in the marmoset. Proc. Soc. exp. Biol. Med. *134:* 410–412 (1970).

WAKSMAN, B. H.: Tissue damage in the 'delayed' (cellular) type of hypersensitivity in mechanism of cell and tissue damage produced by immune reactions; in GRABAR and MIESCHER 2nd Int. Symp. Immunopathology, pp. 146–159 (Grune & Stratton, New York 1961).

Author's address: Dr. B. M. LEVY, The University of Texas at Houston, The Dental Science Institute, *Houston, Tex.* (USA)

Medical Primatology 1972. Proc. 3rd Conf. exp. Med. Surg. Primates, Lyon 1972, part II, pp. 198–203 (Karger, Basel 1972)

The Treatment of Human Teeth *in vitro* for the Replacement of the Original Cells, by Cells from Another Individual

G. LUNDQUIST, P. OTTESKOG and P.-Ö. SÖDER

Department of Oral Surgery, Odontological Faculty, Karolinska Institutet, Stockholm

Autologous tooth transplantation is today a practical clinical procedure well established for replacement of lost teeth [GURALNICK and SHULMAN, 1962; NORDENRAM, 1963]. Unlike autografts, the survival time of the tooth allografts is limited depending on immunological processes [FONG, 1965; SHULMAN, 1968]. The strongest source of antigenicity on the tooth is the periodontal membrane and the pulp [SHULMAN, 1964]. This antigenicity is, however, weaker than that of most other tissues e.g. skin [VALENTE and SHULMAN, 1966]. To overcome the rejection phenomenon either immunosuppressive drugs could be used or the allograft could be altered. Because of the toxicity of the immunosuppressive drugs they are unacceptable for use in transplantation of teeth [GURALNICK, 1970].

It has been our experience that to be successful in transplantation of teeth the cells lining the root have to be viable [SÖDER and LUNDQUIST, 1972].

The aim of the present investigation is to develop techniques to remove the original cells on the root surface, to treat the root, mechanically, chemically and/or enzymatically to remove antigenic material from the surface and to replace the cells originally present by cells from another individual.

Material and Methods

The material consisted of two donor teeth, 17 receiver teeth and ten control teeth.

Donor teeth. The donor teeth were premolars extracted for orthodontical reasons. Immediately after extraction the teeth were carefully washed in Hanks' BSS, put in Eagle's

1 Supported by the Swedish Medical Research Council (Project No. K 71-24X-3345-01).

medium [EAGLE, 1959] and incubated at 37 °C. The donor teeth were incubated in liquid medium for four weeks at 37 °C and examined daily with a Leitz inverted microscope for signs of contamination. At the end of the cultivation period the teeth were trypsinized with 0.25% trypsin in Hanks' balanced salt solution pH 7.5 at 25 °C for one hour with agitation every ten minutes. The cells obtained from the teeth were centrifuged 1,400 *g* for ten minutes at room temperature, resuspended in complete growth medium and counted in a hemocytometer. The trypsinized cells were then cultivated separately in small tissue culture flasks for about one week. When the cells were outgrown to a confluent monolayer, they were trypsinized again and reincubated in a larger vessel.

When the cells should be transferred to a receiver tooth, the cells were trypsinized, centrifuged, resuspended in medium and counted. They were then concentrated and resuspended in a small volume to a density of 1×10^6 cells/ml.

Receiver teeth. The receiver teeth were canines (three), premolars (ten) and third molars (four). The canines and premolars were extracted or surgically removed for orthodontical reasons. The third molars were surgically removed. The teeth were without periodontal disease or crown restorations.

To remove the cells surrounding the root, a method was used which was a combination of mechanical, enzymatic and chemical treatment. Mechanically the root surface was polished with pumice and chalk. To remove residues of antigenic material, the teeth were then incubated for two hours at room temperature in a 2.5-percent solution of trypsin. Finally the teeth were incubated for one hour at room temperature in 80 mM Na-ascorbate and 80 mM H_2O_2 (1:1) to remove polysacharides and proteins. The teeth were then incubated for one week in Hanks' buffer and checked for signs of contamination before the cell-transfer experiments were carried out.

Control teeth. Premolars extracted for orthodontical reasons were incubated in liquid medium for two or four weeks at 37 °C and examined daily. At the end of the incubation time the cells were trypsinized, centrifuged, resuspended in 10 ml complete growth medium and counted in a hemocytometer.

Medium. Eagle's minimum essential medium was used supplemented with 10% calf serum, 4 mM L-glutamine, 1 mM Na-pyruvate, sufficient Na-bicarbonate to make the pH about 7.2, 100 μg/ml streptomycin and 100 IU/ml penicillin. Occasionally between 10 and 100 IU/ml Mycostatin®, 1 μg/ml Fungizone®, Garamycin® 10 μg/ml, Aureomycin® 10 μg/ml and Kanamycin® 10 μg/ml were used together with penicillin and streptomycin in various combinations.

Cell-Exchange Experiments

Nine of the receiver teeth were coated with thiolated gelatin (Sigma Chemical Company, St. Louis, Miss., USA) in a 10-percent solution. The gel on the tooth was then stabilized to form S-S bounds by incubation in a 3-percent solution of H_2O_2 for ten minutes.

All receiver teeth (17) were placed separately in beakers containing a semisolid medium consisting of Eagle's medium and 1% agar [LITWIN *et al.*, 1971]. The tooth was held in an upright position until the agar solidified. When the donor cells were to be transferred, the tooth was temporarily removed, and 5×10^5 cells/0.5 ml were added into the socket in the agar formed by the root. The tooth was then replaced into the agar for 12 h at 37 °C to allow the cells to attach to the surface. After the incubation time in the semisolid medium the tooth was removed and placed in a screw cap tissue culture tube containing liquid Eagle's medium. The receiver teeth were then incubated for different periods of time. After several weeks of incubation time the number of cells on the root was estimated by treating the tooth with 0.25% trypsin. The suspended cells were centrifuged, resuspended in medium and counted in a hemocytometer.

Results

The number of cells trypsinized from the root in the control group after two weeks of incubation varied between 1.8×10^4 and 10.0×10^4 (table I). After four weeks the number of cells from the root surfaces varied between 1.4×10^4 and 14.8×10^4. Cells could also be trypsinized from the receiver teeth with and without gelcoating (tables II and III). After four weeks the number of cells trypsinized from nongel-coated receiver teeth varied from 2.0×10^4 to 12.0×10^4. One tooth was cultivated six weeks and the number of cells trypsinized from the root surface was 10.0×10^4 (table II).

In the gel-coated group the number of cells varied from 0.4×10^4 to 2.8×10^4 two weeks after the cell exchange. In two cases (C and E in table III) no cells were detectable.

All of the cell populations were fibroblasts and were capable of growing to a confluent monolayer in the culture bottles, indicating a division capacity of at least six to seven cell divisions.

Table I. The number of cells that could be trypsinized from the control teeth after two and four weeks of incubation

Teeth	Cell cultivation time, weeks	Total number of cells trypsinized from teeth $\times 10^4$
A_1	2	2.0
B_1	2	3.0
C_1	2	3.2
D_1	2	1.8
E_1	2	10.0
F_1	4	14.8
G_1	4	4.4
H_1	4	4.4
I_1	4	1.4
J_1	4	1.8

Table II. The number of cells that could be trypsinized from the receiver teeth four and six weeks of incubation after cell exchange

Teeth	Cell cultivation time, weeks	Total number of cells trypsinized from teeth $\times 10^4$
A_2	4	6.8
B_2	4	12.0
C_2	4	8.0
D_2	4	6.0
E_2	4	2.0
F_2	4	7.0
G_2	4	3.0
H_2	6	10.0

Table III. The number of cells that could be trypsinized from the gel-coated receiver teeth two weeks after incubation after cell exchange

Teeth	Cell cultivation time, weeks	Total number of cells trypsinized from teeth $\times 10^4$
A_3	2	2.8
B_3	2	2.8
C_3	2	–
D_3	2	0.8
E_3	2	–
F_3	2	1.6
G_3	2	0.8
H_3	2	0.8
I_3	2	0.4

Discussion

This study is an effort to alter the surface of the teeth by removing the antigenic material from the donor and replacing it with cells to be derived from the recipient, thus, permitting the possibility of allotransplantation. The cell cultivation techniques employed permit the possibility of incubating teeth and cells for long periods of time *in vitro* to allow the cells to multiply. Moreover, by propagating cells separately in culture vessels, enough cells are obtained to allow for cell transfer. These techniques make it also possible to select suitable donors and receivers with fewer genetic differences. The microbial contamination could be controlled by the selective use of antibiotics during the cell cultivation period. The cells and much of the proteins lining the root were removed mechanically and enzymatically with trypsin and Na-ascorbate and H_2O_2 to eliminate their antigenic effect and minimize the possibility of immunological rejection. The pulp which also evokes a rejection reaction could be removed and replaced by a rootcanal filling.

The results suggest that fibroblasts obtained from human teeth can attach to a new surface on another root. The number of cells found after four weeks of incubation are of the same magnitude as the number of cells in the control group after the same time of incubation (tables I and II). The attachment of the cells to the surface of the receiver tooth with new periodontal ligament fibers has not yet been further investigated.

Nine of the receiver teeth were coated with thiolated gelatin to get a faster attachment to the surface and possible formation of a periodontal ligament. The number of cells obtained after two weeks' incubation were less than from the noncoated teeth. This could result from the shorter incubation time, but also from differences in attachment between the cells growing on gelcoated surface and those on a noncoated root. The cells growing on the coated surface were older than the cells growing on the noncoated teeth and probably had a slower growth potential.

Acknowledgements

The authors gratefully acknowledge the helpful advice of Dr. J. LITWIN, Department of Applied Microbiology, Karolinska Institutet, Stockholm.

References

EAGLE, H.: Amino acid metabolism in mammalian cell cultures. Science *130:* 432–437 (1965)

FONG, C. C.: Autologous and homologous experimental tooth transplantation. Seminar of dental tissue transplantation, University of California School of Dentistry, San Fransisco 1965.

GURALNICK, W. C.: Autogenous and allogeneic transplantation of teeth. Oral Surg. *28:* 575–577 (1970).

GURALNICK, W. C. and SHULMAN, L. B.: Tooth transplantation. Dent. Clin. N. Amer. 499 (1962).

Litwin, J.; Lundquist, G., and Söder, P.-Ö.: Studies on long-term maintenance of teeth and viable associated cells *in vitro*. Scand. J. dent. Res. *79:* 536–539 (1971).

Nordenram, Å.: Autotransplantation of teeth, a clinical and experimental investigation. Acta odont. scand. *21:* suppl. 33, p. 17 (1963).

Shulman, L. B.: Transplantation antigenicity of tooth homografts. Oral Surg. *17:* 389–394 (1964).

Shulman, L. B.: In Guralnick Textbook of oral surgery (Little, Brown, Boston 1968).

Söder, P.-Ö. and Lundquist, G.: Autotransplantation of teeth with use of cell cultivation technique. Int. dent. J. *22:* 327–340 (1972).

Valente, L. and Shulman, L.: Transplantation immunity to a single subcutaneously implanted tooth in mice (abstract). IADR Progr. Abstr. 64 (1966).

Author's address: Dr. G. Lundquist, Dept. of Oral Surgery, Odontological Faculty, Karolinska Institutet, *Stockholm* (Sweden)

Medical Primatology 1972. Proc. 3rd Conf. exp. Med. Surg. Primates, Lyon 1972, part II, pp. 204–211 (Karger, Basel 1972)

The Facial and Dental Consequences of Nasal Septum Resection in Baboons

A Preliminary Report

M. I. SIEGEL

Faculty of Arts and Sciences, Department of Anthropology, University of Pittsburgh, Pittsburgh, Pa.

Introduction

Students of human evolution have shown considerable interest in the relationship between dental and facial evolution [2, 18]. Although it has been realized that there exists a correlation between phylogenetic shortening of the face and phylogenetic dental reduction [6], there is little agreement on the cause and effect of this relationship. One component of dental reduction is shortening of the roots. It has been shown [9] that orthognathous human populations are characterized by short tooth roots, while prognathous populations usually have long roots. When the situation was investigated in dogs which had been bred for varying facial configurations [13], high correlations were also found between relative facial protrusion and relative root length. The correlations were highest for those teeth in closest proximity to active sites of facial elongation, the maxillary-premaxillary suture and the maxillary-palatine suture. The use of this comparative technique was extended to nonhuman primates in a study of *Papio anubis* [16]. In that species, where there has been a secondary elongation of the face, high correlations were found between facial length and root length for those teeth in closest proximity to the aforementioned sutures.

Techniques of experimental surgery were first applied to the study of facial elongation by FICK in 1857 [5]. He was the first investigator to surgically excise the nasal septum to produce short-faced animals, thereby demonstrating that the septum affects the anterior growth of the face. Although some investigators [10] question the role of the cartilaginous nasal septum as the impetus for midfacial growth, this technique has been utilized to produce facial shortening in the rabbit [15, 19] and the rat [12]. It has also been demonstrated that when facial shortening is surgically produced in rats, the roots of molar teeth are significantly shortened [12]. Since, with regard to root reduction, there was agreement between the results of the comparative and experimental studies, in *Rodentia*, *Carnivora*, and *Primates*, this investigator [16] postulated a general mammalian trend: Within a given population there exists a biomechanical mechanism which transmits changes in ontogenetic growth rates of the facial skeleton to the roots of teeth. The mechanism was defined as biomechanical rather than genetic, since the relationship could be demonstrated by surgical intervention as well as by comparative studies of breeding populations, and regardless of the cause of the change in facial configuration, an influence on the roots could be demonstrated. In an attempt to better understand the operation of this mechanism in primates, an extensive experimental research program was begun. The longitudinal study was designed to determine the effects of nasal septum resection on facial and dentitional development in the baboon *(Papio cynocephalus)*. Such a program, utilizing nonhuman primates, has relevance for evolutionary studies as well as clinical implications for oral biology in man.

Methods

The study was begun in June of 1970 and is still in progress. The research is being conducted at the Laboratory for Experimental Medicine and Surgery in Primates (LEMSIP) of New York University Medical Center, New York. Young baboons *(P. cynocephalus)* are obtained when they become available. For inclusion in the experimental group, three-month-old animals are preferred and none are accepted which are older than six months. There are presently twelve animals in the study; however, only six are, at this time, old enough to be discussed in this paper.

All preoperative animals were tranquilized with a phencyclidine hydrochloride compound (Sernylan) (1.25 mg/kg, i.m.) and injected with pentobarbital sodium (26 mg/kg, i.v.). The bridge of the nose was then infiltrated with a 2-percent solution of lidocaine (Xylocaine) hydrochloride (1.0–2.0 ml). A butanol compound (Revistat) coagulant was used, and the animals were injected (2.0 ml/kg, i.m., each dose) before, during, and after surgery. An endotrachial tube was passed and the fauces packed with gauze to facilitate clot removal. The animals were then secured to the operating table in a prone position and the snout and face were prepared for sterile technique. An approximately 2.5-cm midline incision was made through the skin (fig. 1) from the nasal root to the proximal border of the rhinarium, exposing the nasal bones. Two incisions were then made along the nasal-premaxillary sutures and connected by a transverse incision along the nasal-frontal suture (fig. 2). The bone flap and tissues were then reflected distally, (fig. 3) exposing the nasal cavity (fig. 4). Maximum amounts of the cartilaginous nasal septum were then resected (fig. 5). The bone

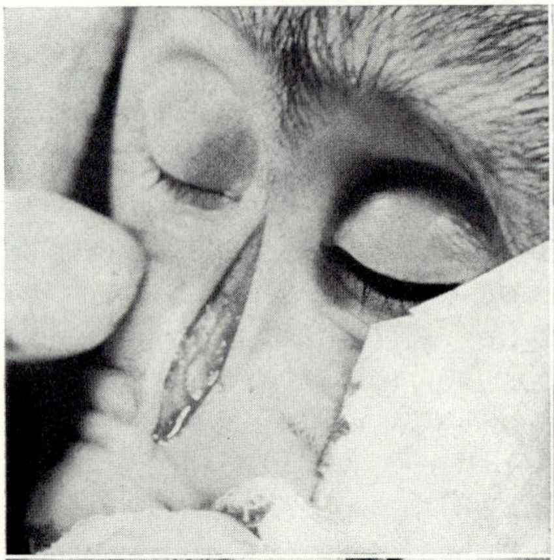
1

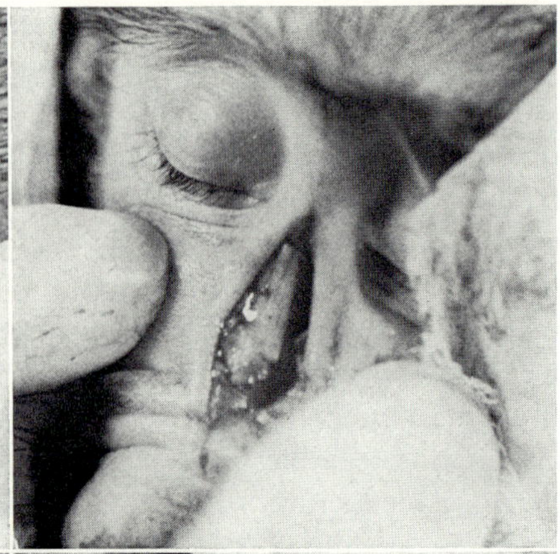
2

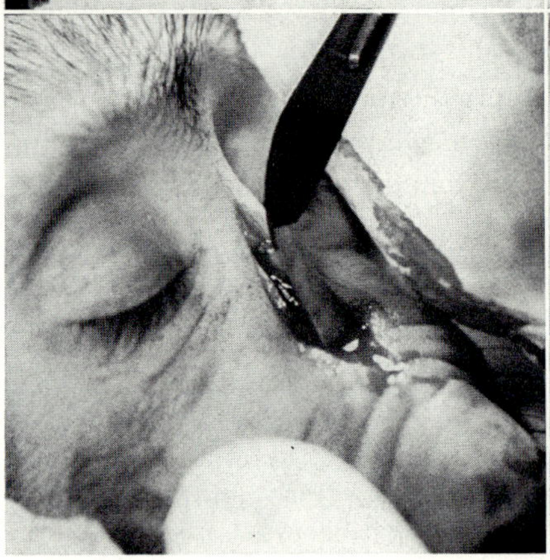
3

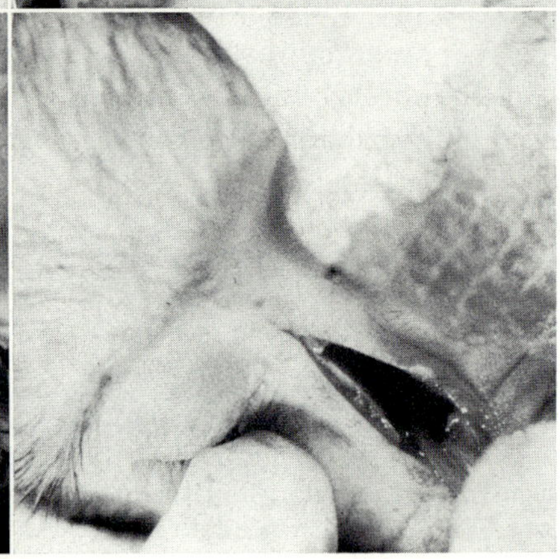
4

Fig. 1. Nasal region of a three-month-old *Papio cynocephalus* showing initial incision.

Fig. 2. Nasal region of a three-month-old *Papio cynocephalus* showing nasal bones with sutural incisions.

Fig. 3. Nasal region of a three-month-old *Papio cynocephalus* showing completed bone flap.

Fig. 4. Nasal region of a three-month-old *Papio cynocephalus* with bone flap reflected, exposing nasal cavity.

flap was replaced and the skin closed with 3-0 nylon sutures. Postoperatively, the animals were given prophylactic doses of tetracycline (100 mg/kg, i.m.), which served a secondary role as a vital stain. The animals were then placed in an incubator for recovery. On the first postoperative day the animals were usually bright and alert. Experimental and control animals were weaned at an age of approximately 24 weeks, and maintained thereafter on a diet of monkey chow (Purina) and supplemental milk. All animals were housed in standard baboon cages and received the normal maintenance of animals at LEMSIP [4]. For the pur-

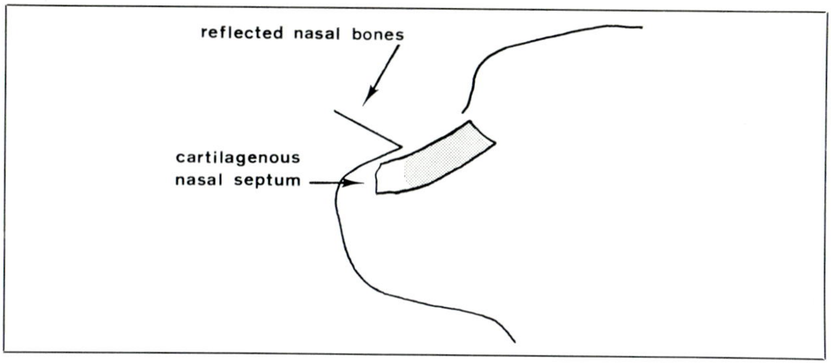

Fig. 5. Extent of cartilaginous nasal septum of a three-month-old *Papio cynocephalus* superimposed on sagittal outline. Stippled area represents surgically-excised portion.

pose of vital staining of growing bone, the animals were injected (100 mg/kg, i.m.) with tetracycline at three-month intervals. At two-month intervals, alginate impressions were taken of maxillary and mandibular dentitions, and dental stone casts prepared. On the same days, lateral cephalograms were made using a specially-designed head holder [11] and a portable x-ray unit.[1] For both of these procedures the animals were tranquilized with a ketamine hydrochloride compound (Ketaject) (10 mg/kg, i.m.).

Data

Assessment of the effects of septum resection on dental roots must await the eruption of the full succedaneous dentition and the closure of the apical diaphragm of the third molar teeth, at an approximate age of six years. At this time, after sacrifice of the animals, histological analysis of alveolar bone as well as growth rate curves based on the results of the tetracycline staining will be prepared. Since the apical third of the root has been shown to be quite plastic [8] and develops within alveolar bone which may have been altered by anterior growth arrest, it is felt that this analysis will provide valuable information on the mechanism of root reduction. The data obtained from the lateral cephalograms will be reported in a separate study now in preparation [17]. From the dental casts obtained thus far, an analysis of premaxillary

1 Two X-ray units were used at various times: A Picker X-ray, using an exposure of 100 mA, 60 kV, 0.1 sec and a Norelco Oralix Dental X-ray unit using an exposure of 7 mA, 50 kV, 3 sec. The results are slightly better with the Picker unit since the short exposure time eliminates problems of slight anterior-posterior movements by the animal.

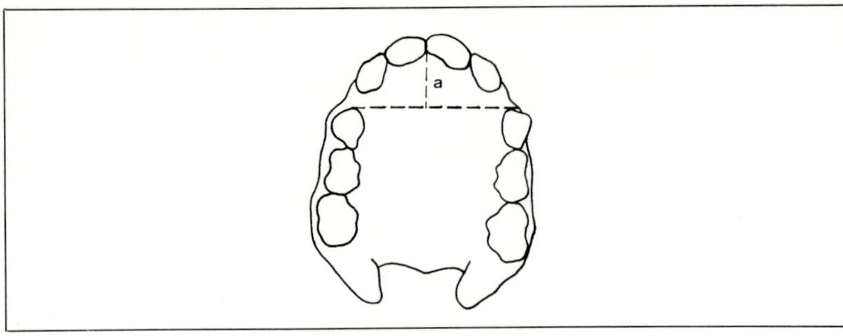

Fig. 6. Dimension a: premaxillary length defined on outline of dental cast of maxillary dentition.

growth has been prepared. Measurements[2] of premaxillary length were taken on the casts with the length defined as the distance from prosthion to a line drawn tangent to the mesial borders of the maxillary canine teeth (fig. 6). These dimensions were then plotted against the age of the animals in days (fig. 7 and 8).

In figure 7, B400 and B406 are operated females and B396 and B398 are unoperated females. The control animals show an increasing premaxillary length which continues until some time after 375 days for B398 and until after 350 days for B396. What appears to be a decrease in length for some of the plotted measurements (for males as well as for females) is actually an artifact of the increased diameter of the canine tooth as it erupts and variation in edema of the gums which produces a decrease in the apparent length of from 0.1 to 0.3 mm. Both experimental females, B400 and B406, show little increase in premaxillary length after 250 days and the ultimate absolute length appears to be less than for controls.

For the males, B103 and B111 (fig. 8) an interesting situation exists. Both animals have been operated on. B111 shows an increasing premaxillary length to 250 days and then levels off, showing no apparent growth to 550 days. B103, on the other hand, shows the type of growth curve on might expect from an unoperated control. The premaxillary length shows a large increase beyond 400 days and does not level off until after 500 days. For this reason the septal region was examined in all experimental animals. For all but B103 a septal defect was present. This, coupled with the fact that B103

2 All measurements were made with a Helios dial caliper read to an accuracy of 0.005 cm.

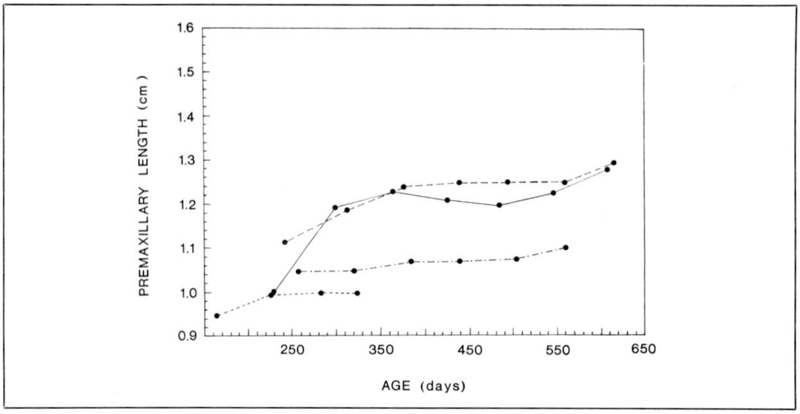

Fig. 7. Graph of premaxillary lengths plotted against age of animal for experimental (●.–.–.–● B400 and ●--------● B406) and control ●——● B396 and ●----● B398) females.

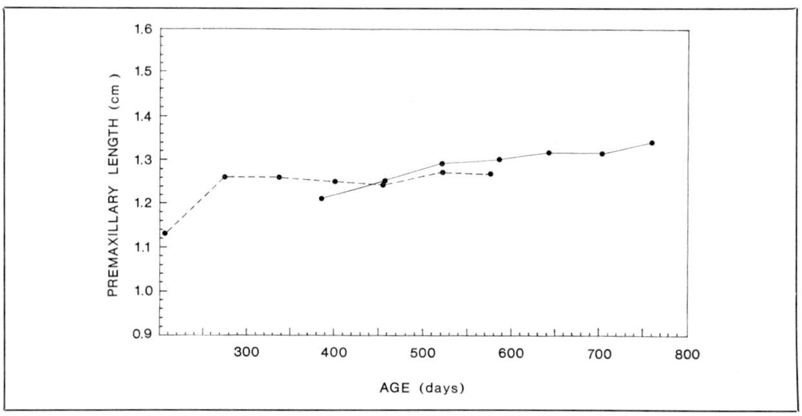

Fig. 8. Graph of premaxillary lengths plotted against age of animal for experimental males (●——● B103 and ●----● B111). (See text for explanation.)

was the first animal to be operated upon, leads me to believe that the surgical technique was not, at the time of surgery, well enough refined and that an insufficient amount of the cartilaginous septum was removed. This agrees well with the results obtained in my laboratory on partial removal of the cartilaginous nasal septum in rats, and with the findings reported by SARNAT [14].

Discussion

While comparative primate studies on the cartilages of the nose [7] and the prenatal morphogenesis of the palate [3] have been carried out, no experimental studies on the role of the cartilaginous nasal septum in mid-facial growth have been reported for primates. All previous studies have relied on nonprimate experimental models for evaluating this mechanism [1, 10, 14]. It should be noted, however, that since WOOD *et al.* [20] have reported the nonexistence of a separate ossification center for the premaxilla in man, even the nonhuman primate model may be imperfect for understanding the human condition.

The results, thus far, demonstrate that when the proximal two-thirds of the cartilaginous nasal septum is surgically excised in growing baboons, arrest of the anterior growth of the premaxilla is produced. Not only is the arrest earlier, but the overall growth of the premaxilla is shown to be less than for unoperated controls. Further refinements of the surgical technique, to allow for the removal of the entire cartilaginous nasal septum, are currently being investigated. While the long-term effects of septum resection on overall facial growth are not yet known, the technique may have clinical applications for the correction of gross disharmony between the face and mandible. Demonstration of the effects of septum resection in primates serves to strengthen the suggestion of SARNAT [14] that care must be taken during cleft palate or septal surgery to prevent trauma to this area which might cause undesirable growth retardation of the upper face.

Acknowledgements

I am grateful to Dr. J. MOOR-JANKOWSKI for his support of this research at LEMSIP and to Dr. W. NIEMANN, LEMSIP, for his advice and technical assistance. This paper was made possible by Dr. J. ROSENBERG, Dean, Faculty of Arts and Sciences, University of Pittsburgh, NIH Biomedical Science Support Grant FR 07084.06. The coagulant, Revistat, was provided through the courtesy of Dr. E. REVICI, Research Director of the Institute of Applied Biology, New York, Photographs were prepared by J. M. STUCKART and illustrations by CAROLYN A. KONISH.

References

1 BABULA, W. J., jr.; SMILEY, G. R., and DIXON, A. D.: The role of the cartilagenous nasal septum in midfacial growth. Amer. J. Orthod. *58:* 250–263 (1970).

2 BIEGERT, J.: The evaluation of characteristics of the skull hands and feet for primate taxonomy; in WASHBURN Classification and human evolution, pp. 116–145 (Aldine, Chicago 1963).
3 BOLLERT, J. A. and HENDRICKX, A. G.: Morphogenesis of the palate in the baboon *(Papio cynocephalus)*. Teratology *4:* 343–353 (1971).
4 DAVIS, J. H.; MCPHERSON, B. R., and MOOR-JANKOWSKI, J.: Maintenance and handling of primate animals for medical research. Ann. N. Y. Acad. Sci. *162:* 329–335 (1969).
5 FICK, L.: Über die Ursachen der Knochenformen. Experimental-Untersuchung (Wigand, Göttingen 1857).
6 HOOTON, E. A.: Up from the ape (MacMillan, New York 1931).
7 JORDAN, J. and SERNICKI, A.: Further investigations on the cartilages of the nose and palate in *Cebus capucinus*. Folia morph. *27:* 447–458 (1968).
8 KOVACS, I.: Contribution to the ontogenetic morphology of roots of human teeth. J. dent. Res. *46:* 865–884 (1967).
9 MOSS, M. L.; CHASE, P. S., and HOWES, R. I., jr.: Comparative odontometry of the permanent post-canine dentition of American Whites and Negroes. Amer. J. phys. Anthrop. *27:* 125–142 (1967).
10 MOSS, M. L. and BROMBERG, B.: The passive role of the nasal septal cartilage in midfacial growth. Plast. reconstr. Surg. *41:* 536–542 (1968).
11 PRATT, J. L.: Three-view x-ray cephalometry of primates. J. dent. Res. *48:* 1312 (1969).
12 RIESENFELD, A.: The effect of environmental factors on tooth development. An experimental investigation. Acta anat. *77:* 188–215 (1970).
13 RIESENFELD, A. and SIEGEL, M. I.: The relationship between facial proportions and root length in the dentition of dogs. Amer. J. phys. Anthrop. *33:* 429–432 (1970).
14 SARNAT, B. G.: The face and jaws after surgical experimentation with the septovomeral region in growing and adult rabbits. Acta oto-laryng. (suppl.) *268:* 1–30 (1970).
15 SARNAT, B. G. and WEXLER, M. R.: The growth of the face and jaws after resection of the septal cartilage in the rabbit. Amer. J. Anat. *118:* 755–767 (1966).
16 SIEGEL, M. I.: The relationship between facial protrusion and root length in the dentition of baboons. Acta anat. *83:* 17–29 (1972).
17 SIEGEL, M. I.: An x-ray cephalometric study of normal and abnormal growth of the face and palate in *Papio cynocephalus* (in preparation).
18 WEIDENREICH, F.: The brain and its role in the phylogenetic transformation of the human skull. Trans. Amer. phil. Soc. *31:* 321–442 (1941).
19 WEXLER, M. R. and SARNAT, B. G.: Rabbit snout growth after dislocation of nasal septum. Arch. Otolaryng. *81:* 68–71 (1965).
20 WOOD, N. D.; WRAGG, L. E.; STUTEVILLE, O. G., and OGLESBY, R. G.: Osteogenesis of the human upper jaw. Proof of the non-existence of a separate premaxillary centre. Arch. oral Biol. *14:* 1331–1341 (1969).

Author's address: Dr. MICHAEL I. SIEGEL, Department of Anthropology, University of Pittsburgh, *Pittsburgh, PA 15213* (USA)

Medical Primatology 1972. Proc. 3rd Conf. exp. Med. Surg. Primates, Lyon 1972, part II, pp. 212–216 (Karger, Basel 1972)

Application of Osseous-Dental Homografts to Monkeys and Men

IOAN SPIRESCU

Autologous dental transplants can yield satisfactory results, but homologous dental grafts usually fail, partly because of tissue incompatibility [3]. In order to create essentially normal conditions for the transplanted tooth, we decided to experiment with osseous-dental transplants. These transplants have the advantage of preserving the morphofunctional bond between the teeth and the alveolus through the periodontal ligament. Since lyophilized homologous osseous transplants have been successfully applied in bone surgery, we decided to conduct research utilizing lyophilized homologous osseous-dental transplants. It was thought that if the osseous tissue was accepted, it would lead to acceptance of the tooth [1, 4].

Material and Methods

Osseous-dental transplant jaw sections were taken from two rhesus monkeys; one contained three anterior teeth, and the other contained one molar. The osseous transplants were removed, leaving 0.5–2 mm of spongey osseous tissue surrounding the dental roots. The material was taken 20 min following the death of the animal. The grafts were washed five times in a nutrient solution containing amino acids and antibiotics (penicillin 1,000 units/ml, streptomycin, 250 μg/ml, mycostatin, 500 μg/ml) and then frozen slowly to –52 to –54 °C in the chamber of the lyophilization apparatus (SMJ Usiroid). Primary desiccation was carried out at –40 °C, and the secondary desiccation at +30 °C. The lyophilization process required 44–48 h. The osseous-dental transplants were then preserved for 30 days in sealed bottles containing argon [2].

In view of the spongey osseous structure of the human's jaw, we attempted to introduce the grafts into bones of a similar structure, namely, the long bones of monkeys (fig. 1). The upper tibial epiphysis was selected as a recipient site for the graft. A transverse incision was made over the upper third of the tibial crest, exposing the boney surface. A cavity of

1, 2

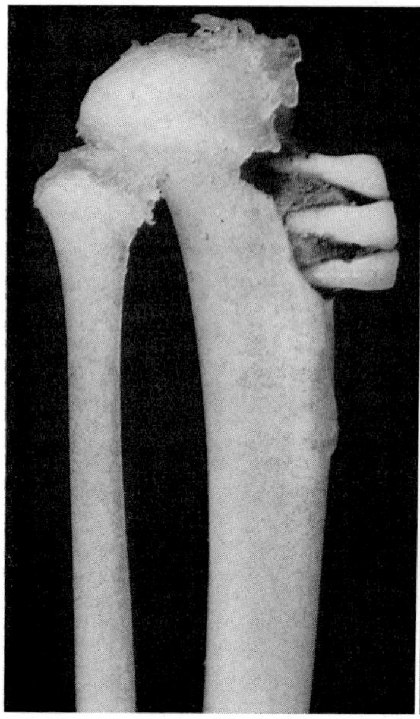

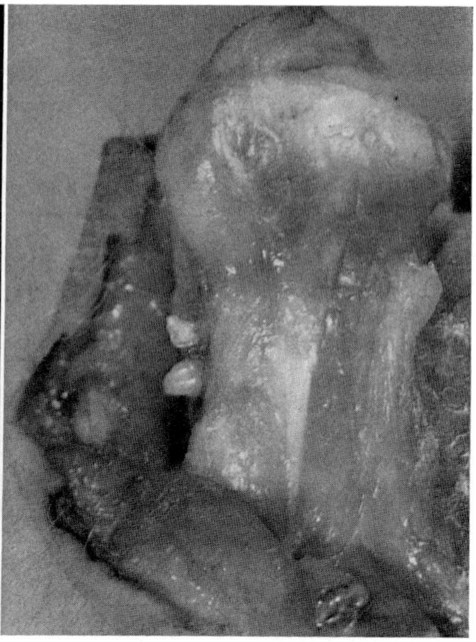

Fig. 1. Experimental pattern in monkeys. Note osseous-dental transplant and cavity made in upper tibial epiphysis.

Fig. 2. Tibial epiphysis of monkey containing homogenous dental transplant two months after insertion.

the same size as the graft was created, and the graft was introduced in such a manner that the crowns of the teeth remained outside the bone (subcutaneous). The graft was fitted without pressure to avoid traumatic bone necrosis, and the skin was sutured in place.

Results

The postoperative course was uneventful. No inflammatory phenomena were observed, and the wounds healed normally. The animals were sacrificed after two months. We observed that a consolidation of the osseous-dental transplant to the host bone had occurred, without any mobility of the teeth. Hard ossified tissue was present around the teeth, and the junction between the osseous-dental transplant. The host bone was also ossified. The crowns of the teeth were undamaged, with the exception of one tooth which was fractured (probably by trauma in the cage). The color of the teeth remained unchanged (fig. 2).

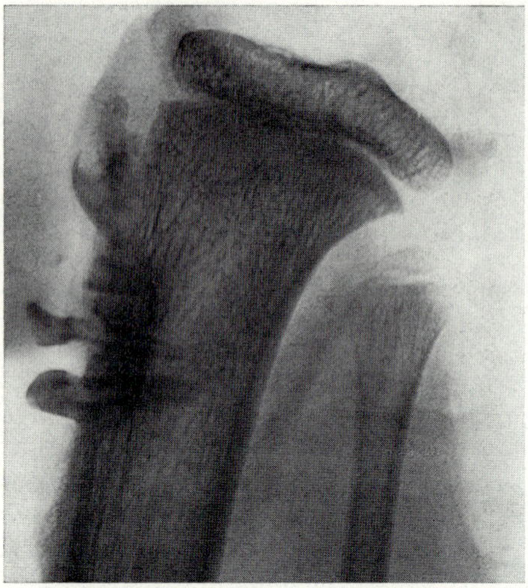

Fig. 3. Roentgenograms of the tibial epiphysis in monkeys. Note the ossification surrounding homogenous osseous-dental transplant and the integrity of the dental roots two months after implantation.

Roentgenograms revealed osseous compatibility around the osseous-dental transplant. The dental roots remained normal, with no areas of root resorption or narrowing of the periodontal space (fig. 3).

Discussion

The successful general surgical application of lyophilized homologous bone grafts for the replacement of osseous voids gave us the idea of employing lyophilized osseous-dental transplants. To test the rejection, homologous lyophilized osseous-dental transplants were embedded in large bones of similar animals. No dental root resorption and no graft rejection was observed. The transplants were firmly attached to the recipient sites; there was no tooth mobility and no dental root resorption.

Application in Man

Starting from these experimental data, lyophilized homogenous osseous-dental transplants were performed in man. The aim of these trans-

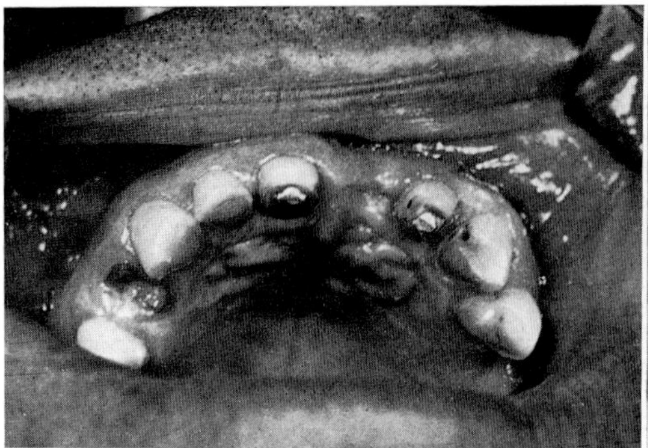

4a

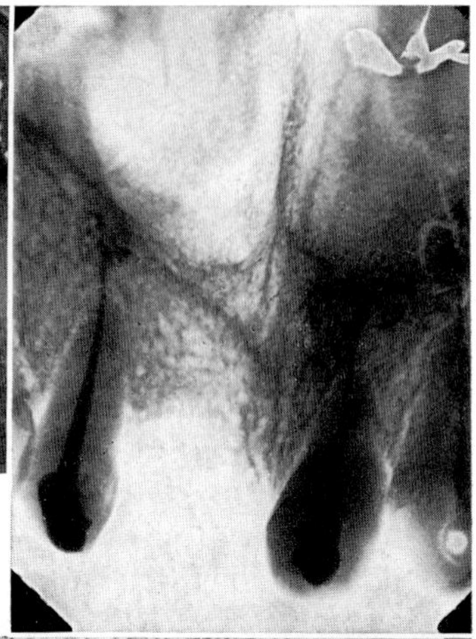

4b

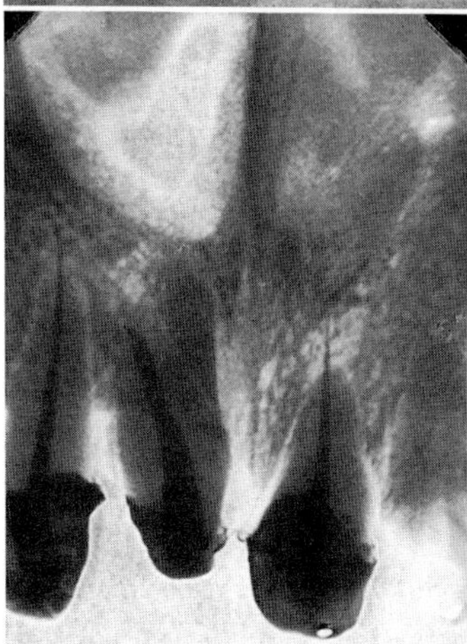

5a

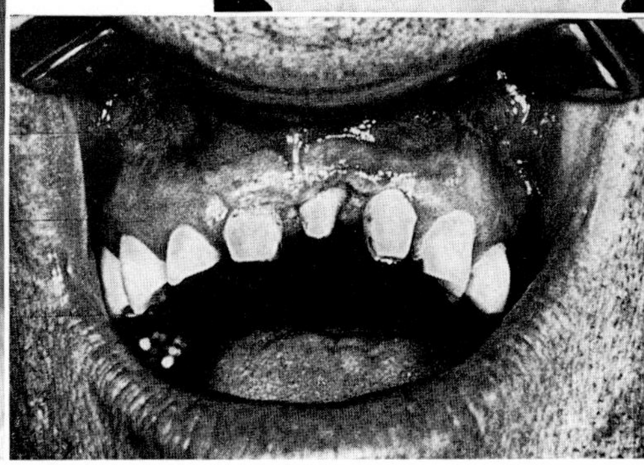

5b

Fig. 4. Osseous-dental transplant in man. *a* Clinical and *b* radiological appearance prior to transplantation.

Fig. 5. Osseous-dental transplant in man. *a* Clinical and *b* radiological appearance 15 months after transplantation.

plants is the support of dental bridge and replacing missing teeth, especially in young patients. Under aseptic conditions, a jaw block graft was taken from the jaw of a 22-year-old male at approximately ten hours after death from an accident. The transplant was composed of the tooth with a thin layer (1–3 mm) of alveolar bone surrounding the root.

The pieces were prepared and lyophilized according to the method described above. Some of the boney pieces were used for bacteriological control. After reflecting a mucoperiosteal flap, we created a cavity in the alveolar bone using a surgical bur. This cavity was of the same size and shape as the osseous-dental homograft which was then inserted into the cavity. The transplant was stabilized by means of a bridge cemented to the neighboring teeth.

After three months of immobilization the bridge was removed. The homotransplant appeared firmly in place. The gingiva was firmly and tightly attached to the tooth transplant. In all cases after this immobilization period the teeth were used in mastication. Periodic radiographs taken postoperatively from three to 18 months, and clinical examination performed from three to 16 months, revealed a progressive restoration of the bone continuity, without any signs of osteolytic or tooth root resorption defects. According to this technique, we performed nine transplants on eight patients (fig. 4 and 5).

In spite of the brief follow-up period and the impossibility of drawing definitive conclusions, the osseous-dental homografts in our cases suggest the possibility of using this method in edentulous areas [5].

References

1 ANDREW, C. and BESSETT, L.: Tissue banks; in REY Aspects théoriques et industriels de la lyophilisation, p. 431 (Hermann, Paris 1964).

2 HYATT, G. W.: La banque des tissus d'origine humaine; in REY Aspects théoriques et industriels de la lyophilization, p. 279 (Hermann, Paris 1964).

3 SHULMAN, L. B.: The transplantation antigenicity of tooth homografts. Oral Surg. *17:* 389 (1964).

4 SPIRESCU, I.; THEODORESCU, M., and VASILIU, D.: Studii asupra antigenicitatii tesuturilor dentare. Rev. Stomatologia (Romanian) *4:* 301 (1971).

5 SPIRESCU, I.; SPIRESCU, GH., and CHIRITA, S.: Biologia homotransplantului osos. Personal commun. USSM Stomatologie, Bucuresti 1972.

Author's address: Dr. IOAN SPIRESCU, Asistent Universitar, Clinica de Chirurgie Buco Maxilo Faciala, Facultatea de Stomatologie, Str. Plevnei 19, Sect. 7, *Bucaresti* (Romania)

Medical Primatology 1972. Proc. 3rd Conf. exp. Med. Surg. Primates, Lyon 1972, part II, pp. 217–231 (Karger, Basel 1972)

Pharmacological and Dental Filling Materials Testing in Nonhuman Primates

H. TRIADAN

Clinic of Dental Medicine and Department of Scanning Electron Microscopy, University of Bern, Bern

I. Introduction

It is becoming increasingly apparent in dental medicine that new drugs and materials must be methodically investigated [2–5, 8, 9, 23]. Well-defined specifications for physical, chemical and pharmaceutical *laboratory* testing have been available for some time and improved standardized test procedures are continually being developed.

For practical and ethical reasons, however, the *biological* testing of drugs and materials in dentistry is hard to deal with. It is difficult to extract human teeth in large numbers for experimental purposes, and the resulting paucity of human data prevents investigators from reaching statistically valid conclusions. In particular, the comparison of various materials in the same individual is practically impossible. For these reasons the use of experimental animals is necessary. The anatomy and physiology of the monkey (*Macaca mulatta* or other species) make it the most suitable of all the species for testing of drugs and materials. A maximum of information

Table I. Scheme

First monkey		Second monkey	
A	B	B	A
$M_3M_2M_1P_2P_1CI_2I_1$	$I_1I_2CP_1P_2M_1M_2M_3$	$M_3M_2M_1P_2P_1CI_2I_1$	$I_1I_2CP_1P_2M_1M_2M_3$
$M_3M_2M_1P_2P_1CI_2I_1$	$I_1I_2CP_1P_2M_1M_2M_3$	$M_3M_2M_1P_2P_1CI_2I_1$	$I_1I_2CP_1P_2M_1M_2M_3$
B	A	A	B

'A' denotes the test treatment and 'B' the control (standard or placebo).

can be obtained with suitable experimental design and the frequently-expressed objection that the results of animal experiments are not transferable to the human loses a large part of its validity (for caries research, however, the small rodents have been shown to be most suitable).

II. Histological Examination of Dental Pulp in the Monkey

A. Testing of Substances for Direct Capping

(1) Extensive literature on the problem of covering open pulp and stimulating dentinogenesis exists [20, 24]. It is particularly difficult to promote the formation of tertiary dentine by dental pulp when the pulp has been exposed by the excavation of carious dentine.

(2) *Material and methods.* The same method in each experiment is employed. Whenever possible each monkey receives two treatments, one treatment of the upper jaw on one side and the other treatment of the lower jaw on the opposite side. This cancels out differences between upper and lower jaw and right and left side in the statistical analysis.

Variation remains possible according to the purposes of the experiment and external influences.

The preparations listed in table II were tested by direct application to the dental pulp.

(3) *Results.* Measurement of the degree of inflammation of the pulp and dentinogenesis showed test preparation No. 2 to be inferior (at a 0.03 significance level using the sign test).

B. Model for Testing Preparations Used in the Treatment of Pulpitis

(1) There is considerable literature on the subject of healing inflamed dental pulp [18, 24]. Such treatment, if successful in retaining the vitality of

Table II.

No.	Number of teeth	Antiphlogistic	Antiinfectious	Dentinogenetic agent	
1	6	triamcinolone	neomycine	calcium hydroxide	standard
2	18	triamcinolone	neomycine	lead chloride + calcium phosphate	test
3	6	–	neomycine	diphenylhydantoine	test

the pulp, offers an alternative to its removal and replacement by a root-canal filling. In addition to clinical observation of the effects of drugs on naturally-occurring pulpitis in human teeth, it is advantageous to observe histologically the effect of these drugs on monkey pulp in which one has artificially stimulated a pulpitis. The author has described the need for this evidence more fully in another article [21] and is currently testing the following model:

(2) *Material and methods.* Stimulants of pulpitis: *physical* (drill trauma); *chemical:* histamine dihydrochloride, extractum capsici fluidum concentratum Ph H (Pharmacopoea Helvetica V), phenolum Ph H, 50% orthophosphoric acid (Bio-Trey fluid®) (De Trey, London); *bacterial.*

The substances were applied to dental cavities which were prepared in a standardized manner and were left *in situ* for 24 h, 72 h and, in some cases, 14 days. Standardized bacterial suspensions containing known counts of selected strains (bacteroides) were placed in the cavities in direct contact with the pulp for three, 13 and 42 days.

(3) *Results.* The experimental series is not yet complete, but preliminary results show that changes in the dental pulp occurred in 24 of 28 control teeth. Approximately half of these pulps showed acute inflammation which was caused by trauma from the drill [21]. The bacterial preparations were more reliable as stimulants of pulpitis, regularly producing acute inflammation after only three days.

This *finding* is not surprising, since it has already been established by TRIADAN [19–21] and MEYER [11] that the pulp of *Macaca* monkeys is more sensitive to infection than human pulp, but more resistant to chemical toxins. It can be concluded that standardized drill trauma and carefully-dosed infection with the bacteroides strain both produce a pulpitis model which can be used for the histological testing of antiinflammatory agents. It has yet to be determined whether the pulpitis treatment recommended by TRIADAN *et al.* [1, 19, 22] on the basis of systematic clinical studies in humans and histological observations in healthy monkey teeth is the most efficient. This treatment involves the use of a preparation containing DMSO (dimethylsulfoxide), corticosteroid (triamcinolone) and antibiotic (minocycline) (Lederle Laboratories, Pearl River, USA).

C. Testing of Substances for Protection of Cavities

(1) The testing of the ability of various cements and varnishes to protect the pulp from physical (mechanical and thermal) and chemical (acids and

Table III.

No.	Number of teeth	Pulp protection	Observation period, days	
1	12	none	14	control
2	8	Dropsin®[1]	14	test
3	8	Rapid-Lasul®[2]	14	test

1 Wera Karl, Bisingen, Germany.
2 Svedia, Enköping, Sweden.

resin components) damage from filling materials can be carried out advantageously in the monkey.

(2) *Material and methods.* A student working for his thesis under the supervision of the author prepared standardized cavities in monkey teeth and protected the pulp from the toxic effects of silicate cement using the preparations listed in table III.

(3) *Results.* Cavities treated with protective varnish showed less pulp damage due to the phosphoric acid component of the silicate cement than those treated with Dropsin® cement or the controls [11].

III. Testing and Comparison of Two New Composite Filling Materials

(1) In recent years various synthetic resins have been introduced for the filling of dental cavities. These materials contain approximately 25% organic and 75% inorganic component. The organic fraction is either a resin based on the *Bowen* formula

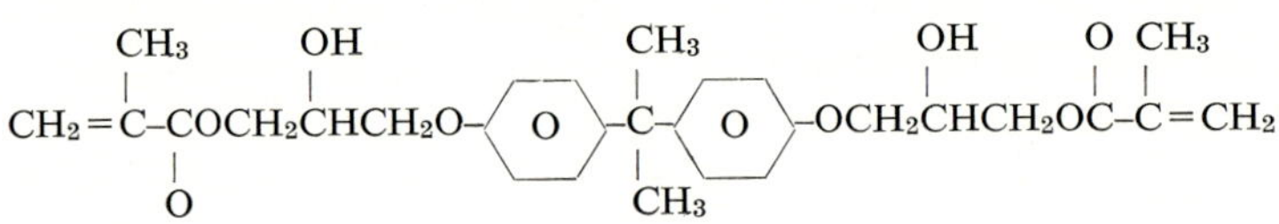

or pure methacrylate. The inorganic filler, which also has a reinforcing function, consists of glass, quartz particles or other substances. These filling materials have four advantages over conventional amalgam: They allow satisfactory color blending without the physical, chemical and biological disadvantages of silicate cement; the thermal conductivity is low; the filling can be stressed within one hour; the patient does not have to return for polishing of the filling. These composite materials have also been discussed in a large numer of publications [12, 16]. The method of testing filling materials which has gained most popularity in recent times is that of scanning electron microscopy. A basic publication on the working principles and potential of the scanning electron microscope is that of FROMME *et al.* [7]. Their article describes the advantages of the continuously variable magnification (20–30,000 times), the considerable depth of focus, an optical resolution of 200 Å and a step discrimination of 30 Å, in studying the behavior of surfaces and structures of tooth and filling materials and the margins between them. Other articles on this subject describe the failure of glass fibers and glass spherules to bind with the matrix when they are used as reinforcement in silicate cements [13, 14].

This failure of filler and matrix to bind together has also been reported in certain composite materials [6, 14] and in addition, stressing of the filling has been shown to produce a surface roughness. RIEDEL *et al.* [15] investigated the surface and interface properties of another composite material whose matrix consisted of pure methacrylate (Palakav®) (Kulzer, Bad Homburg v.d.H., Germany) and which was placed in teeth for 69–80 days. Moderate interface fissures of 12–30 μm occurred (maximum 72 μm, minimum 6 μm), these values being less than those obtained from amalgam, silicate or gold fillings. LEE *et al.* [10] published the results of a scanning electron microscope study on surfaces and sections of fillings from extracted human teeth. Four composite materials (Adaptic® [Johnson & Johnson, New Brunswick, USA], Addent 12®, Addent 35® [3M, Minnesota Mining, St. Paul, Minn., USA], Dakor® [Caulk, London]) were investigated in addition to pure methacrylate, silicate and amalgam. Adaptic® showed the best adaptation to the dental cavity. The size of the interface fissure could be accounted for by the degree of shrinkage occurring during polymerization. The authors mention the possibility of displacement of the filling in the dental cavity which occurs during section of the tooth. We have also observed this phenomenon.

The following investigation represents an attempt on the part of the author to improve and systematize the testing and comparison of dental materials using the scanning electron microscope. Once again, the monkey is used as the experimental animal.

(2) *Material and methods.* The two new *Bowen* composite material Adaptic® (inorganic filler – approximately 75% quartz) and Concise® (3M, Minnesota Mining,

Table IV.

First monkey – 5 months		Second monkey – 15 months		
adaptic®	concise®	concise®	adaptic®	
× × ××	××× ×	× × ××	××× ×	Class 1
+ + +++++	++++++ +	+ + +++++	++++++ +	Class 5
$M_3M_2M_1P_2P_1CI_2\ I_1$	$I_1\ I_2\ CP_1\ P_2M_1M_2M_3$	$M_3M_2M_1P_2P_1CI_2\ I_1$	$I_1\ I_2\ CP_1\ P_2M_1M_2M_3$	
$M_3M_2M_1P_2P_1CI_2\ I_1$	$I_1\ I_2\ CP_1\ P_2M_1M_2M_3$	$M_3M_2M_1P_2P_1CI_2\ I_1$	$I_1\ I_2\ CP_1\ P_2M_1M_2M_3$	
+ + +++++	++++++ +	+ + +++++	++++++ +	Class 5
× × ××	××× ×	× × ××	××× ×	Class 1
concise®	adaptic®	adaptic®	concise®	

St. Paul, Minn., USA) (inorganic filler – approximatly 72% borosilicate) were placed in the teeth of two monkeys in the manner shown in table IV.

It was possible in this manner to test a large number of fillings with two different materials in the same individual and under the same physiological conditions over a long period of time. (This is not possible in the case of human patients, due to the variability in alcohol consumption, and the eating and tooth cleaning habits.) Each premolar and molar contained two fillings: The occlusal filling was stressed and abraded by its antagonist while the buccal filling was not under direct pressure. Thus at the end of the experimental period, it was possible to observe both types of filling in one and the same section.

Following sacrifice of the monkeys, the dental crowns were carefully cut out of the jaw and, beginning with the apex, the pulp was extracted. The tooth was cemented to the carrier grid using Justis® resin cement, shadowed *in vacuo* with carbon and gold, and examined in the Stereoscan microscope (Stereoscan, Cambridge, Department of Scanning Electron Microscopy [Prof. ALLEMANN], Operator: F. ZWEILI) at various magnifications. The width of the fissure at various points could be directly read off on the cathode ray display. The means and standard deviations of the measurements can be seen in tables V and VI. In the case of monkey No. 2 (*in situ* for 15 months) one tooth was always examined *in toto* in order to determine the contours of restoration border and surface, while the next tooth was cut through the center and the plane of the section was then polished and examined.

(3) *Results*. It can be seen from tables V and VI that the t-tests ($a<0.05$) showed no statistically significant difference in the widths of the fissures between the buccal and occlusal fillings or between the two filling materials.

The increased mean width of the tooth/restoration interface fissures in the five-month experiment can be ascribed to technical causes; owing to a breakdown of the stereoelectron microscope, it was necessary to keep the teeth longer in physiological saline. For this reason we did not attempt to correlate the width of the fissure with the length of time that the filling was *in situ*. However, the two variables appeared to be statistically independent.

Because of the small numbers in some of the individual groups in the 15-month test, caused by tooth fractions and by displacement of the restoration (fig. 21), all the values from Adaptic® and Concise® (excluding the

Table V. Tooth/restoration interstice during a 5-month observation period

Adaptic®						Concise®					
occlusal (class 1)			vestibular (class 5)			occlusal (class 1)			vestibular (class 5)		
n	x	s	n	$\bar{x}$	s	n	$\bar{x}$	s	n	$\bar{x}$	s
5	10.6 μm	7.13	8	11.2 μm	6.84	5	11.3 μm	8.24	10	6.6 μm	2.53

Table VI. Tooth/restoration interstice, observation period: 15 months

Adaptic®						Concise®					
surfaces											
occlusal			vestibular			occlusal			vestibular		
n	$\bar{x}$	s	n	$\bar{x}$	s	n	$\bar{x}$	s	n	$\bar{x}$	s
3	6.7 μm	2.05	7	5 μm	2.44	4	3.7 μm	0.47	6	4 μm	3.66
sections											
5	6 μm	6.34	10	1.7 μm	1.91	2	3.6 μm	0.95	7	2.7 μm	1.14
surfaces+sections											
N	$\bar{x}$	s				N	$\bar{x}$	s			
25	4.08 μm	3.77				19	3.4 μm	2.17			

values from the bases of the cavities) were pooled for comparison (table VI, bottom line). Once again there was no statistically significant difference between the two filling materials.

Examination of the surfaces and, particularly, of the sections regularly showed the presence of round bubbles in the interior of the filling and loss of inorganic particles from the surface (fig. 22).

(4) The following *conclusions* can be drawn from the results shown:

(a) The anatomy and mode of mastication of the monkey is basically the same as that of the human, and the information derived from the experiments described is thus of direct use in human dental medicine. The design of the experiment in which each monkey received 44 fillings (28 buccal, 16 occlusal – 22 Adaptic® and 22 Concise®) allows comparison between the various materials used while taking account of the variability due to anatomical site, time and mechanical influences.

(b) The surface of fillings which had been *in situ* for five months were examined and the interface fissures between the tooth and the filling were measured. Absolute mean values of 11 μm were observed in the Adaptic® fillings and 9 μm in the Concise® fillings. The difference was statistically insignificant (t-test, $a<0.05$) (fig. 1–4). The corresponding figures in the

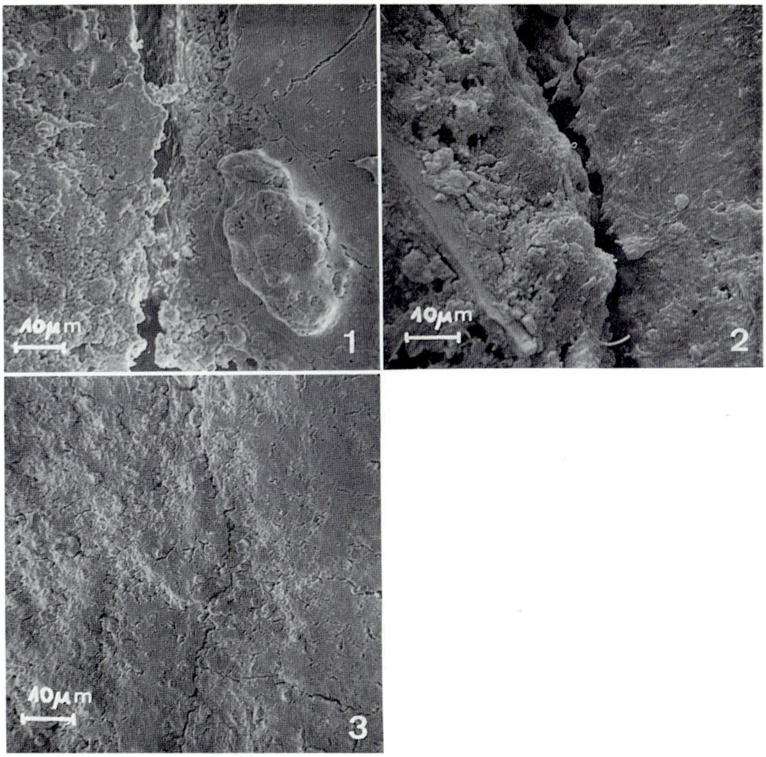

Fig. 1–3. Surfaces, five-month observation period. *1* $+P_1$, Concise®. *2* $+I_1$, Concise®. *3* $+I_1$, Adaptic®.

15-month test were 6 μm in the case of Adaptic® and 4 μm in the case of Concise®. Again, the difference was statistically insignificant (t-test, $a < 0.05$) (fig. 5–23).

Technical problems which have been described by LEE and SWARTZ [10] make evaluation of the sections difficult. During the cutting of the teeth an unavoidable displacement of the filling in the direction of surface occurs. For this reason, fissures at the base of the cavity were not evaluated. We are currently attempting to improve our method (embedding of the teeth in methacrylate, Microslice circular saw) (Metals Research Ltd., Melbourn [Royston], Great Britain).

Examination of the sections showed a mean fissure of 4 μm size in the case of Adaptic® and 2 μm in the case of Concise®. Combining examinations

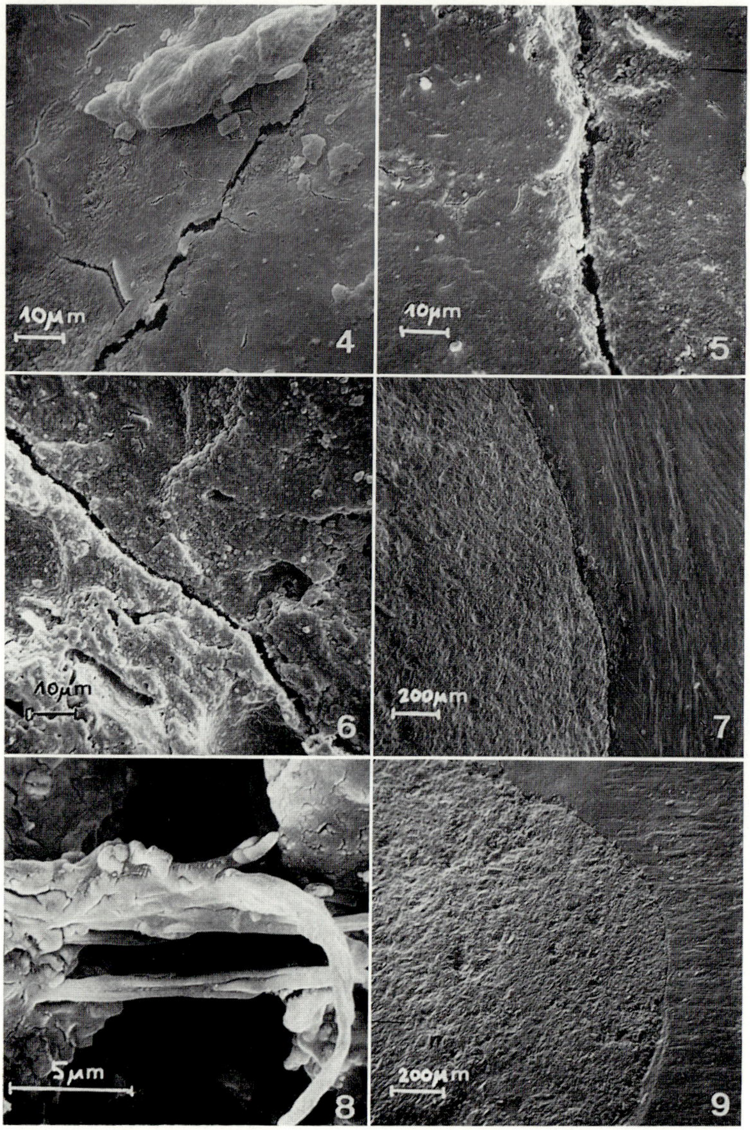

Fig. 4–12. Surfaces, 15-month observation period. *4 and 5* $+I_1$, class 5, amalgam, for comparison. *6* $+I_1$, class 5, Concise®. *7* C^+, class 5, Concise®. *8* $-M_1$, class 5, Concise®. *9* C– C^-, class 5, Adaptic®, *10* $+P_2$, class 1, Adaptic®. *11* $+P_2$, class 5, Adaptic®. *12* M_2– M_2, class 5, Adaptic®.

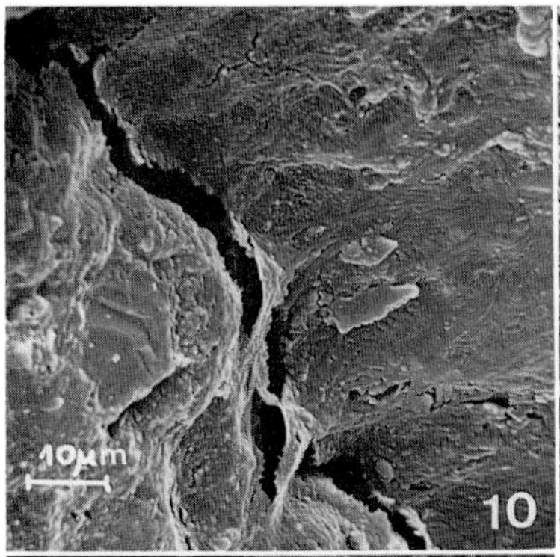

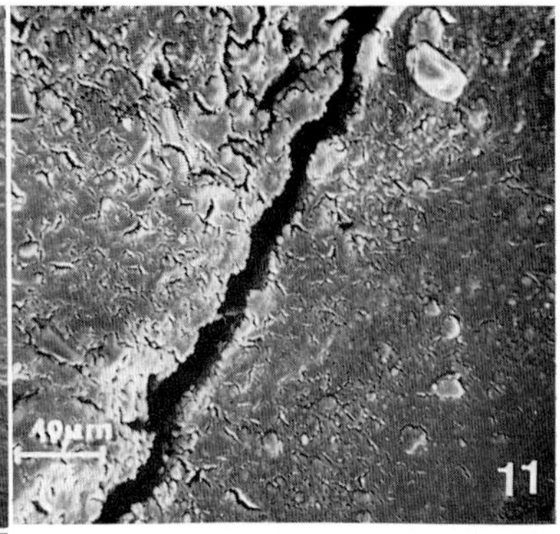

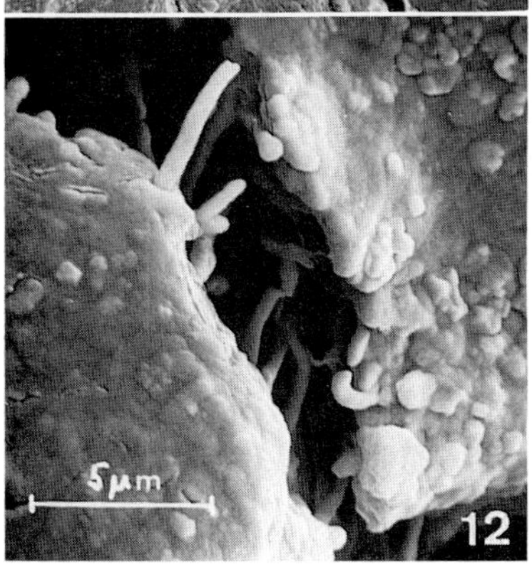

of surface fissures and cross-sectional fissures showed them to be approximately 4 µm in width in the Adaptic® fillings and 3.5 µm in the Concise® fillings. The difference was statistically insignificant (t-test, $\alpha < 0.05$).

The fissures observed by us between tooth and filling material were considerably smaller than those described by RIEDEL *et al.* [15] in the case of Palakav® fillings (36 µm mean value). They were, however, larger than those reported by LEE and SWARTZ [10] who recorded mean values of 1 µm in Adaptic® restorations in extracted human teeth which had not been physiologically stressed.

(c) Unlike STRASSBURG and KNOLLE [17] we could not find any correlation between the width of fissure and localization of the filling.

(d) The bubbles within the filling material and the breaking off of surface inorganic filler could not be related to localization of the filling or to the material used.

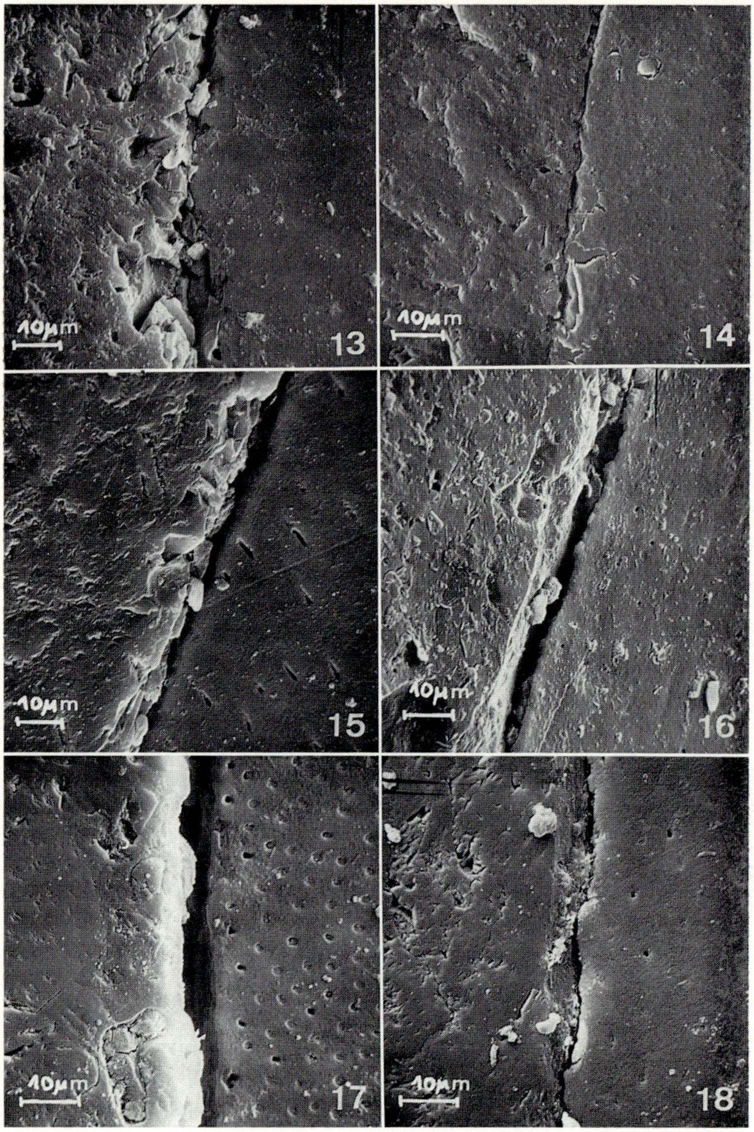

Fig. 13–23. Sections, 15-month observation period. *13* $+P_1$, class 5, Adaptic®, side of cavity. *14* $+P_1$, class 5, Adaptic®, opposite side of cavity. *15* $+P_1$, class 5, Adaptic®, base of cavity. *16* –C, class 5, Concise®, side of cavity. *17* –C, class 5, Concise®, opposite side of cavity. *18* –C, class 5, Concise®, base of cavity. *19* $-M_1$, class 5, Adaptic®. *20* –M, class 5, Adaptic®, higher magnification. *21* $+M_1$, class 1, Adaptic®, artifact by section. *22* $+I_2$, class 5, Concise®. *23* $+P_1$, class 5, Concise®.

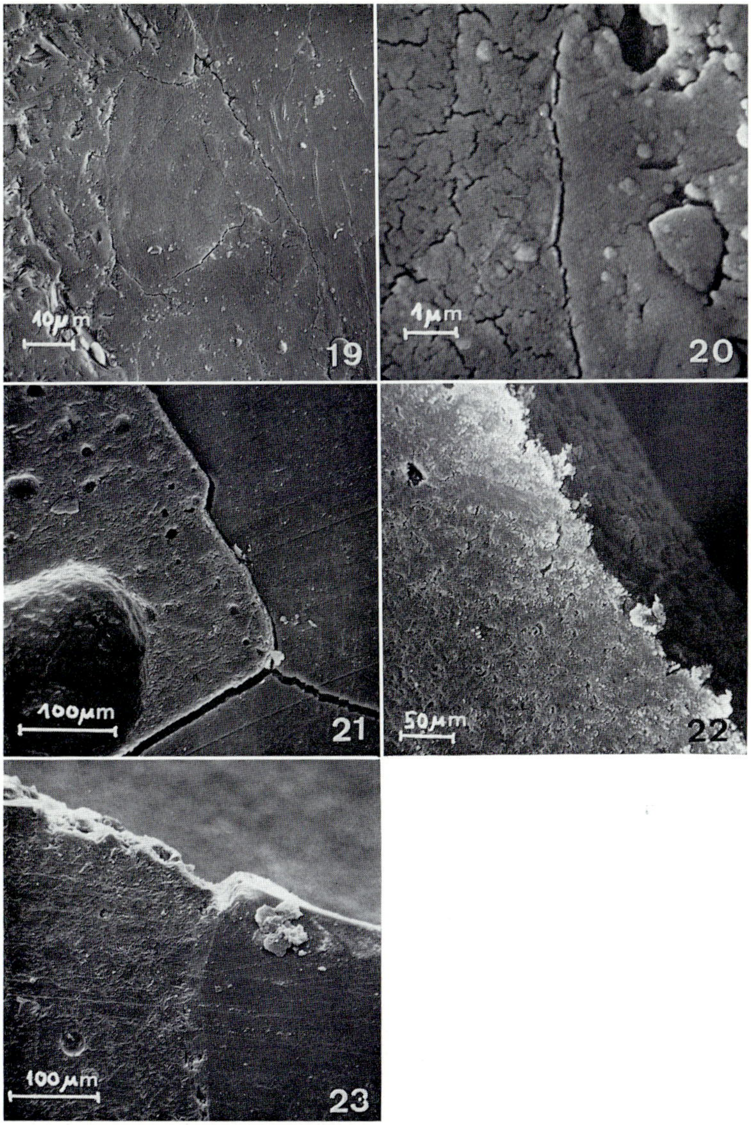

References

1 ATTENHOFER, R.: Kurzfristige klinische Untersuchung über die Wirkung eines Minocyclin-Kombinationspräparates auf den pulpitischen Schmerz. Med. Diss. Bern (1971).

2 BAUME, L. J. and FIORE-DONNO, G.: Response of the human pulp to a new restorative material. J. amer. dent. Ass. *76:* 1016–1022 (1968).

3 BAUME L. J. et FIORE-DONNO, G.: Biologie de la pulpe et matériaux dentaires. Schweiz. Mschr. Zahnheilk. *80:* 962–984 (1970).

4 BAUME, L. J.; FIORE-DONNO, G., and HOLZ, J.: Biological pulp testing of restorative materials. Brit. dent. J. *131:* 9–16 (1971).

5 BAUME, L. J.: La réaction pulpaire à l'égard de l'Addent XV et sa prévention. Schweiz. Mschr. Zahnheilk. *81:* 1099–1120 (1971).

6 CHANDLER, H.; BOWEN, R., and PAFFENBARGER, G.: Method for finishing composite restorative materials. J. amer. dent. Ass. *83:* 344–348 (1971).

7 FROMME, H. G.; RIEDEL, H. und VAHL, J.: Beobachtungen zum Gefügeaufbau der Zahnhartsubstanzen mit Hilfe des Raster-Elektronen-Auflichtmikroskopes. Dtsch. zahnärztl. Z. *22:* 395–400 (1967).

8 HARNDT, R.: Biologische Testmethoden. Dtsch. zahnärztl. Z. *26:* 316–326 (1971).

9 LANGELAND, K. und KLOETZER, W. T.: Verfahren zur Prüfung der biologischen Eigenschaften zahnärztlicher Werkstoffe. Dtsch. zahnärztl. Z. *26:* 298–315 (1971).

10 LEE, H. J. and SWARTZ, M. L.: Scanning electron microscope study of composite restorative materials. J. dent. Res. *49:* 149–158 (1970).

11 MEYER, U.: Untersuchungen über ein neues chemisches Dentin- und Pulpenschutzmittel (Rapid-Lasul®). Med. Diss. Bern (1972).

12 PHILLIPS, R. V.; SWARTZ, M. L., and NORMANN, R. D.: Materials for the practicing dentist (Mosby, St. Louis 1969).

13 RIEDEL, H. und VAHL, J.: Experimentelle Untersuchungen über neuzeitliche Füllungsmaterialien mit Hilfe des Raster-Elektronen-Auflichtmikroskopes. Dtsch. zahnärztl. Z. *22:* 476–486 (1967).

14 RIEDEL, H.; VAHL, J. und HAHNE, D.: Untersuchungen über die Mikromorphologie von 3M Addent 35 mit Hilfe des Oberflächen-Raster-Elektronenmikroskopes. Dtsch. zahnärztl. Z. *23:* 199–204 (1968).

15 RIEDEL, H.; VAHL, J. und WINKELMANN, G.: Untersuchungen über die Mikromorphologie getragener Palakav®-Füllungen mit Hilfe des Raster-Elektronenmikroskopes. Dtsch. zahnärztl. Z. *26:* 267–273 (1971).

16 SCHROEDER, A.: Stand der Erfahrungen auf dem Gebiet der neuen Kunststoff-Kompositionsmaterialien auf Basis der Bowen-Formel. Schweiz. Mschr. Zahnheilk. *81:* 999–1010 (1971).

17 STRASSBURG, M. und KNOLLE, G.: Füllungstherapie mit Palakav®. Dtsch. zahnärztl. Z. *26:* 247–252 (1971).

18 TRIADAN, H. und SCHROEDER, A.: Die pharmakologische Heilung der Pulpitis (1. und 2. Teil). Schweiz. Mschr. Zahnheilk. *70:* 724–731 (1960) und *71:* 94–100 (1961).

19 TRIADAN, H.: Histologische Untersuchungen über die Wirkung von Dimethylsulfoxid (DMSO) auf die Affenzahnpulpa. Schweiz. Mschr. Zahnheilk. *78:* 480–489 (1968).

20 TRIADAN, H.: Histologische Untersuchungen über die Wirkung einiger potentieller direkter Überkappungsmittel auf die Affenzahnpulpa. Schweiz. Mschr. Zahnheilk. *80:* 716–731 (1970).

21 TRIADAN, H.: Tierexperimentelle Erzeugung einer Pulpitis. Modellversuch. Schweiz. Mschr. Zahnheilk. *81:* 629–644 (1971).

22 TRIADAN, H.: Histologische Untersuchung über die Wirkung eines neuen DMSO-Triamcinolon-Minocyclin-Kombinationspräparates auf die Affenzahnpulpa. Schweiz. Mschr. Zahnheilk. *81:* 1011–1023 (1971).
23 VIOHL, J.: Labortestmethoden und FDI-Spezifikationen. Dtsch. zahnärztl. Z. *26:* 290–297 (1971).
24 WILHELM, S.: Das Problem der Vitalerhaltung der Pulpa, betrachtet an der geschichtlichen Entwicklung der konservativen Therapie der Pulpaerkrankungen. Med. Diss. Bern (1969).

Author's address: Prof. Dr. H. TRIADAN, University of Bern, Clinic of Dental Medicine, Freiburgstrasse 7, *CH–3000 Bern* (Switzerland)

Medical Primatology 1972. Proc. 3rd Conf. exp. Med. Surg. Primates, Lyon 1972, part II, pp. 232–241 (Karger, Basel 1972)

Perfusion of Fixatives via the Carotid-Jugular Route for the Histopathological Study of the Dental Pulp in the *Macaca fascicularis*

G. Zeilig, J. Nitlich, P. Chamodot and J. P. Fortier

U. E. R. d'Otontologie, Université René Descartes de Paris

Introduction

The dental pulp is composed of a connective tissue which is basically made up of vascular and nervous elements and is surrounded by a wall of differentiated cells. This soft tissue is moulded within a highly-calcified hermetic casing. The complete unit is itself embedded in bony tissues.

In order to make a histopathological study of the dental pulp, it is necessary to remove the tooth and then, rapidly, saturate it in a fixative. This procedure is usually carried out after first sectioning the root, and due to the fact that it is a lengthy and delicate procedure, it can modify the structure of the studied tissues. In addition to vascular variations due to infiltration of local anesthetic, it has been seen that lesions of the peripheral layer of the odontoblasts can occur in the areas compressed by the forceps. Fixation of the pulpal mass is often incomplete, even after enlargement of the foramina, as only a small area of the tissue has been exposed to the fixative. It was in order to avoid these difficulties that we used the method of prefixation by a perfusion of formalin via the carotid-jugular route in the *ante mortem* state. This method was practiced on the *Macaca fascicularis* as part of a study of dentin-pulp reactions following trauma.

Principles of the Method

Our first objective was the elimination of the blood mass from the cephalic end and then the perfusion of formalin in physiological conditions [4].

Anatomical Account

In the upper region of the neck the common carotid arteries (primitive carotid) divide into two branches. The external carotid insures the irrigation of the maxillae, the mandible, the gingiva, the periodontal membrane and the teeth, by its terminal branches [10].

The distribution of vessels which irrigate the mandible and the maxillae was studied in the *M. fascicularis*. As a general rule, the irrigation of the mandible follows the same pattern as that in man [8]. However, the incisory region of the mandible is supplied by the sublingual artery so, therefore, the distribution of the lower alveolar artery is limited to the body of the mandible [3]. On the other hand, the distribution of the lower alveolar artery and of its ascending dentoalveolar branches, is similar to that seen in man. These vessels branch out into fine collaterals to the interalveolar and interradicular septa towards the periodontal membranes and the dental pulp [2]. One or two arterial branches penetrate into the pulp tissue of each dental root. These results concord with those recorded in man [7].

With regard to the maxillary, the study of the *M. fascicularis* has revealed the absence of the upper, anterior alveolar artery which, in man, where it is a terminal branch of the suborbital artery, supplies the alveolars and the upper incisors and canines. Indeed, in the *M. fascicularis* this region is irrigated as far as the anterior region of the maxillary bone by an extension of the palatine artery. The apical artery or arteries divide into two or three arteries. Sometimes an additional apical artery penetrates the root wall of the tooth, starting from the periodontal membrane [5]. The pulp vessels fan out upon approaching the coronal pulp chamber to terminate in a subdentin capillary plexus. In the molars, they form a vascular bridge above the floor of the pulp chamber. The aperture of the subdentin capillary loops is on the order of 7–10 μm and it increases as it approaches the venous capillaries which drain them. The diameter of the veins decreases in the direction of the apex indicating a slow venous return, favoring exchanges at the capillary level.

Physiology of the Perfusion

When performing a perfusion, a constant physiological pressure has to be maintained because the control of vascular functioning is patterned by a series of homeostasis mechanisms. These mechanisms tend to maintain numerous physiological variables at a constant level. One of these variables, the dominant circulatory reflex, is brought about by the carotid sinus. The nerve endings of the depressor nerve are receptors which are specifically sensitive to the rise of arterial blood pressure and they are located at the junction of the common carotid artery. When the carotid sinus is perfused under a constant pressure, near to the threshold level (approximately 40 mm Hg), the discharge is very slight compared to that noticed when a pulsation pressure of identical average value exists. This reflex mechanism of blood pressure control is brought about by a local release of noradrenalin and by a distant effect of this noradrenalin at the level of the symphathetic nerve endings, producing the contraction of the smooth vascular muscle. In addition, a venomotor tone exists which, while preventing any displacement of blood volume, is just as important for the carotid sinus reflex for compensating any blood pressure change. The veins are richly innervated by the sympathetic nervous system [1].

Circulation Time

Knowledge of the circulation time is equally important. We obtained this circulation time by performing a simple equation, taking two points, A and B, of a vascular bed:

$$\text{Circulation time (in seconds)} = \frac{\text{volume of vascular bed between A and B (in ml)}}{\text{volume of flow from A into B (in ml/sec)}}.$$

The circulation time in the *M. fascicularis* is on the order of one min. Under conditions of cephalic perfusion, the required time to obtain a satisfactory fixation will be more or less the same as total circulation time.

Fixation

The liquid fixative is brought into direct contact with the cells which are to be fixed and this considerably reduces the artifacts due to a prior autolysis. In fact, as soon as blood circulation is interrupted, changes occur which modify the balance between cells and environment [6–9]. However, fixation by perfusion can lead to histological changes of the vascular walls, which are not negligible, as well as edematous reactions of the precapillary regions. The authors advise reducing the amount of wash fluid in order to avoid, as far as possible, these latter reactions which occur at the time of the wash preceding the injection of the fixative. For fixation of dental tissues, 10% formalin is used. Penetration into the tissues is quite fast, and preservation of cell structures is good after their coagulation.

To obtain satisfactory results in our experiment, we took into account the following factors:

1. Technical Factors

(a) Intervention time must be as short as possible (blood stasis time).

(b) Absence of pain or twisting of vascular walls provoking peripheral spasms.

(c) Importance of the hydrostatic pressure which must be maintained within the limits of the normal arterial pressure.

(d) Necessity of a lavage with a solution of heparinized physiological serum prior to the fixative perfusion.

(e) Prolonged perfusion is unnecessary. It is apparent within the first few seconds whether the fixation is a success or a failure. Furthermore, more than 1 or 2 min can cause perivascular edema.

2. Composition of the Fixative

The osmolarity of the fixative plays a role in the quality of the fixation.

Efficient hydrostatic pressure depends on the fixative used. For instance, when using formalin, this pressure can be lower than the arterial pressure, whereas for other fixatives, such as glutaraldehyde or Bouin's solution, it is preferable that the pressure be higher.

Finally, the organ reacts and is fixed differently according to which fixative is perfused. It is for this reason that formalin can provoke an increase of the capillary diameter with intercellular edema.

Material and Methods

The animal is laid on a horizontal plane made up of a stainless steel rack that permits the run-off of fluids.

The material needed for the perfusion comprises:

(a) a water bath vat used to maintain the perfusion fluids – heparinized Ringer's solution and 10% formalin – at a constant temperature of 38–39 °C;

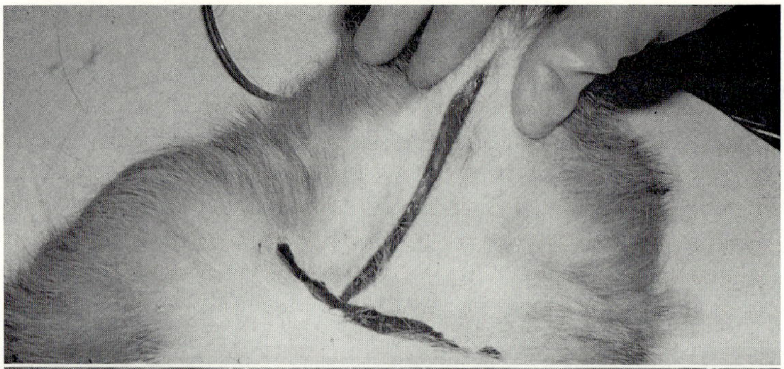

1

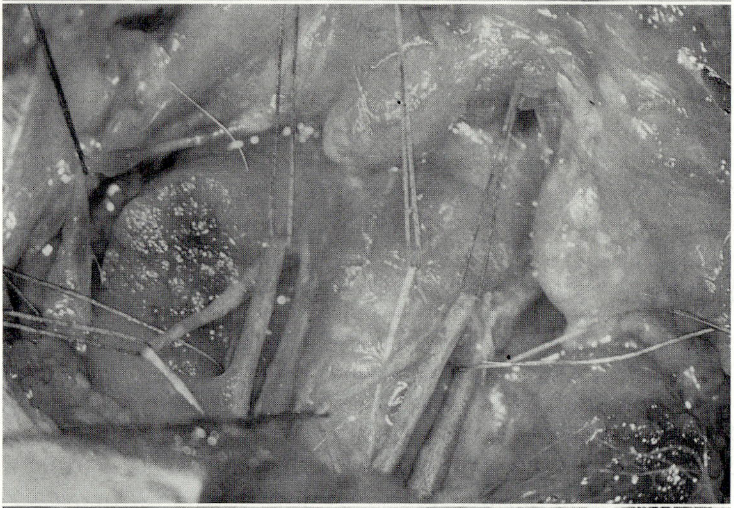

2

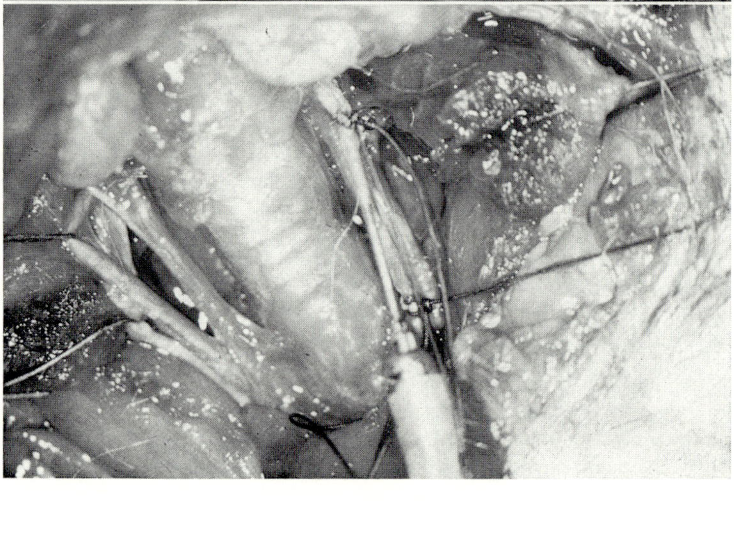

3

(b) the diluted solution which is connected by two transparent, plastic tubes to a digital pressure pump (Lillehei type) enabling to check on the absence of air bubbles (this pump ensures a flow and a constant pressure of 80 mm Hg);

(c) two mercury manometers and two Brooks' flow meters allowing a control on the flow and the pressure of the circulation; any excess pressure may cause an edema;

(d) two cannulas complete the circuit (one for each carotid) equipped with stopcocks which are used to interrupt the perfusion when a change of perfusion liquid is required.

Following the general anesthetic, an inverted T-shaped incision is made. The first median incision extends from the lower part of the mandible symphysis to the upper part of the sternal manubrium. The second transverse incision follows the upper edge of the clavicles (fig. 1). The incision marks off two surface strips which will be dissected and folded back and upwards. The important anterolateral muscle group can then be clearly seen as well as the sternoclavicular mastoid, on the surface of which, in its upper part, runs the external jugular. The jugular will be dissected and isolated by a ligature thread (fig. 2). At this stage everything is ready to start the perfusion, which should be carried out fairly rapidly, and in the same order for each side, according to the following procedure:

(a) The lower ligature of the carotid is closed in order to interrupt the arrival of the arterial blood.

(b) A bevel (slant) incision is made in the vascular wall using small scissors. The vessel is less easy to handle due to the suppression of the blood flow.

(c) The cannula is inserted from top to bottom into the axis avoiding, as far as possible, any stretching or twisting of the vessel.

(d) With the aid of the upper ligature, the vascular wall is solidly secured over the cannula (fig. 3).

Perfusion is then commenced by opening the stopcocks which control the flow of the heparinized solution. The internal and external jugulars are then cut, and the elimination of the blood mass is carried out. The heparinized fluid will replace the blood in the cephalic circulation and the flow is observed through the jugulars. At the required moment, the heparinized solution is replaced by the fixative solution by changing the bottle, having first clamped the plastic tubes to avoid air penetration into the cannulas.

When the fixative solution starts flowing, several contractions in the neck region can be seen. The animal quickly presents a state of extreme rigidity and continued perfusion is unnecessary. Samplings may then be made while taking into consideration that the rigidity of the fixed tissues prohibits extractions by means of forceps alone. If necessary, X-rays are taken. Each sample is placed in an individual, labelled bottle containing the liquid fixative.

Fig. 1. T-shaped incision.

Fig. 2. Under the reflected central muscular mass one can see the vasculoneural bundles of the neck, which immediately border the trachea. Primitive carotid; internal jugular; pneumogastric nerve; external jugular; trachea.

Fig. 3. Bevel (slant) incision. Insertion and block of the cannula by means of top ligature. The walls of the carotid deprived of the blood flow which distended them make the carotid difficult to grasp and to cut. External jugulars sectioned. Internal jugulars sectioned. Cannulas in place in the carotids. The perfusion arrangement is in place, and the flow of the lavage fluid is controlled at the sectioned ends of the internal and external jugulars.

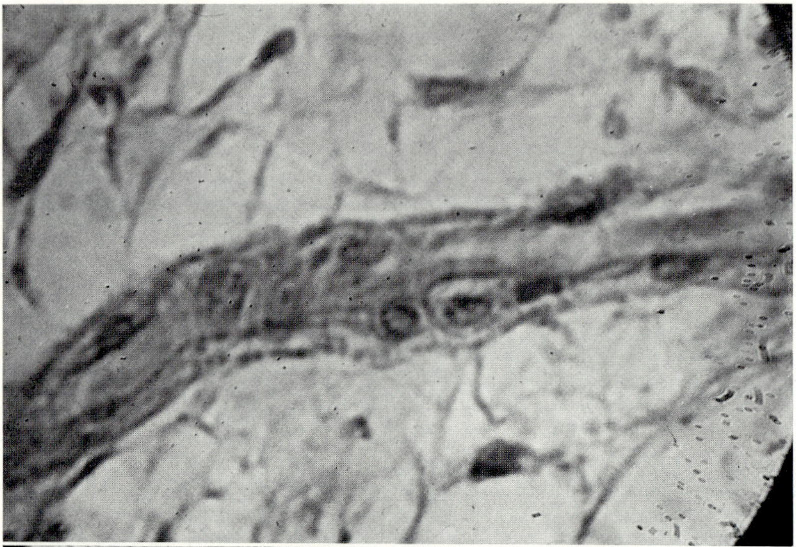

4

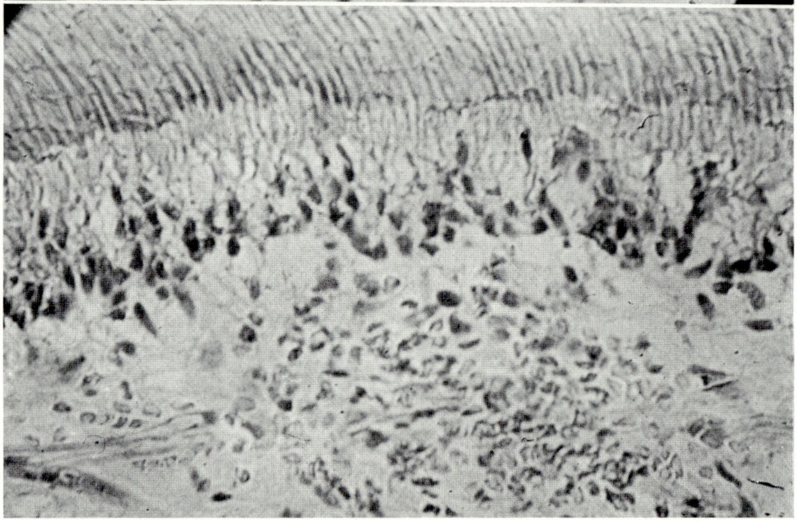

5

Fig. 4. Arterioles and precapillaries. The vessels are not damaged by the perfusion. They present an empty lumen. Note the pericytes and the muscle cells in the walls.

Fig. 5. Inflammatory sites located in the subodontoblast region.

6 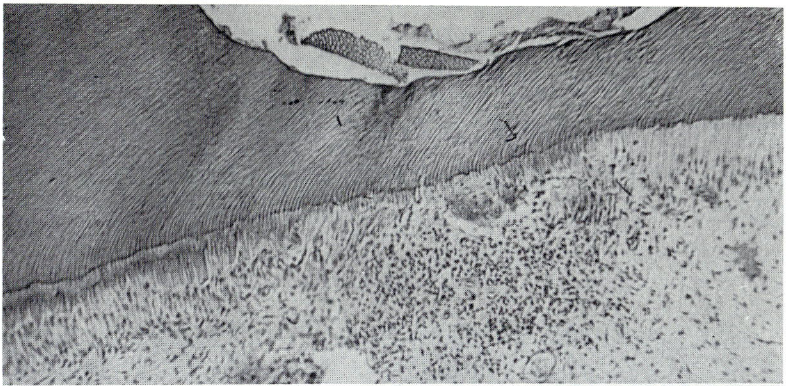

7 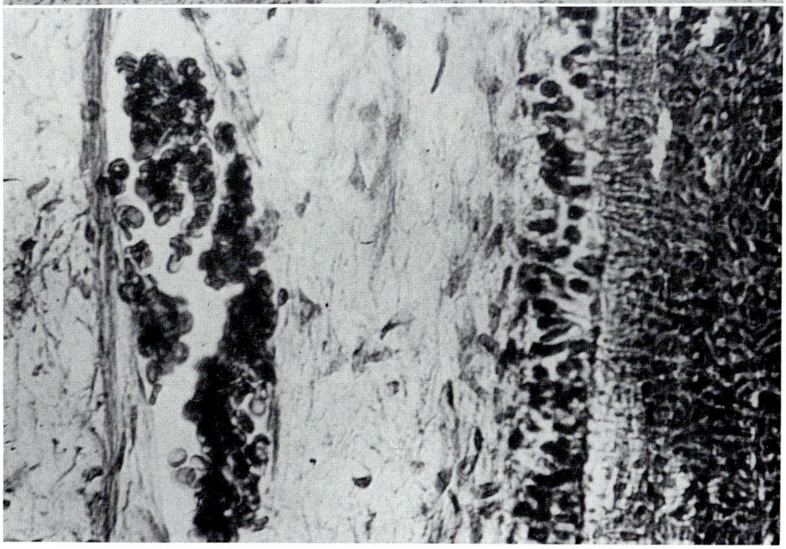

Fig. 6. Site of necrosis under a deep cavity.

Fig. 7. Images of vascular thrombosis. Lavage perfusion fluid. Though it involves all the elements figured, it is without effect on the intravascular thromboses. The organized clots remain caught on the walls.

Results

The method described above was used as part of a study to observe dentin-pulp reactions following trauma. Some examples of the results of histopathological studies give an idea as to the effectiveness of this experimental procedure. They illustrate the quality of the tissue responses to trauma on the

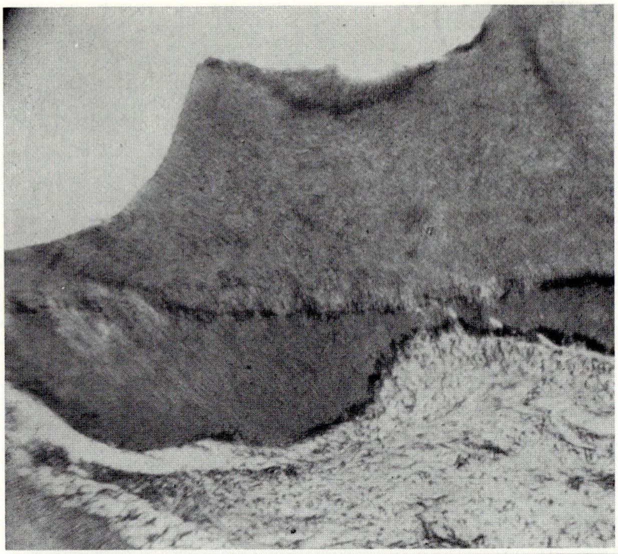

8

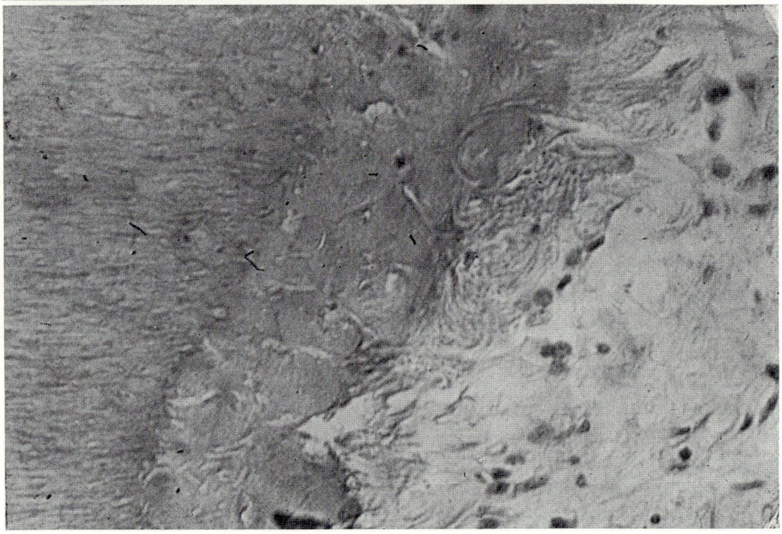

9

Fig. 8 and *9*. These two illustrations show different aspects of dentin reaction deposits.

dentin, obtained by the *ante mortem* perfusion method (fig. 4–9). It should be noted that in almost all sections, the pulp-dentin membrane was not dissected. The odontoblast body remains joined to the predentin. The pulp parenchyma appears well fixed, and the vessels do not seem to have suffered from the perfusion. The thrombosed elements are not carried away by the lavage of the perfusion, but remain fastened to the walls. The perfusion does not remove vascular sequelae which are due to considerable traumata. Finally, the different aspects of reactional dentin are clearly individualized in terms of the traumata to which they responded.

Conclusions

The method described here enables the elimination of delays and traumatisms which are otherwise incurred when classic methods for the histological study of dental pulp are practiced. While avoiding numerous artefacts, this method permits the observation of pulp tissues that have preserved an appearance close to their natural biological state.

References

1 Beveridge, E. E. and Brown, A. C.: The measurement of human dental intrapulpal pressure and its response to clinical variables. Oral Surg. *19:* 655–668 (1965).

2 Butcher, E. O. and Taylor, A. C.: The blood vascular supply of the dental pulp of the monkey and its alteration by tooth reaction. J. dent. Res. *31:* 239–242 (1952).

3 Castelli, W. C.: Vascular structure of dental pulp in cat. J. dent. Res. *41:* 213–220 (1962).

4 Chamodot, P.; Fortier, J. P.; Nitlich, J. et Zeilig, G.: Contribution à l'expérimentation animale en endodentie. Sci. Rech. Odonto-Stomat. *5:* 19–24 (1971).

5 Kramer, I. R. H.: The vascular architecture of the human dental pulp. Arch. oral Biol. *2:* 177–189 (1960).

6 Nouet, R.; Villa-Porcile, E. et Olivier, R.: Fixation de l'hypophyse par perfusion aortique. Micr. electron. *8:* 72a–73a (1969).

7 Provenza, D. V.: The blood vascular supply of the dental pulp with emphasis on capillary circulation. Circulat. Res. *6:* 213–216 (1958).

8 Saunders, R. L. de C. H.: X-ray microscopy of the periodontal and dental pulp vessels in the monkey and in man. Oral Surg. *22:* 503–518 (1966).

9 Villa-Porcile, E.; Nouet, R.; Smiechowska, I. et Olivier, R.: Problèmes posés par les techniques de fixation du lobe distal de l'adenohypophyse. Bull. Ass. Anat. *149.*

10 Walter, R.; Castelli, W. C., and Huelke, D. F.: The arterial system of the head and neck of the rhesus monkey with emphasis on the external carotid system. Amer. J. Anat. *116:* 149–170 (1970).

Author's address: Dr. G. Zeilig, Université René Descartes de Paris, U.E.R. d'odontologie, 1 rue Maurice Arnoux, *F–92 Montrouge* (France)

Neurophysiology and Experimental Psychology

Medical Primatology 1972. Proc. 3rd Conf. exp. Med. Surg. Primates, Lyon 1972, part II, pp. 244–249 (Karger, Basel 1972)

Sleep and Environment

J. BERT and V. PEGRAM

Département de Neurophysiologie appliquée, Institut de Neurophysiologie et de Psychophysiologie, CNRS, Marseille, and Department of Psychiatry, University of Alabama Medical Center, Birmingham, Ala.

It appears that three basic factors determine the organization of sleep and shape its course: a genetic factor, a learning factor, an environmental factor. It is difficult to separately analyse each one of them because they obviously interact together. If we compare the sleep of different species of monkeys, we can study the influence of genetics; if we study the ontogeny of sleep within a species we can study the influence of learning; and if we study monkeys of the same species submitted to different milieu conditions we may define the influence of environmental factors. In an attempt to better define the effect of environment on sleep this report deals with the EEG sleep of adult baboons *(Papio papio)* under two different conditions, in a laboratory at Marseille (France) and in their natural environment in the African bush.

Material and Methods

The subjects used in both studies were adult baboons *(Papio papio)* implanted in a similar manner. In-dwelling stainless steel electrodes were surgically implanted. The cortical electrodes located on the surface of the dura mater were used for the EEG; the periocular electrodes were used for recording of eye movements, and braided wire was implanted in the nuchal muscles to determine muscle tone. In the wild, a removable plastic box capable of holding five FM transmitters was attached with screws to the dental acrylic covering the implantation to transmit the EEG to a nearby telemetry receiver and polygraph. In the laboratory, the baboons were recorded directly onto a polygraph while seated in a chair. They were placed in an isolation room with the lights on at 7:00 a.m. and off at 6:00 p.m. Several times a week the EEG was recorded during the night (between 6:00 p.m. and 7:00 a.m.). The monkey's environment was disturbed only for cleaning of the room and for distribution of food.

In Africa we worked in two regions of Eastern Senegal: in 1968 at Sansande in the Faleme river valley, and in 1971 at Damantan inside the National Park of Niokolo-Koba. All the baboons were captured by using a baiting procedure with peanuts laid out in a trail leading into large wire mesh cages located by a frequently used water hole. While the animals fed in the cages, they were trapped by a guillotine type door dropped upon command from a distance of 30–40 m. In both expeditions the baboons were recorded close to the site of their capture and set free after the sleep experiments were terminated. At Sansande the animals were kept inside an open enclosure surrounded by an electrified fence with a radius of 10 m placed around a small sleeping tree. At Damantan the animals were recorded inside a totally enclosed wire mesh cage (3 × 3 × 2.5 m), with a tree branch set horizontally inside the cage at 1.3 m to provide a sleeping perch. At Sansande four baboons were recorded for a total of six nights of acceptable records and at Damantan three baboons were recorded for a total of six nights. It should be noted that many nights of sleep recordings in Africa had to be discarded due to dust, heat, and at times an unreliable generator.

Results

Laboratory Experiments

Newly acquired baboons placed in a restraining chair showed a succession of three well-defined periods with an increase of total sleep time and a modification of stage distribution (see table I).

During the first period, immediately following restraint of the animal, sleep occupied a little more than half of the recording, i.e. 55.2%. Stage 1 was predominant, stage 2 was variable and stage 3 was completely absent. REM sleep was apparently severely disrupted and represented only 1.8% of sleep. The length of this first period varied greatly from one monkey to another, ranging from one to 15 days. During the second period of adjustment to the laboratory and restraint, the amount of sleep was notably longer,

Table I. Percentages of laboratory recordings (n=4)

	% awake TRT	% sleep TST			
		stage 1	stage 2	stage 3	REM
Period 1	44.8±16.6	59.9±17.7	40.8±16.4		1.8±2.9
Period 2	23.2± 6.9	29.4± 9.0	57.4± 6.0	5.6±5.0	7.4±2.0
Period 3	18.3± 3.6	20.5± 4.5	57.8± 2.7	8.8±2.9	12.5±2.7

TRT=Total recording time; TST=total sleep time.

76.8% of the total recording time. The proportion of stage 1 decreased, while stage 2 increased. Stage 3 was observed, but remained scarce and very variable, while REM increased significantly. The third period, which was reached towards the end of the third week, was marked by a further increase in the total length of sleep and a continued decrease in stage 1. There was a considerable increase in stage 3 and REM sleep. Also during this period the variability between nights decreased markedly.

Experiments in the Natural Environment

The recordings in the wild at Sansande and Damantan showed some evident differences (table II). At Damantan the total percentage of sleep was less and REM sleep was more evident than at Sansande. Stage 2 was approximately the same in both regions, while stage 3 was quite evident at Sansande and nearly absent at Damantan. One point not evident from table II was the consistency of sleep between nights, especially at Damantan. This variability was much less than that observed during the first two adjustment periods in the laboratory.

Comparison of Recordings in the Laboratory and in the Natural Environment

The recordings of the first adjustment period in the laboratory were clearly different from the recordings in the natural environment. The laboratory recordings were distinguished by frequent awakenings and a prevalence of stage 1. If we consider the sleep percentages recorded in the

Table II. Percentages of African recordings (n = 6)

	% awake TRT	% sleep TST			
		stage 1	stage 2	stage 3	REM
Sansande	7.9 ± 3.0	24.5 ± 7.3	64.0 ± 11.2	7.5 ± 8.0	3.8 ± 3.2
Damantan	18.2 ± 1.4	32.2 ± 6.2	58.8 ± 7.9	0.5 ± 0.5	8.3 ± 2.2

TRT = Total recording time; TST = total sleep time.

natural environment, we find them close to the second period in the laboratory. The recordings in the African bush never showed the length of total sleep time and the simultaneous rise of stage 3 and REM sleep which characterized the third period of laboratory recordings.

The morphology of the EEG recordings from the baboon studies was essentially the same as had previously been recorded in the laboratory except for two adult animals at Sansande which showed a slow wave activity classifiable as stage 4 sleep (1–3 Hz of high amplitude EEG without transitional elements). This brief appearance of stage 4 was rather significant since stage 4 had never been observed in previous laboratory studies using adult baboons and was not seen at Damantan. Certainly, our attempts at recording 'sleep in the wild' did not represent the sleep patterns of a free-ranging animal. However, the results do provide evidence that stage 4 is not a necessary component of *adult* baboon sleep, but may occur when the animal is fatigued. This lack of stage 4 in the adult baboon is even more striking since infant baboons show stage 4 comprising as much as 20% of the total sleep record by three to four months of age which declines to almost zero by two years. Perhaps a better understanding of the function of stage 4 in primates might be obtained by examining how or under what conditions the baboon is able to eliminate a stage of sleep which is apparently critical in its early development. For example, the young baboon sleeps while being held in its mother's arms. Perhaps the stage 4 decrease occurs about the time the young baboon 'acquires' its new sleeping position, that of sitting on a branch.

Discussion

A number of interesting points resulted from the comparisons reported in this paper. One of them was certainly the differences in sleep patterns found between the two capture sites in Africa. While the data were not sufficient for many firm conclusions, they certainly presented some interesting possibilities. For example, the percentage of REM sleep was higher at Damantan and might be explained by the size and type of sleeping trees *(Ceiba pentandra* and *Elaeis guinensis)*. These trees were quite large and provided a relatively safe shelter for the sleeping baboons. The Sansande region was a much dryer country and the sleeping trees were smaller. Observing large baboons sleeping on narrow branches was often puzzling, for the total suppression of postural tonus during REM sleep seemed impossible. In fact, two of the animals recorded at Sansande showed very fragmentary periods of REM with the muscle tone persisting well into many of these REM periods. While the size and relative safety of the sleep trees might explain the difference in REM time, this would not explain why the animals at Damantan had less total sleep time and less of stage 3 than the animals at Sansande. We did observe more awakenings during the night at Damantan and there were probably more causes for these awakenings. Being an area of denser vegetation, the sleeping trees of adjoining troups at Damantan were closer together and loud barks often penetrated the night for as long as 10–25 min. It was never possible to find direct cause for these outbursts and such episodes were not observed in the Sansande region.

A second interesting point is the similarity found between the laboratory animals and those recorded in the wild. Previous studies on sleep in primates have been carried out under laboratory-type conditions or with animals which have lived in captivity for months or years. These animals have often been deprived of their normal physical activity and, in many recording conditions, have also had to adapt to a modified sleep position. Further, they have been exposed to an essentially human environment including climatic conditions and the elimination of predators. Such procedures certainly raise the possibility of distortions of the normal organization of sleep under laboratory conditions. Thus, one of the findings in these African studies was in fact the similarity between laboratory animals and newly captured animals. As previously mentioned, the African recordings were similar to the second period of habituation in the laboratory. This makes sense as we recorded the African animals about ten days after capture and thus missed the dramatic adjustment to captivity which might be comparable to the first experience in restraint chairs of the laboratory animals. As noted, we also never observed as much of stage 3 and REM in the African recordings as occurred in the laboratory animals. No doubt the restrained baboons in the laboratory were exposed to sensory restriction. Once they were habituated to the restraint chairs, there were very few stimuli to disturb their sleep and they lived in a monotonous environment. This type of environment has been shown to influence the sleep of man as well. RAYBACK and LEWIS [1971] studying the sleep of healthy young adults confined to bed during five weeks, observed an increase of sleep stages 3 and 4. Actually these young men, even with TV and books available, were submitted to a monotonous environment, for they were deprived for a large part of the stimuli which normally occur in the daily routine. WEBB and SWINBURNE [1971] reported an increase of sleep duration in the aged who spent about 12 h in bed while in the hospital ward. They suggested that this increase was due to boredom.

By comparison the night environment in the bush was quite different. The amount of moonlight, wind and noises was constantly changing. But above all, the night was full of various noises: rustle of leaves, branches cracking, cries of nocturnal animals (lion, hyena, birds, etc.), cries of other baboons. In this way the baboons in the wild were exposed to a number of stimuli. A large part of these stimuli were probably meaningless and provided a neutral background noise, but some had a powerful alerting effect (i.e., the roar of the lion).

Thus, we feel that these studies have shown that some tight relations link sleep and environment. Environmental pressure can control sleep duration and modify its pattern. In animal survival imperatives inside a specific ecological niche command its course. In man it is probably social constraints (cultural patterns).

Therefore, sleep should not be studied as an independent and autonomous phenomenon. It is submitted to environmental influences which shape its pattern inside the limits determined by two other factors, learning and genetics.

References

Ryback, R. S. and Lewis, O. F.: Effects of prolonged bed rest on EEG sleep patterns in young, healthy volunteers. Electroenceph. clin. Neurophysiol. *31:* 395–399 (1971).

Webb, W. B. and Swinburne, H.: An observational study of sleep of the aged. Percept. Mot. Skill. *32:* 895–898 (1971).

Authors' addresses: Dr. Jacques Bert, INP 3, CNRS, 31, ch. Joseph-Aiguier, *F-13274 Marseille, Cédex 2* (France); Dr. Vernon Pegram, Department of Psychiatry, University of Alabama, 1919 Seventh Avenue South, *Birmingham, AL 35233* (USA)

Medical Primatology 1972. Proc. 3rd Conf. exp. Med. Surg. Primates, Lyon 1972, part II, pp. 250–259 (Karger, Basel 1972)

Group Social Tension as a Determinant of Alcohol Consumption in *Macaca mulatta*

T. E. CADELL and R. CRESSMAN[1]

Department of Psychology, University of Waterloo, Waterloo, Ontario

In recent years a number of investigators have studied changes in social behavior following alcohol consumption by simians. The behavioral change which occurs when the animal is intoxicated appears to depend upon the species of animal studied, the nature of its early experience, and its current stage of development. In adolescent male feral rhesus monkeys *(Macaca mulatta)* the outstanding behavioral result of alcohol consumption is an increase in contact play [CRESSMAN, 1972; CRESSMAN and CADELL, 1971]. The data of PERETTI and LEWIS [1969] appears to support this observation although they report some aggression and dominance shifts as well. Rhesus monkeys which were socially deprived during their early life indulged in self-aggression and self-mutilation under the intoxicating effects of alcohol in a social group [CHAMOVE and HARLOW, 1970].

In adult chimpanzees, alcohol consumption differentially increased the frequency of aggressive and sexual behaviors in males and females [FITZ-GERALD and FOX, 1969]; males evidenced more aggressive behavior than females and females displayed more sexually related behavior than males. In the above research, attention was not directed at those factors in the social experience which might affect the consumption of alcohol by individual members of the group. More recently, CRESSMAN's [1972] work has shed light on some of the social determinants of alcohol consumption and led to the present research.

CRESSMAN [1972] used four groups of four monkeys and housed each group in a separate enclosed social room where they lived continuously except when removed for cleaning of the room. Once each day, over a period of

1 This research was supported by a Research Grant and a Research Fellowship from the Addiction Research Foundation, Toronto, Ontario.

several weeks, groups were observed for 100 min while they were provided access to either alcoholic or nonalcoholic beverages. Home-cage alcohol consumption was recorded for each monkey both before and after group social living. Dominance order within each group was assessed under water deprivation conditions at the beginning and end of a series of tests.

A comparison of the average amount of alcohol consumed by animals in each dominance position for the four groups revealed that those animals occupying the intermediate dominance positions consumed significantly more alcohol per kg body weight than those in the extreme dominance positions. The same monkeys did not differ in terms of their home-cage consumption at the start of the experiment. It appeared that the social situation produced an overall decrease in alcohol consumption during these short-term observations, but this reduction was due mostly to one group of monkeys.

To assess whether dominance was a major determinant of alcohol consumption, the groups were recombined by placing the two most dominant monkeys from each of two groups into one new group and the remaining subordinate monkeys into a second new group. In this way at least two monkeys would be forced to change dominance position.

The results of the above experiment were equivocal; the correlation of expected changes with observed changes was zero and, overall, the consumption of the monkeys according to their new dominance positions did not differ. However, one observation was of great interest. One of the original groups, as mentioned above, drank much less alcohol than the other three. Further, this group had an unstable dominance hierarchy, had a high level of contact play, and showed a minimum of competition for the automatic drinking device. The animals of this group, when recombined into new groups, vastly increased their consumption. It appeared as though the original group differed from the new groups in terms of the quality of social interactions. The present study is an extension of this observation and compares the alcohol consumption of two groups of monkeys which have been selected on the basis of group 'social tension'.

Method

Subjects

The subjects for this experiment were eight male rhesus monkeys *(Macaca mulatta)* maintained on Rockland Primate Diet plus a daily supplement of fruits, nuts, or vegetables. The monkeys were quarantined for a period of 90 days before being used. The animals were housed in individual home-cages (71.1 × 86.4 × 71.1 cm in height), each equipped with an an automatic drinking device in the center back of the cage. The weight range of the subjects at the start of the experiment was 3.4–4.4 kg.

Apparatus

Social observation room. A line drawing of the social observation room may be seen in figure 1 and a more detailed description of the room may be found in CRESSMAN and CADELL [1971]. Basically this room measured 316 cm long × 244 cm wide × 244 cm high. It had a grey Neotex floor, three white epoxied cement block walls, and a wire mesh ceiling. The fourth wall, referred to as the observation wall, was made up of steel panelling and included a guillotine door on the right side to introduce or remove the monkeys.

A 183 × 244 cm anteroom, which shared the observation wall, was utilized for observing the monkeys. This room was equipped with a white noise generator which was in 24-hour operation to mask the laboratory noise throughout the duration of the experiments. The anteroom and the social test room were both air-conditioned. The former was maintained at a temperature of 21.1 ± 1 °C and the latter was about 4 °C higher because of the incandescent lighting.

Monkey detection drinking system. This is a radiofrequency controlled device which permits monitoring of consumption of individual monkeys in groups. The transmitter consisted of a multivibrator tuned to a particular frequency in the range of 375–450 Hz. Each monkey carried one attached to a leather vest [CRESSMAN and CADELL, 1970] and the high-frequency components of the signal were undetectable unless the monkey contacted the drinking device (Hardco No. 5540). The signal was detected by a standard AM car radio as a quasi-sine wave and this signal was used to drive a tuned reed relay which in turn operated a standard relay. In this way each monkey operated a separate standard relay by contacting the drinker and the relay opened a fluid solenoid, connecting the monkey's bottle to the drinker. An event recorder was used to record the operation of the relays and thus the sequence and timing of drinking of the monkeys. A detailed description of this device is given in CADELL and CRESSMAN [1972].

General Procedures

Measurement of dominance. For the purpose of this study, dominance was measured under conditions of water deprivation. The monkeys were deprived of water for 20 h and then allowed access for a period of 4 h. The order in which the monkeys took their longest drink at the beginning of the 4-hour access period was recorded as the dominance order.

Measurement of group social tension. Group social tension was measured in terms of mean interindividual distance between monkeys in a group under conditions of water deprivation. For this purpose the room was divided into a total of 35 areas based upon distance from the drinking device which supplied the water. The area of the floor nearest the drinking device was divided into six semicircular areas with each succeeding one having a

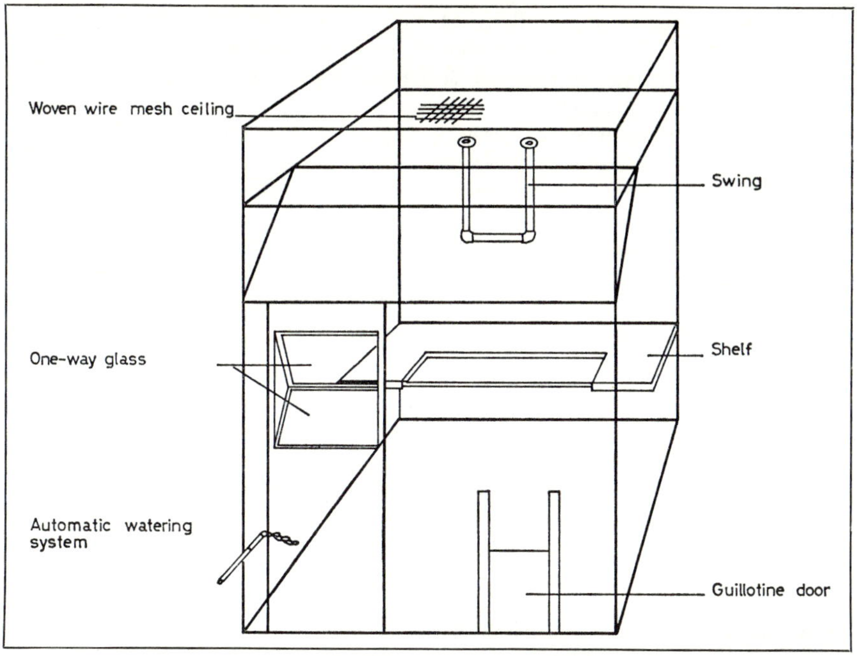

Fig. 1. Outline drawing of the social observation room showing the relative positions of the automatic watering system, the shelf, the swing, the observation window and the guillotine door.

radius of 15.2 cm greater than the last. Beyond the first 91.2 cm, the semicircular areas laid out on the floor occurred at intervals of 61.0 cm resulting in three additional areas. The floor was further divided into four sectors by three lines radiating from the drinker at 45, 90 and 135° to the wall. The shelf was divided into two parts: (1) the iron pipe and the side of the shelf nearest the drinker, and (2) the back of the shelf and opposite side together. Two additional areas in which the monkeys could be located were the swing and the ceiling.

The location of all animals was recorded every 10 sec in terms of their location in the 35 areas demarcated in the room. The scoring continued in this manner for a period of 10 min following introduction of the water drinker. The mean distance between the dominant monkey, when at the drinker, and the other three monkeys represented the group social tension; large values indicated high social tension and small values indicated low social tension.

Measurement of social behavior. Basically, the procedure for recording a social behavior session was the same as that described by CRESSMAN and CADELL [1971], utilizing a modified time sample. In each test session, which was of 20-min duration, the animals were observed individually for 1 min in a random sequence selected for that session. Each monkey was observed in turn for 1 min until the sequence was complete and the sequence was repeated until the end of the session. Within each 1-min period the behavior occurring at the end of each 10-sec interval was recorded. Social interaction behaviors were scored further in terms of which animal was the initiator and which the recipient of the behavior.

A number of behaviors were recorded during the social observations but only contact play was of importance to the present paper and it was defined as: mauling, biting and wrestling, which rarely results in even mild injury to either partner.

Preparation of alcohol solution. The alcohol consumed by the monkeys in this experiment was prepared from a Canadian Sauterne diluted to 5% alcohol with grape juice and water. The concentration of grape juice was 40%. Solutions were prepared every other day and stored in a well-stoppered bottle in a refrigerator.

Experimental Procedures

It was important for the purposes of this study that two groups of monkeys were selected in such a way that they differed greatly in terms of interindividual spacing and that they did not differ initially in terms of home-cage alcohol consumption.

Individual home-cage consumption. Once the eight monkeys had been removed from quarantine they were transferred to individual cages and given access to a 5-percent alcohol solution 100 min each day, Monday through Friday, for a period of two weeks. The data on individual consumption collected during the second week was used as one of the bases for comparing the various groupings of monkeys selected by the procedure outlined below. Alcohol was accessible through a drinking device attached to the front of the cage. Water was continuously available from a similar device at the back of the cage.

Group selection. Following the last day of home-cage consumption, all monkeys were put on daily 20-hour water deprivation. At the end of the 20-hour deprivation in the home cage the monkeys were divided into two groups of four and interindividual distance was measured for each group according to the method outlined above. Following testing, each group of monkeys was returned to their home-cages and allowed access to water for the remainder of the 24-hour period. At the end of this time the water was withdrawn again.

The groups of monkeys were selected by first ranking the monkeys by body weight and systematically selecting various groupings to vary monkey sizes within the groups. At the end of the first week (five days) one grouping was selected which produced the greatest difference in interindividual distance between the two groups. This particular grouping was tested for an additional week to insure that the difference was reliable. At the end of this time the monkeys were allowed three days free access to water in the home-cage to reduce the effect of repeated 20-hour water deprivation periods on subsequent consumption of alcohol.

Alcohol consumption. Once the groups had been selected and had recovered from water deprivation, they were reintroduced to their respective social rooms and allowed access to the alcohol solution 24 h per day for the period of the experiment. The only time that the

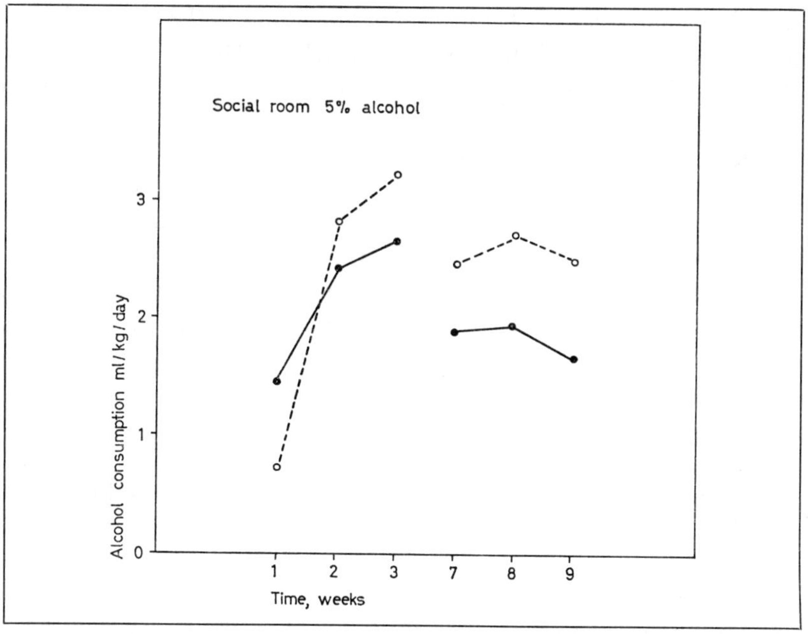

Fig. 2. Daily alcohol consumption per kg body weight for the high (○) and low (●) social tension groups. The reasons for omitting the data of weeks 4 through 6 are given in the text.

monkeys were removed from the social room was for cleaning which occurred twice each week for a period of 4 h. During the cleaning period the monkeys had access to alcohol in their individual home-cages.

Alcohol consumption was recorded, and when necessary, bottles filled at five different times each day: 9:30 a.m., 12:30 p.m., 3:30 p.m., 6:30 p.m., and 9:30 p.m. Behavioral observations were recorded in three sessions daily: 9:40 a.m., 12:40 p.m. and 3:40 p.m.

There were occasional breakdowns in the monkey detection system. For example, the most common problem was one monkey removing the vest of another. Whenever such breakdowns occurred the data for both groups were discarded for that day. Of the total of 42 days covered in this experiment five were discarded owing to failure of the detection system.

Results

Two groups of monkeys were established on the basis of their overall interindividual distance scores. Group E, the low social tension group, had an average distance score of 171.4 cm, whereas group F, the high social tension group, had an average distance score of 261.1 cm. These differences were statistically reliable as determined by a simple analysis of variance ($p<0.05$).

The home-cage consumption for the E and F groups was 1.25 and 1.67 ml/kg, respectively, for the 100 min allowed. These values were not found to differ statistically.

Initially, as may be seen in figure 2, the E group consumed more alcohol per unit body weight than the F group. However, after the first week the high social tension group consumed significantly more alcohol per kg body weight than the group selected for low social tension ($p<0.05$). The statistical comparison, a simple analysis of variance, was based on repeated measures of average daily group consumption rather than on a between subjects comparison. Between subjects variation in alcohol consumption was already known to be increased by differences in dominance position as outlined above.

As may be seen in figure 2, data are missing for weeks 4 through 6. This omission is due to failure of the air-conditioning system and a temperature elevation of about 8 °C. As a result of the change in temperature, both groups increased their consumption of alcohol by 20–25% and differences between the groups were diminished. Although the F group retained its higher consumption rate, the differences were not statistically reliable.

The pattern of alcohol consumption for the four dominance positions in group F revealed that the monkeys in dominance positions 2 and 3 had a higher consumption rate than those in dominance positions 1 and 4. The consumption values for dominance positions 1 through 4 were 1.79, 2.86, 2.45, and 2.39 ml/kg/day, respectively. The pattern of consumption for the E group differed sharply from that of the F group, with dominance position 1 having the highest consumption rate. The consumption values for the four dominance positions were 2.64, 2.38, 1.48 and 1.64 ml/kg/day, respectively.

The main concern of this paper was with the amount of alcohol consumed by each of the groups. The data concerning the social behavior of the monkeys will be treated elsewhere. However, two observations are noteworthy here. First, the frequency of contact play for the two groups differed. In the first week following the selection of the groups the average frequency of contact play for the E and F groups was 4.24 and 0.88 observations per 20-min session. In other words, under the influence of alcohol and given that the E group had a lower 'social tension', the E group indulged in contact play more than four times more frequently than the F group.

The second observation was that, for both groups, contact play decreased from the first week onward as the amount of alcohol consumption continued to increase. The correlation between contact play and alcohol consumption for the E and F groups was –0.40 and –0.26, respectively.

Discussion

Over a period of eight weeks the differences between the two groups of monkeys were maintained in the direction suggested by CRESSMAN [1972]. This indicates that group social tension, measured in terms of interindividual distance, is an important determinant of alcohol consumption by monkeys in social groups. This difference in alcohol consumption is not attributable to initial differences between the groups since they did not differ significantly in the home-cage test and the social consumption for the low tension group was greater than that for the high tension group during the first week.

The significance of social tension in determining alcohol consumption may be questioned in terms of its ability to raise consumption above levels recorded in the home-cage. Certainly the data of CRESSMAN [1972] appear to indicate that alcohol consumption in the social situation is somewhat less than that recorded for the same monkeys in the home-cage. Preferably, however, a comparison of solitary and social consumption should be based on measurements taken while the monkey is alone or with others, respectively, in the social test room. Since this kind of test was made impossible by a limited amount of testing space, this comparison has not yet been made. Furthermore, in the present research the home-cage consumption was measured during a 100-min period instead of a 24-hour period as in the social situation.

The data of this study support only in part the observation of CRESSMAN [1972] that the intermediate dominance positions consume the greatest amount of alcohol. The pattern of drinking in the F group, in which case the dominant monkey consumed the least, corresponded well with the pattern recorded for CRESSMAN's groups. However, the E group showed a sharp departure from this pattern with the dominant monkey having the highest consumption rate. Since, according to CRESSMAN's data [1972], the pattern does not re-emerge after recombination, differential consumption according to dominance position may be a rather delicate effect. It may interact with other factors such as experience in social consumption of alcohol, developing taste preferences or changes in the dominance hierarchy toward a nonlinear hierarchy.

Both of the above social determinants of alcohol consumption are related to dominance interactions among the monkeys. It is not surprising that the occurrence of play behavior is negatively correlated with group social tension; the highest values of play are achieved by the low social tension group and the lowest by the high tension group. What is surprising is the result showing a negative correlation between the amount of play occurring and the amount of alcohol consumed on a daily basis. CRESSMAN [1972] confirmed the finding of CRESSMAN and CADELL [1971] that the short-term effect of alcohol is enhanced contact play. Furthermore, the occurrence of contact play was highly positively correlated with the amount of alcohol consumed during the same session. However, when the amount consumed by each of his four groups was correlated with the mean frequency of contact play for each group, the correlation was –0.84. Although this correlation is only based on an 'n' of 4, it nevertheless supports the observations of the present research.

The simplest explanation of these conflicting correlations is one which assumes that alcohol consumption relieves or reduces social tensions within the group. In other words a monkey which finds itself in a group having high social tension will not normally indulge in a great deal of contact play. This monkey at the same time will consume a great deal of alcohol and its frequency of contact play will thereby be increased. Even under the influence of alcohol the frequency of play will still be less than that for a monkey in a low tension group. As the data reveal, the latter monkey will have a high rate of contact play and consume little alcohol. It is unlikely that the latter effect is due to a competition of activities; that is, more play produces fewer opportunities for drinking, since overall the highest rates of contact play do not represent a large proportion of total activity.

The above interpretation is consonant with studies which have shown a temporary increase in alcohol consumption for monkeys subjected to a shock-stress situation [CLARK and POLISH, 1960; MELLO and MENDELSON, 1966]. However, more work needs to be done on the social determinants of alcohol consumption both to uncover further variables influencing consumption and to replicate and extend the present results.

References

CADELL, T. E. and CRESSMAN, R. J.: A contact detection system to identify individual animals in groups. Physiol. Behav. (in press).

CHAMOVE, A. S. and HARLOW, H. F.: Exaggeration of self-aggression following alcohol intake in rhesus monkeys. J. abnorm. soc. Psychol. *75:* 207–209 (1970).

CLARK, R. and POLISH, E.: Avoidance conditioning and alcohol consumption in rhesus monkeys. Science *132:* 223–224 (1960).

CRESSMAN, R. J.: Consumption of alcohol. Social behavioral determinants and behavioral consequences in monkeys; unpubl. doct. diss., Waterloo, Ontario (1972).

CRESSMAN, R. J. and CADELL, T. E.: Drinking and social behavior in groups of rhesus monkeys *(Macaca mulatta)*. Quart. J. Stud. Alcohol *32:* 764–774 (1971).

CRESSMAN, R. J. and CADELL, T. E.: A monkey vest to carry light weight equipment. Lab. anim. Care *20:* 120–122 (1970).

FITZ-GERALD, F. L. and FOX, V.: Effects of alcohol upon sexual and aggressive behavior in chimpanzees. Paper read at the Meetings of the American Association for the Advancement of Science, Boston 1969.

MELLO, N. K. and MENDELSON, J. H.: Factors affecting alcohol consumption in primates. Psychosom. Med. *28:* 529–550 (1966).

PERETTI, P. O. and LEWIS, B. R.: Affects of alcohol consumption on the activity patterns of individual rhesus monkeys and their behavior in a social group. Primates *10:* 181–188 (1969).

Authors' address: THEODORE E. CADELL and REGINALD J. CRESSMAN, Department of Psychology, University of Waterloo, *Waterloo, Ontario* (Canada)

Medical Primatology 1972. Proc. 3rd Conf. exp. Med. Surg. Primates, Lyon 1972, part II, pp. 260–273 (Karger, Basel 1972)

A Comparison of EEG Changes by Drugs of Social Abuse[1]

K. F. KILLAM and J. E. GEHRMANN

Department of Pharmacology, National Center for Primate Biology, University of California-Davis, Davis, Calif.

Earlier studies from these laboratories compared the electroencephalographic changes observed following the administration of 11 benzodiazepines using spectral techniques [JOY *et al.*, 1971a and b]. Those studies indicated that time series analysis is a powerful tool to examine the relationships between chemical structure of particular substances and changes in CNS function and characteristics. This report represents findings from an extension of those studies and is part of a continuing program to study centrally acting drugs with particular reference to drug dependency.

Methods

Under barbiturate anesthesia, stainless steel screws were implanted in the calvaria of adult *M. mulatta*. The lead wires from the recording sites were exteriorized and attached to a multipin receptacle subsequently affixed with dental acrylic to the skull. The neocortical sites for recording were selected to represent the international 10×20 system of electroencephalography. In all cases the electrodes were epidural (fig. 1).

Subsequent to recovery from the surgical procedure, the animals were placed in restraining chairs for periods of 72 h. During these periods water was available *ad libitum;* and food, on a fixed period basis. Multichannel records were made in parallel on an FM magnetic tape recorder (Ampex FR100b or 1300) and on chart recorders (Grass Model IV or Offner Model T). A binary coded decimal time code generator (IRIG-B) was used to synchronize the two recording media. General comportment of the animals was described by one or more observers. Vital signs were assessed periodically. Bioelectric activity was quantitized using Linc II and Time/Data 100 Computers.

1 Supported by USPHS Grant No. MH 17471 and Department of Justice Contract J-70-37.

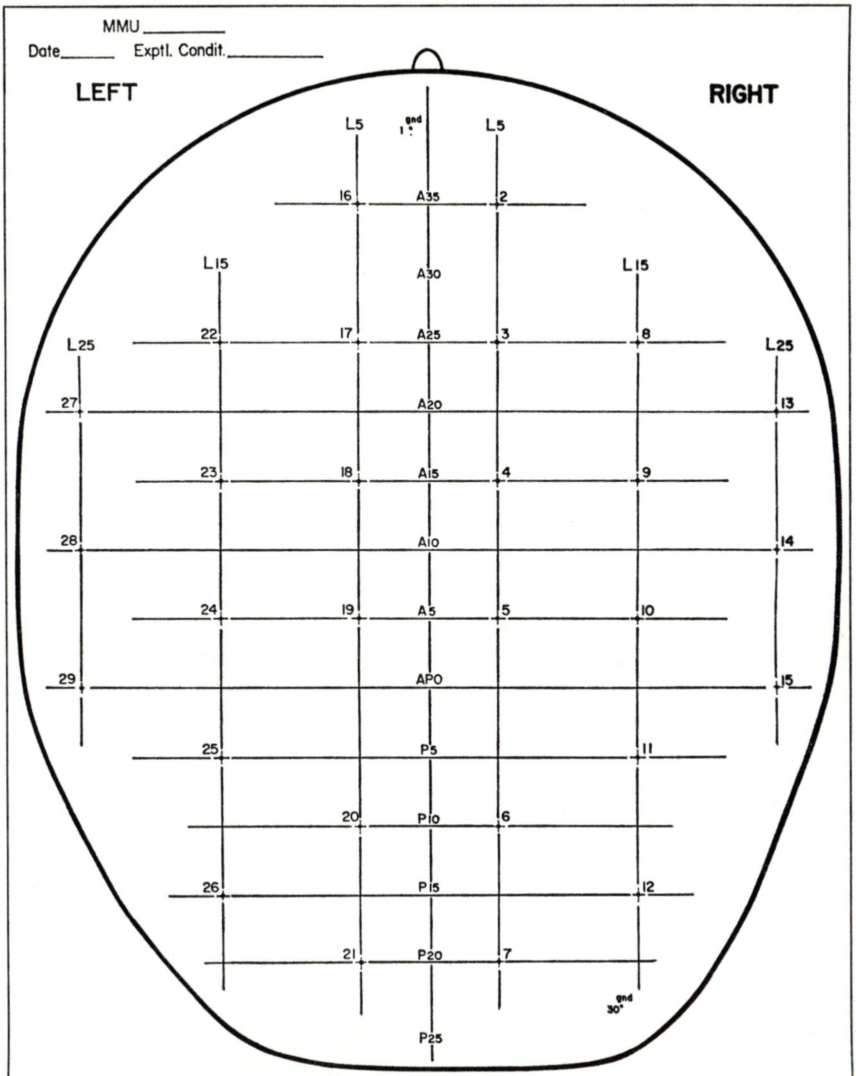

Fig. 1. Schema of electrodes placement.

Drugs were administered i.m. using two schedules: fixed doses and progressive increments. The latter schedule was used to establish dose-range sensitivities for individuals and for the species. Fixed dose assessments were guided by data from the progressive increment schedule. In all cases estimates of onset and duration of effects were assessed.

No attempt was made to relate CNS effects to body fluid levels. The injection of vehicle solutions served as control.

At least one week intervened between each drug session. Amphetamine sulfate (supplied by Smith, Kline and French Labs.), fenfluramine (supplied by A. H. Robbins Co.) and pentobarbital sodium were used in physiological saline. Propylene glycol served as the vehicle for delta 9, tetrahydrocannabinol (supplied by the National Institutes of Mental Health).

The FM magnetic tapes were digitized and autospectra calculated using a Time/ Data 100 Computer controlled by a Linc II Computer. Each 4-sec interval was used to calculate serial autospectra from 0 to 64 Hz. The autospectra were stored in Linc digital tapes and plotted in an isometric format on a Calcomp Digital Plotter. Subsequent calculations of average spectra, total power and variability were performed on the Linc and plotted. Synchronization of events was achieved by decoding the binary decimal time code record.

Results

Pentobarbital

Pentobarbital was examined over a dose range of 2.5–30.0 mg/kg. Figure 2 illustrates conventional EEG recordings describing the effects of progressive increments to anesthetic levels and a recovery record. Many have observed similar data in a variety of species. However, the marked degree of so-called burst suppression seen in the cat was not obvious. Note the appearance of increased voltage and decreased frequency in all leads. Figure 3 represents averaged autospectra over 20-min periods derived from the serial spectra depicted immediately adjacent. Note that spectral changes following 5.0 mg/kg pentobarbital are confined to the parietal and occipital regions. At anesthetic levels of the drug the spectral peaks diminish in the parietal and occipital regions with the appearance of selected lines in frontal regions.

Amphetamine

Amphetamine was studied over dose ranges of 0.1–4.0 mg/kg. Typical signs of psychomotor stimulation were evident beginning at 0.5 mg/kg. These changes in comportment were accompanied by tachycardia. At 2.0 and 4.0 mg/kg dose range two animals died following severe paroxysmal tachycardia. These side effects precluded further exploration of dose ranges.

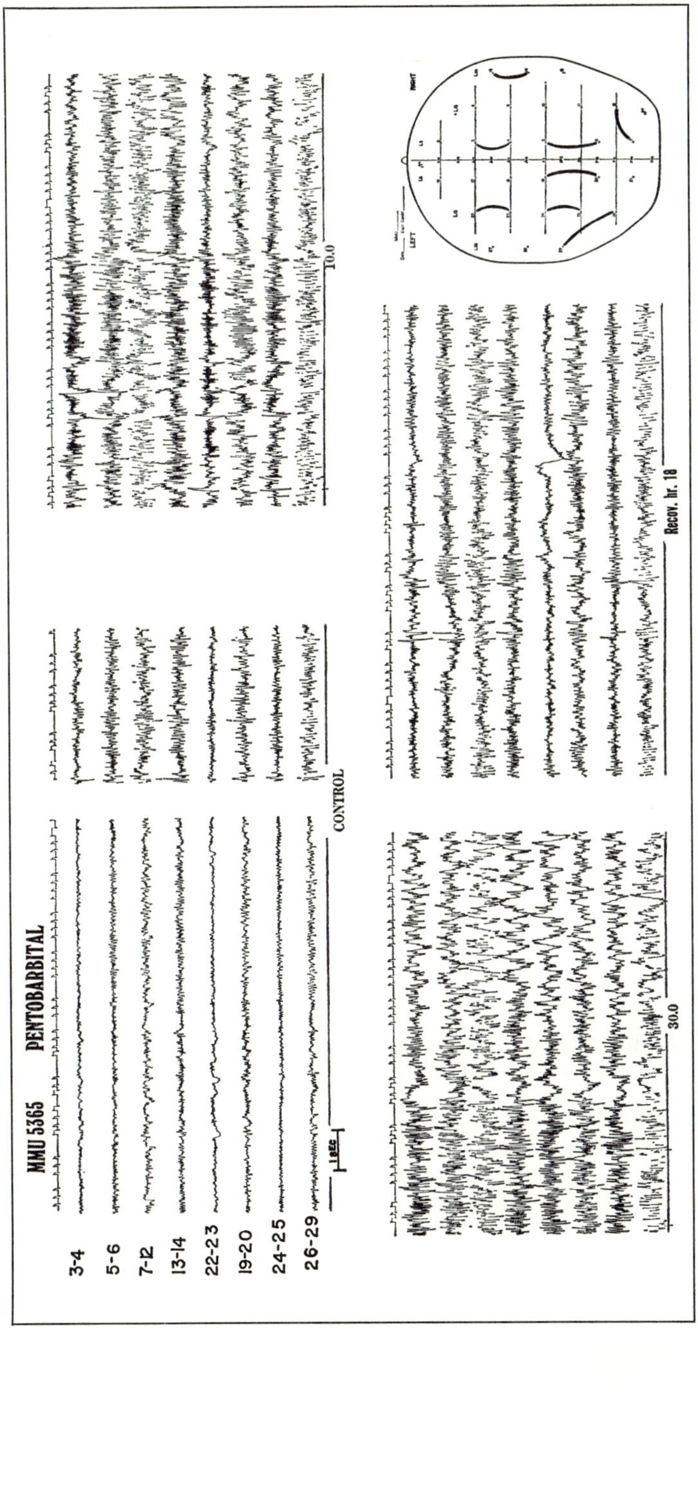

Fig. 2. Spontaneous EEG records under pentobarbital. All records from neocortical derivations. The upper four channels from the right side, the lower four from the left with the exact sites indicated on the inset. Time base as indicated. The control records reflect alert and drowsy states, respectively, doses indicated are in mg/kg. Note overall increase in energy under drug conditions.

Chronic administration regimens may be needed to explore higher dose ranges commonly seen in the syndrome of drug dependency where the amphetamines are the reinforcing substances.

Figure 4 illustrates conventional EEG recordings describing the effects of progressive increments of dosage of amphetamine. No evidence is seen of low frequency hypersynchrony. It is not possible with a single illustration to describe the overall fluctuations in state. In all control recordings as the animals change from maximal alertness to drowsiness to alertness, spectral characteristics change in parallel. The control portion of the illustration is from a period wherein the animal was alert. The effects of amphetamine appear not to be different from such periods. Further, the state fluctuations are not present.

Figure 5 displays averaged spectra following amphetamine. As before, the corresponding serial spectra are juxtaposed. Note that spectral changes are confined to the posterior parietal regions and the medial temporal regions. The spectral changes become more evident as the dose is ncreased.

Fenfluramine

Fenfluramine, a congener of amphetamine, reported to have little psychomotor stimulant properties, was studied over a dose range of 0.6–10.0 mg/kg. No significant change in comportment was observed except a more general adaptation to restraint. No evidence of sympathomimetic activity could be seen.

Figure 6 illustrates EEG tracings taken during progressive increments in the dosage of fenfluramine. Note the absence of slow activity. The control record was selected to illustrate the control EEG state during maximal alertness. Under fenfluramine high frequency activity was noted in frontal areas. These changes were not related to obvious changes in muscle tone or patterns of eye movements.

Fig. 3. Composite of spectra calculated from the spontaneous EEG in a pentobarbital experiment. Both serial spectra and averages are from the indicated derivation. Each line in the serial block represents the spectrum calculated over successive 4-sec time intervals. The samples shown were selected from the later part of a treatment epoch which usually lasted 35 min. The averaged autospectra were calculated from the serial spectra representing 20–25 min of EEG, under the specific condition. The total power has been normalized.

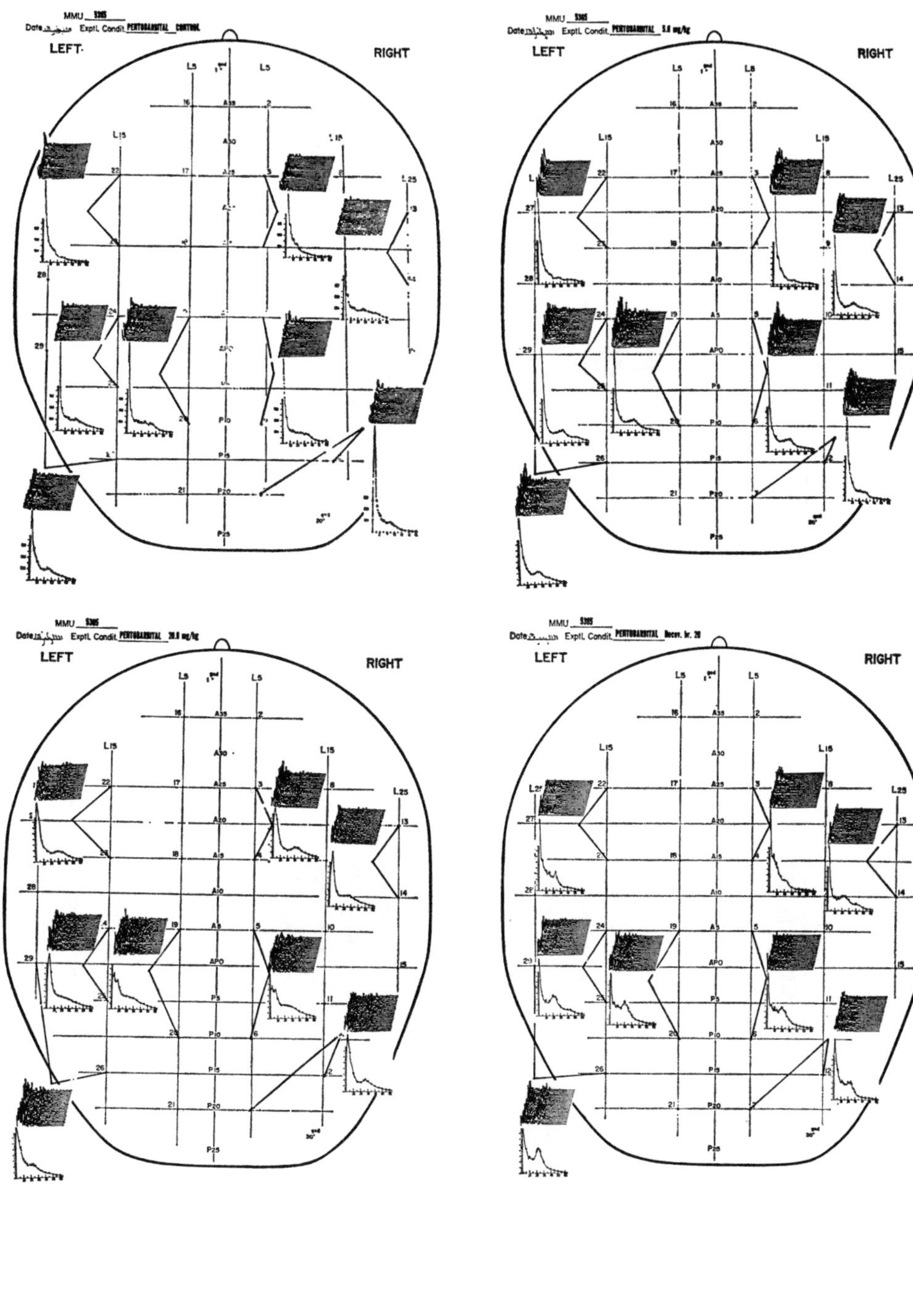
MMU 5365
Date Exptl. Condit.
LEFT
RIGHT
MMU 5365
Date Exptl. Condit.
LEFT
RIGHT
MMU 5365
Date Exptl. Condit.
LEFT
RIGHT
MMU 5365
Date Exptl. Condit.
LEFT
RIGHT

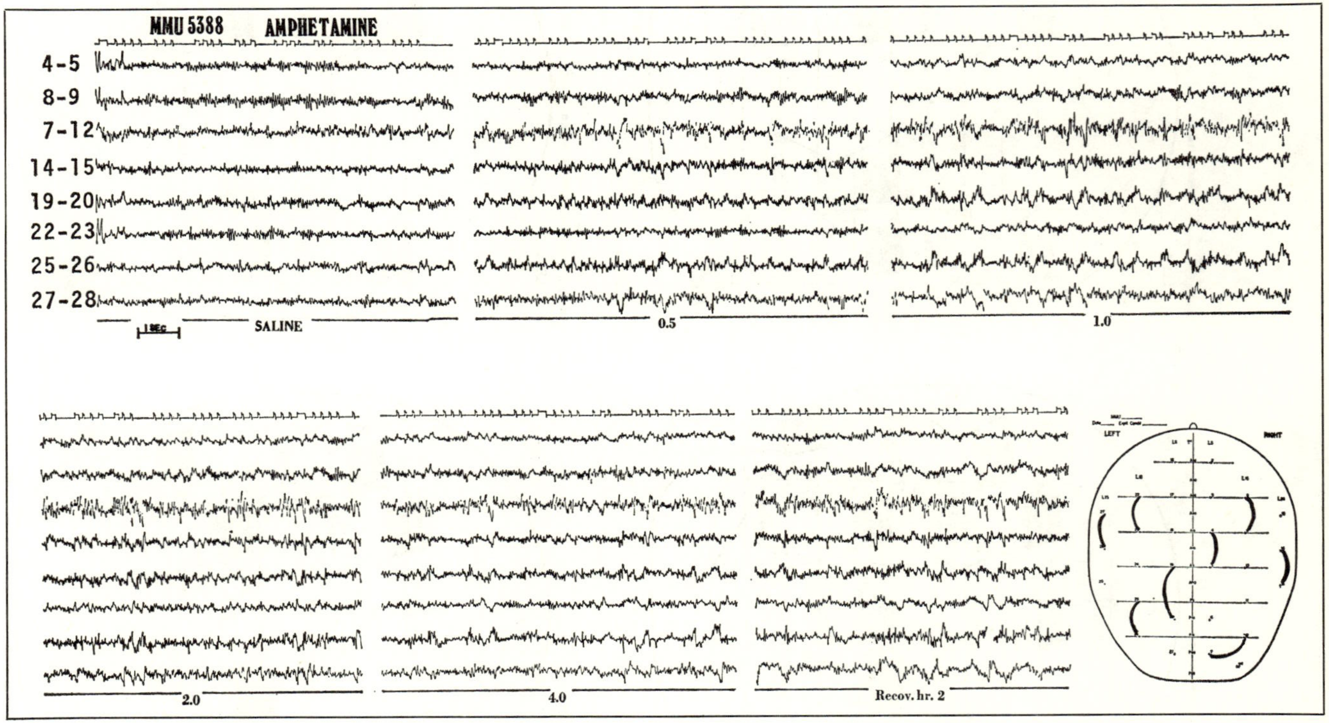

Fig. 4. Representative examples of the spontaneous EEG associated with amphetamine at the indicated doses (in mg/kg). Derivations indicated as in figure 1 and on the inset.

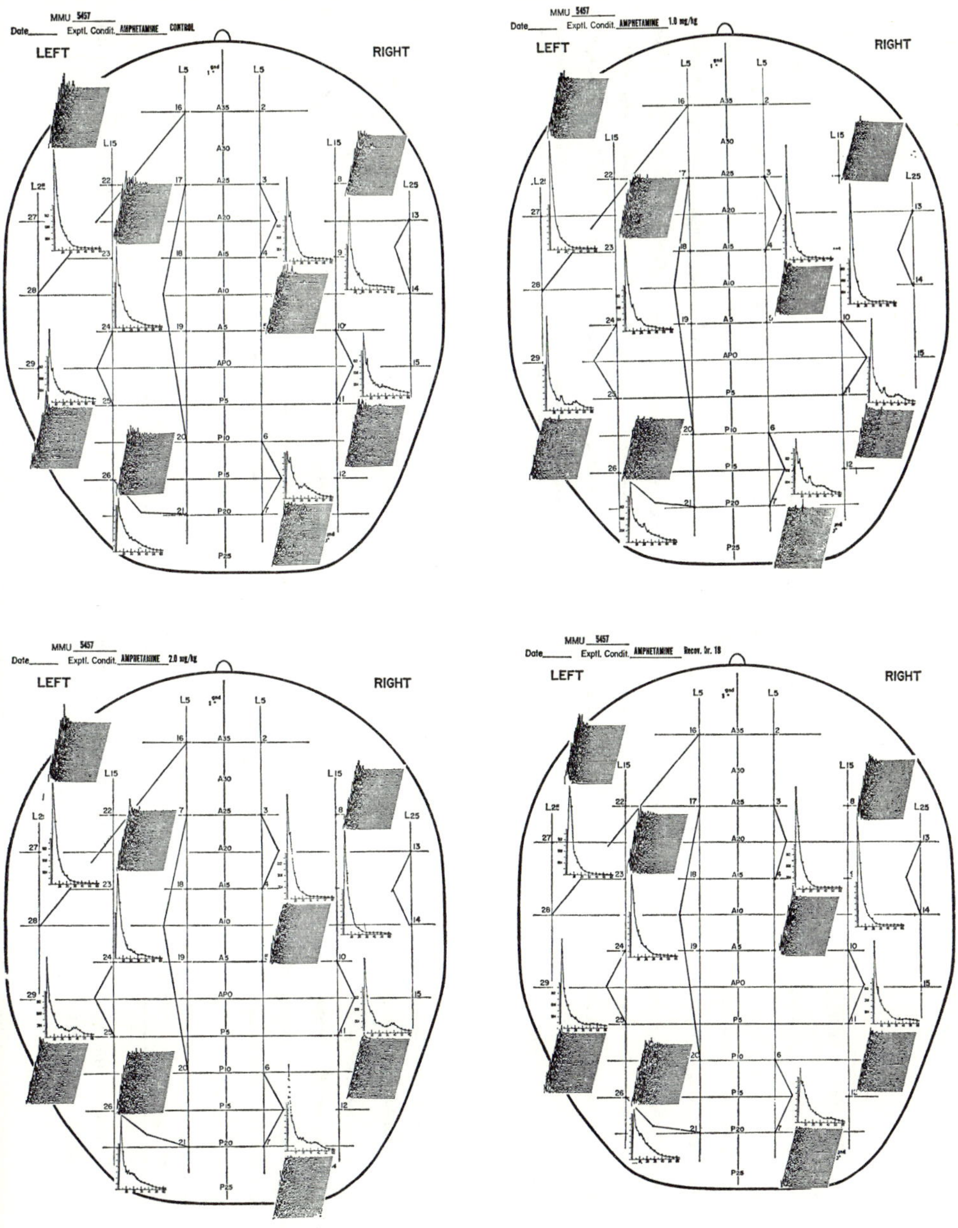

Fig. 5. Composite of spectra calculated from the spontaneous EEG in an amphetamine experiment. The representation is as for figure 2. See text.

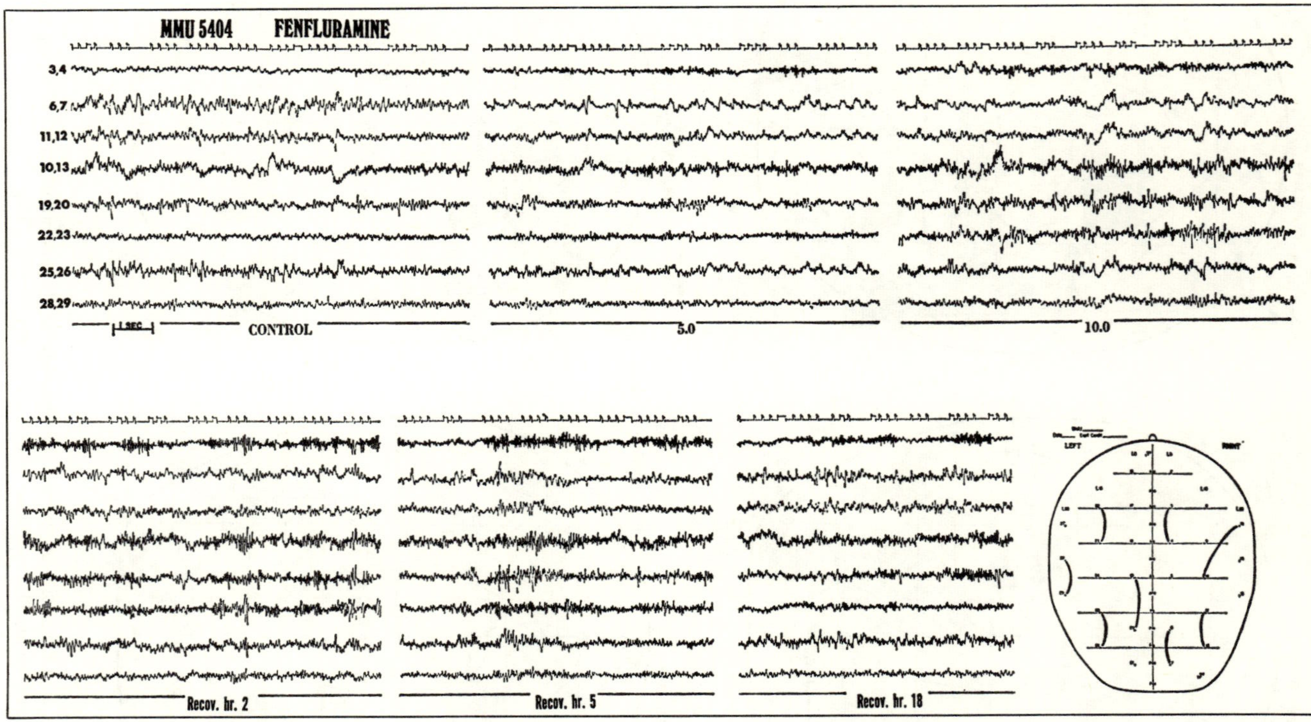

Figure 6. Conventional spontaneous EEG recordings with progressive dosages (in mg/kg) of fenfluramine. Recovery is in hours following the highest dose. Derivations as indicated.

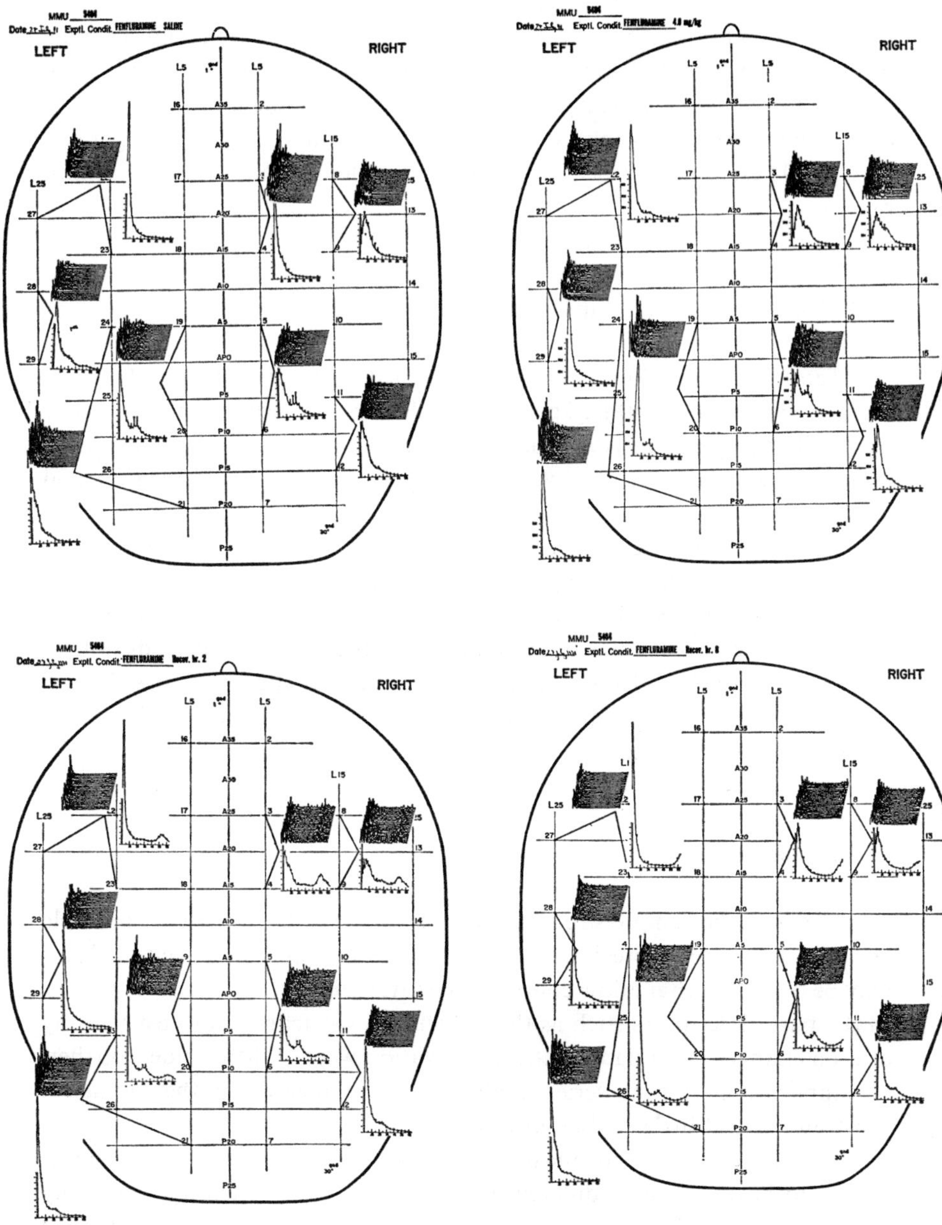

Fig. 7. Composite of spectra calculated from the fenfluramine experiment. Representation as in figure 2. The anatomical distribution is distinct and characteristic.

Figure 7 characterizes spectral changes after progressive increments in the dose of fenfluramine and recovery. Note that changes are confined to the frontal region with a distinct loss of very low frequencies. Further, as the dose increases high frequency spectral lines appear in these same areas with concomitant spectral peaks in the alpha band in the posterior parietal regions. These changes persist up to 36 h after the last dose of fenfluramine.

Δ^9-Tetrahydrocannabinol

Tetrahydrocannabinol (THC) was studied over a dose range of 20–520 mg/kg. Beginning at 160 μg/kg signs of sedation were obvious accompanied by a mild tachycardia. At the high dose employed, 520 μg/kg, behavioral nonresponsiveness was observed accompanied by perfuse salivation and mild tachycardia.

Figure 8 illustrates the EEG changes observed after progressive increments of THC. Note the appearance of synchronous activity in a variety of leads. These changes persist for 12–18 h depending upon the dosage used.

Figure 9 compares spectral estimates over the same time periods following THC. Note the selective change in the spectral characteristics derived from the posterior parietal region. These changes parallel the observable behavioral perturbances.

Discussion

These studies emphasize the point of view that by the use of multiple levels of discourse or precision, the characteristic changes induced by drugs may be described that allow continued research but with different emphases. In the past, descriptors such as the induction of synchrony or desynchrony or even changes in semiautomatic quantization were of little value. Further, the application of time series analysis without regard for the biological substrate yielded little usable information.

In this study two chemical analogs, amphetamine and fenfluramine, appear to have distinctly different effects on the EEG. Such data not only support behavioral findings but also focus on specific changes and time frames for changes allowing more in-depth exploration of the neuropharmacological mechanism of action for these drugs. Since the drugs have both

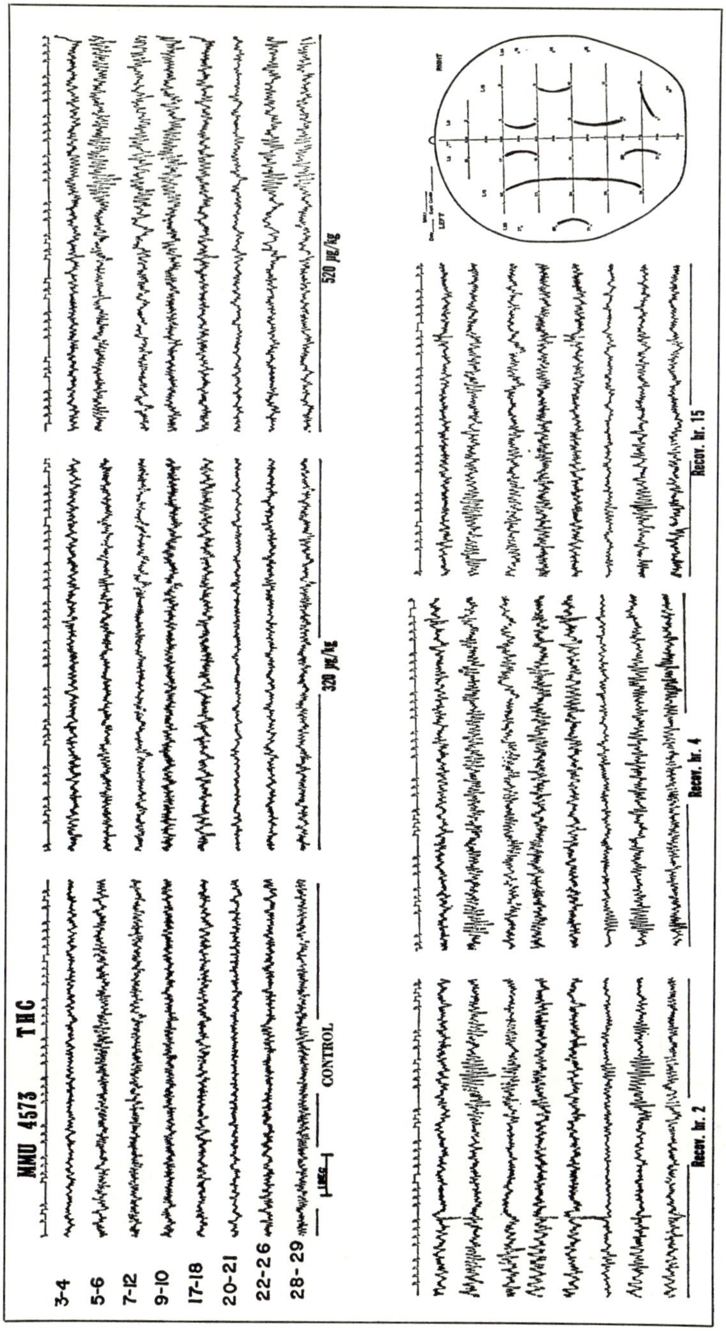

Fig. 8. Spontaneous EEG observed with Δ^9-THC at the indicated doses. Recovery measured in hours following 520 μg/kg. Derivations as indicated.

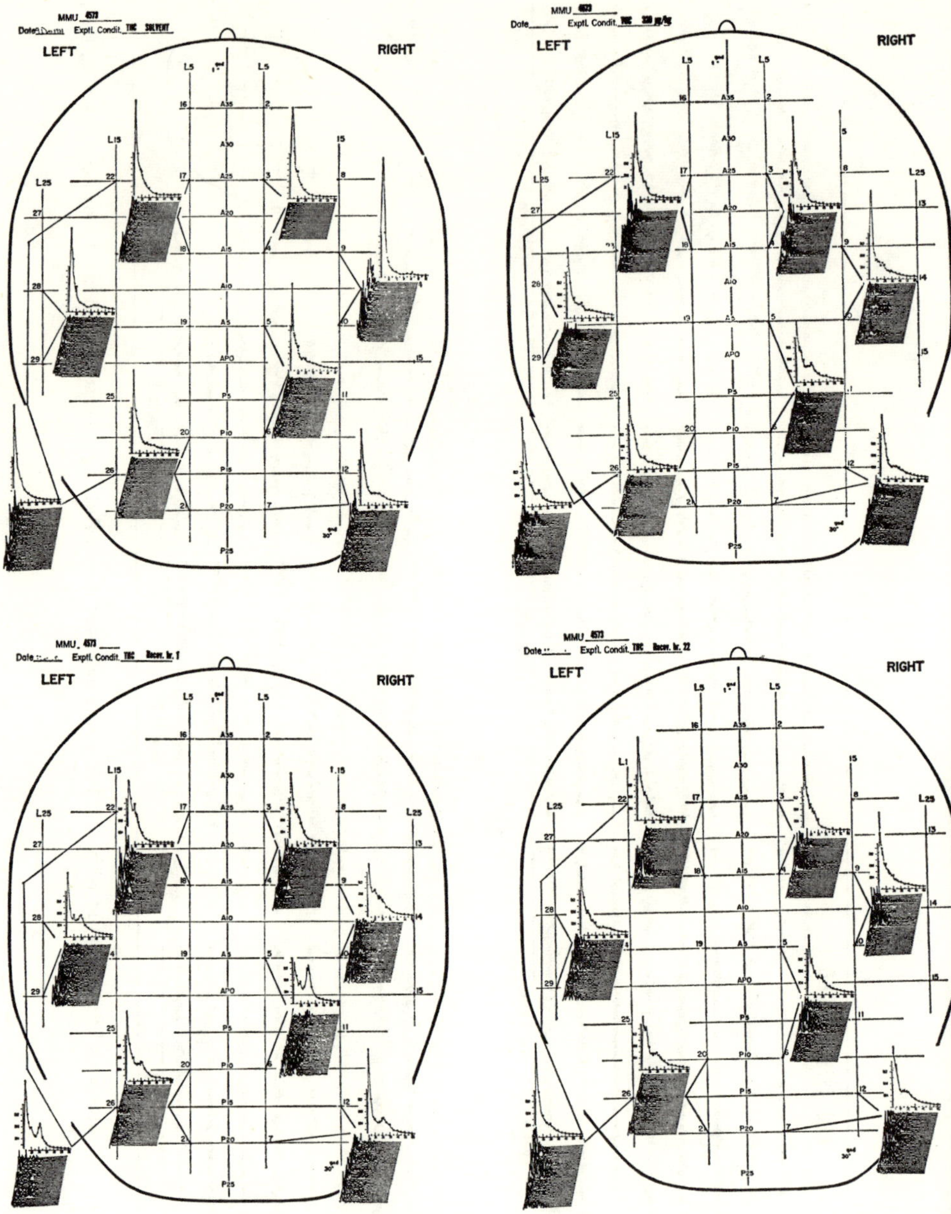

Fig. 9. Composite of spectra calculated from the THC experiment. Note the clear development and distribution of synchronous activity. See text.

been reported to have anorexogenic properties but differing psychomotor effects, further studies should emphasize neocortical *versus* hypothalamic differences.

The data following pentobarbital were unexpected relative to the semi-selectivity of effects. Considering the well-established effects of barbiturates on systems affecting the cortex diffusely, the difference in EEG patterns may reflect either thalamocortical circuits or neocortical differences.

The selective changes in spectral characteristics following THC are the most delimited changes we have observed to date for any centrally-acting drug. The spectral changes parallel changes in comportment. Further experimentation will be needed to determine whether these data are useful in the study of changes induced by the administration of THC.

In considering the data from our previous studies, it would appear that the neocortical activity of the nonhuman primate may be selectively altered from site to site. Whether these data mirror local sensitivities or alterations in functional connectivity from other structures is not obvious.

References

JOY, R. M.; HANCE, A. J., and KILLIAM, K. F.: Neuropharmacology *10:* 471–481 (1971a).
JOY, R. M.; HANCE, A. J., and KILLIAM, K. F.: Neuropharmacology *10:* 483–497 (1971b).

Author's address: Dr. K. F. KILLAM, Department of Pharmacology, National Center for Primate Biology, University of California-Davis, *Davis, CA 95616* (USA)

Medical Primatology 1972. Proc. 3rd Conf. exp. Med. Surg. Primates, Lyon 1972, part II, pp. 274–281 (Karger, Basel 1972)

Activity of Cerebellar and Lower Brain Stem Neurons in Monkeys with Harmaline-Induced Tremor[1]

Y. LAMARRE and M. DUMONT

Centre de Recherche en Sciences neurologiques, Université de Montréal, Montréal, Québec

The central mechanisms underlying the motor abnormalities characteristic of Parkinson's disease are difficult to investigate in humans. However, the drug harmaline, a short-acting monoamine oxidase inhibitor, provides an interesting experimental model of the Parkinsonian tremor. This substance induces a coarse, Parkinson-like tremor at 4–6 cps in monkeys with midbrain and/or lateral cerebellar nuclei lesions [1, 8, 14]. Moreover, in normal monkeys [14] and cats [10, 11, 15], harmaline produces a fine, generalized tremor at 8–12 cps in which the cerebellum is implicated [10, 11]. The present experiments in the monkey indicate that, as in the cat [7], harmaline induces rhythmic activity in neurons of the cerebellar cortex, inferior olive and vestibular nuclei and also in the bulbar reticular formation. This is of particular interest since it has been suggested that lesion of a rubro-olivo-cerebello-rubral loop was responsible for the genesis of postural tremor induced by harmaline in the monkey [12].

Material and Methods

Fourteen monkeys *(Macaca mulatta)* were used in this study. The tremor-producing lesions were stereotaxically placed under pentobarbital anesthesia. The animals were frequently examined and received harmaline (3–5 mg/kg i.v.), an alkaloid of harmala, at regular intervals. Electromyograms (EMG) of the trembling muscles were recorded with intramuscular needle electrodes and stored on analog magnetic tape.

Central unit recording was performed in six unanesthetized monkeys, five unparalyz-

1 This research was supported by Grant MA-2863 from the Medical Research Council of Canada.

ed and one paralyzed with Flaxedil. The technique used for unit recording in awake animals has been described [9]. Briefly, the animals are prepared in the following way: A few days before the first recording session, and under general anesthesia, a plastic chamber is stereotaxically attached to the skull at the margins of a trephine hole above the cortical or subcortical structure to be explored. The animal is later on placed in a special restraining chair and the chamber is attached to a metal bridge supporting a microdrive. The animal is free to move its body and limbs, but its head is immobilized in a stereotaxic position. A steel or tungsten microelectrode can be vertically inserted into the brain. For recording in paralyzed, unanesthetized monkeys, the same technique is used but in this case, the animal is intubated and artificially ventilated. Paralysis is maintained by continuous i.v. infusion of gallamine.

The precise frequency of central unit activity and muscle tremor was determined with the help of a PDP9 digital computer from the autocorrelograms of the impulse signals.

Histological controls were carried out and recording sites were identified from electrolytic lesions left in the electrode tracks.

Results

Tremor Frequency and the Effect of Cerebellectomy

Combined lesions of the ventromedial tegmentum on one side and of the brachium conjunctivum on the other side (caudal to its decussation) were performed in eight monkeys. In all these animals, the lesions did not result in tremor. However, the injection of harmaline repeatedly induced postural tremor in the limbs contralateral to the tegmental lesion.

The range of frequency of the tremor induced by harmaline in all these animals varied from 4.5 to 7.5 cps (average: 5.7 cps). At least in one animal two distinct tremor frequencies could be clearly demonstrated in the same muscles (fig. 1): a slower tremor at about 4.5 cps which was at times abruptly replaced by a faster tremor at 7.0–7.5 cps. The reciprocality of the bursts in antagonistic muscles was unchanged at the two different tremor frequencies (fig. 1). During some periods of time, the faster tremor was predominant while, at other times, the slower tremor episodes were dominant. In some animal a still faster EMG bursting could be sometimes observed at 10–12 cps in muscles that were otherwise trembling at about half this frequency. This faster tremor rate corresponds to the so-called 'shivering' induced by harmaline in normal monkeys and cats and observed also in monkeys with lateral cerebellar lesions [13].

In two other monkeys bilateral cerebellar nuclei lesions were performed. These lesions involved mainly the dentate and part of the interpositus nuclei

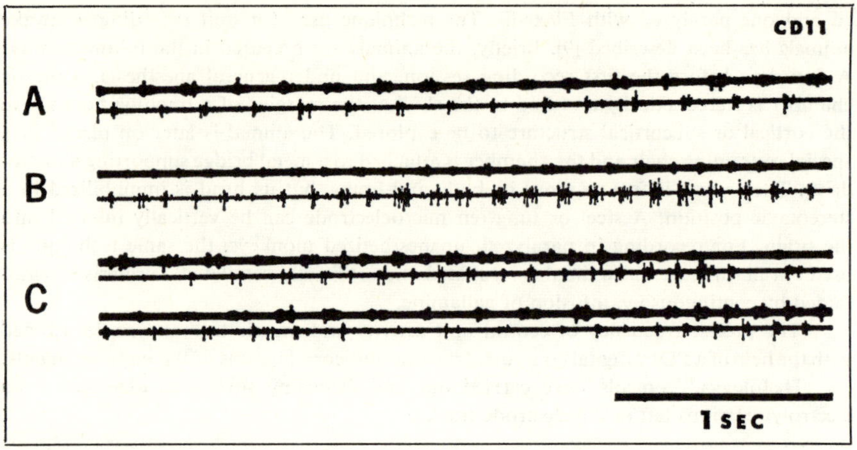

Fig. 1. EMG recordings showing tremor at two frequencies in the left biceps (upper trace) and left triceps (lower trace) in a monkey with right brain stem lesions and harmaline. The frequency is 4.5 cps in A and 7.0 cps in B. The two recordings in C are continuous and show abrupt change from slow to fast tremor.

and surrounding fibers. Harmaline induced bilateral tremor in these animals at a frequency of 5 cps. Moreover, one of these monkeys developed a spontaneous tremor at 6–7 cps, five weeks after the cerebellar lesions.

Four monkeys had complete cerebellectomy including removal of all cerebellar nuclei. Harmaline induced bilateral tremor in all these animals some three to five days after the operation. The average frequency of the tremor was 4.9 cps (range 4.3–6.0). In none of these animals could we detect any fast tremor episodes at frequencies higher than 6 cps. Moreover, in these four monkeys, cerebellectomy did not prevent the appearance of spontaneous tremor associated with a midbrain tegmental lesion preformed either before or after the cerebellectomy. Indeed, the removal of the cerebellum seemed to increase the occurrence of spontaneous postural tremor.

In conclusion, postural tremors induced by harmaline in monkeys either by brain stem tegmentum or by lateral cerebellar lesions could not be distinguished on the basis of their frequency and slow and fast rhythmic EMG bursting could occur in both groups of animals. However, cerebellectomy seems to prevent the occurrence of fast rhythmic muscle activity (6–12 cps) after harmaline treatment and at the same time to favor the appearance of spontaneous tremor associated with midbrain tegmental lesions.

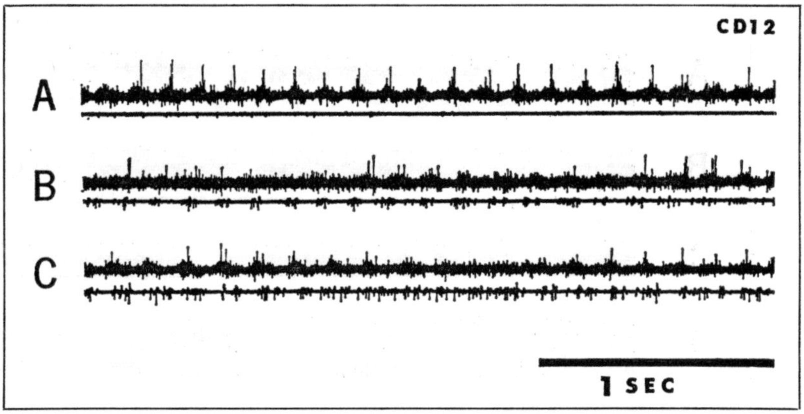

Fig. 2. Recordings in the lateral part of the left principal olive (upper trace) and contralateral triceps EMG (lower trace) in a monkey with left brain stem lesion and harmaline. The olivary spikes are superimposed on large rhythmic field potentials at about 7.0 cps. In A the olivary discharge is sustained with relatively no peripheral tremor. In B the olivary rhythm is not sustained even though muscle tremor is present at about 7.0 cps. Tracings C show a short arrest of olivary rhythmic activity associated with a similar arrest of the peripheral tremor. Positivity is downward in this and subsequent figures.

Central Unit Activity

One outstanding effect of harmaline in the decerebrate cat is the induction of sustained rhythmic activity in the olivo-cerebello-bulbar system [7] with corresponding muscle tremor at 7–12 cps.

The same observations were made in the monkey. Figure 2 shows olivary rhythmic activity and contralateral muscle tremor recorded in a monkey with harmaline. Following midbrain tegmental lesions, harmaline induced a tremor at a frequency of about 7 cps in this animal. Rhythmic firing induced in the olive by harmaline was also at about 7 cps (fig. 2). The inferior olive is a major source of cerebellar climbing fibers. In three animals we have recorded rhythmic climbing fiber responses in the cerebellar vermis at the same frequency as the muscle tremor following harmaline administration. One of these animals was paralyzed, thus indicating that the rhythmic cerebellar activity was not evoked from the trembling limbs.

Rhythmic firing was also recorded in the lateral and descending vestibular nuclei and in the region of giant neurons of the bulbar reticular

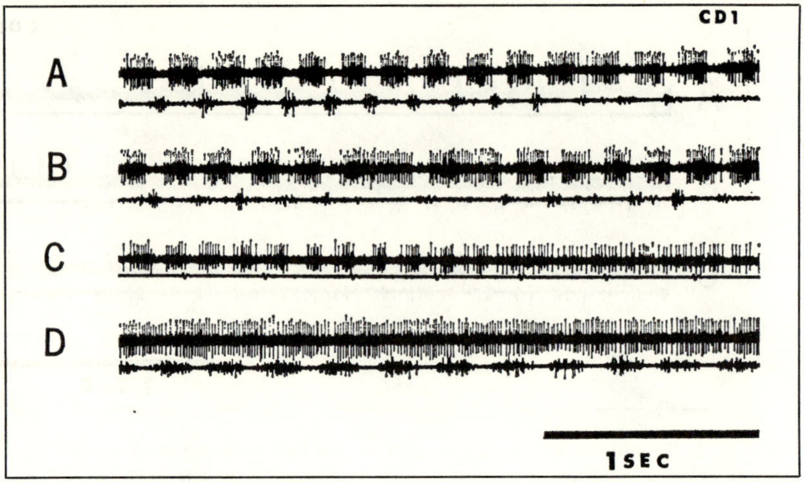

Fig. 3. Recordings in the left Deiters nucleus (upper trace) and left biceps EMG (lower trace) in monkey with right tegmental lesion and harmaline. Each tremor bursts at 5 cps occurs during the interburst intervals of the unit (A and B). In C, in the absence of peripheral tremor the central rhythmic activity increases frequency to 6 cps before desynchronizing. Tracings D show that rhythmic manipulation of the limb at 3.4–3.5 cps has little effect on the vestibular unit.

formation. Figure 3 shows a cell recorded in Deiters nucleus which was firing at the same frequency as the tremor recorded in the homolateral biceps (A and B), even in the absence of peripheral tremor (C). Passive rhythmic movements of the same limb (D) had little or no effect on the activity of that cell. Figure 4 shows the autocorrelograms of three different units recorded in the medial portion (lateral 2) of the bulbar reticular formation in another animal. The three units fired at different frequencies between 5 and 7 cps which was the range of frequency of the peripheral muscle tremor.

All these data indicate that the harmaline-induced tremor can possibly involve the olivo-cerebello-bulbar system as it was demonstrated in the decerebrate cat [7, 10]. However, as we have said earlier total cerebellectomy in the monkey does not prevent either the spontaneous or the harmaline-induced 5 cps postural tremor. We therefore have to postulate that some extracerebellar systems are also involved in the genesis of this type of tremor. Parkinson-like tremor produced or exaggerated by harmaline in monkeys with CNS lesions is already thought to be generated by a central mechanism

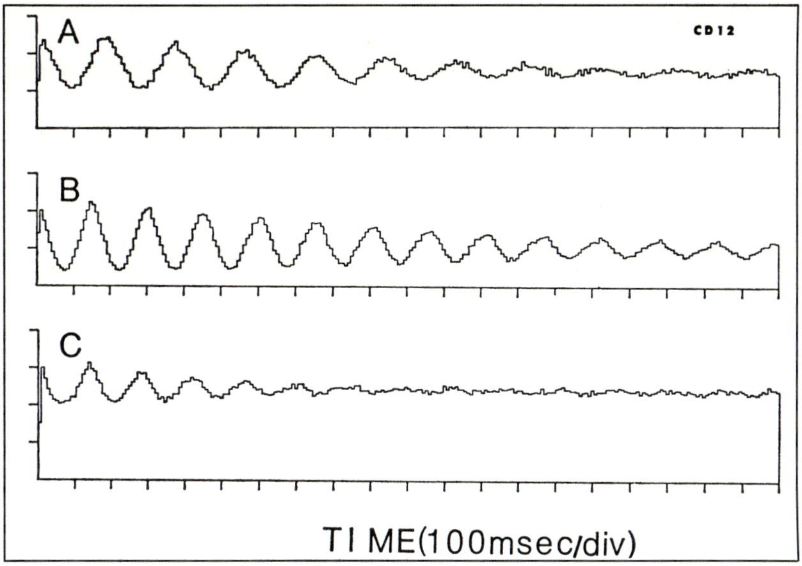

Fig. 4. Autocorrelograms of three units (A–C) recorded in the giganto-cellularis nucleus of a monkey with tegmental lesions and harmaline. The frequencies are 5.3, 6.7 and 7.1 cps, respectively. The bin width is 8 msec.

involving mainly the thalamus and the motor cortex where 'tremorogenic activities' have been recorded [2, 5, 6, 8]. In this series of experiments recordings were obtained in one animal after complete removal of the cerebellum. Tremor synchronous units were found in the bulbar pyramid and in the medial lemniscus (fig. 5A and B respectively). These facts suggest that the cerebellar system may have only an ancillary role in the genesis of tremor and that dien- and telencephalic structures are in fact the real generators of the tremor. In accordance with this hypothesis we have observed that the 5-cps tremor induced bilaterally by harmaline in a cerebellectomized monkey was abolished permanently on the left side after removal of the right sensorimotor cortex.

Discussion

These experiments have demonstrated that harmaline induces rhythmic activity in the olivocerebellar system and that this mechanism, similar to the one described in cats [7], is most probably responsible for the 7- to 12-cps tremor that harmaline induces in the

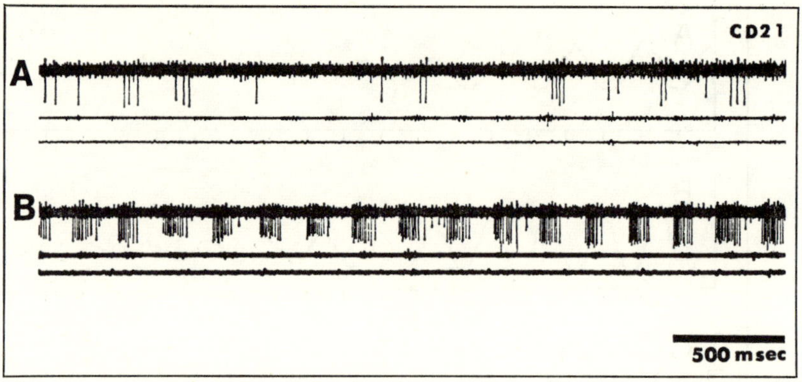

Fig. 5. Monkey with right brain stem lesion, total cerebellectomy and harmaline. A = Recordings in the right pyramid (upper trace) and EMG in left and right triceps (middle and lower trace, respectively). B = Recordings in left medial lemniscus (upper trace) and EMG as in A.

normal monkey. However, such a mechanism alone cannot be responsible for the slower, alternating postural tremor at 4–6 cps observed in monkeys with ventromedial tegmental lesions, since such a tremor persists after total cerebellectomy.

These results are in agreement with the findings that cooling of the motor cortex produces rapid arrest of the 4- to 5-cps Parkinson-like tremor without any effect on the rapid, 8- to 12-cps harmaline tremor [4]. Also, BATTISTA *et al.* [1] were able to demonstrate that lesions of the n. ventralis lateralis of the thalamus served to arrest the Parkinson-like tremor in animals with tegmental lesions, but failed to affect the more rapid tremor induced by harmaline alone. These facts indicate that there are two forms of tremor induced by harmaline through different neuronal mechanisms. The faster harmaline tremor is not dependent upon intact cortical function as is the Parkinson-like tremor [4, 5], but depends primarily on structures lower in the brain stem and cerebellum [7, 10, 11].

It then becomes evident that the action of harmaline in the central nervous system is widespread and that it produces abnormal neuronal activity mainly at the cortico-thalamic level [5, 8] and in the lower brain stem and cerebellum [7]. In intact animals, harmaline induces a rapid, shivering-like tremor at 8–12 cps as a consequence of the predominent influence of cerebellar and lower brain stem control system which tend to correct for the abnormal cortical motor activity induced by the drug. In animals with brain stem or partial cerebellar lesions, such a compensating mechanism would be less effective, resulting in a slower and more severe tremor with a mixture of slow and fast frequencies (4–12 cps). Animals injected with harmaline after total removal of the cerebellum display only a massive tremor at 4–6 cps, which is dependent upon an intact sensori motor cortex. Some of our direct recordings in the pyramidal tract suggest that this system may be involved in the transmission of abnormal rhythmic activities to segmental levels, either directly or indirectly through some lower brain stem relay nuclei.

References

1 BATTISTA, A. F.; NAKATANI, S.; GOLDSTEIN, M., and ANAGNOSTE, G.: Effect of harmaline in monkeys with central nervous system lesions. Exp. Neurol. *28:* 513–524 (1970).

2 CORDEAU, J. P.; GYBELS, J.; JASPER, H., and POIRIER, L. J.: Microelectrode studies of unit discharges in the sensorimotor cortex. Investigations in monkeys with experimental tremor. Neurology *10:* 591–600 (1960).

3 CORDEAU, J. P. and LAMARRE, Y.: Further studies on patterns of central unit activity in relation with tremor. J. Neurosurg. *24:* 213–218 (1966).

4 JASPER, H.; LAMARRE, Y., and JOFFROY, A.: The effect of local cooling of the motor cortex upon experimental Parkinson-like tremor, shivering, voluntary movements, and thalamic unit activity in the monkey; in Corticothalamic projections and sensorimotor activities 4th Symp. Parkinson's Disease Research Center (in press).

5 JOFFROY, A. J. and LAMARRE, Y.: Rhythmic unit firing in the precentral cortex in relation with postural tremor in a deafferented limb. Brain. Res. *27:* 38–389 (1971).

6 LAMARRE, Y. et CORDEAU, J. P.: Etude du mécanisme physiopathologique responsable, chez le singe, d'un tremblement expérimental de type parkinsonien. Actualités neurophysiol. *7:* 141–166 (1967).

7 LAMARRE, Y.; MONTIGNY, C. DE; DUMONT, M., and WEISS, M.: Harmaline-induced rhythmic activity of cerebellar and lower brain stem neurons. Brain. Res. *32:* 246–250 (1971).

8 LAMARRE, Y. and JOFFROY, A. J.: Thalamic unit activity in monkey with experimental tremor; in BARBEAU and MCDOWELL L-Dopa and Parkinsonism, pp. 163–170 (Davis, Philadelphia 1970).

9 LAMARRE, Y.; JOFFROY, A.J.; FILION, M., and BOUCHAUX, R.: A stereotaxic method for repeated sessions of central unit recording in the paralyzed or moving animal. Rev. canad. Biol. *29:* 371–376 (1970).

10 LAMARRE, Y. and MERCIER, L.-A.: Neurophysiological studies of harmaline-induced tremor in the cat. Canad. J. Physiol. Pharmacol. *49:* 1049–1058 (1971).

11 LAMARRE, Y. et MERCIER, L.-A.: Etude neurophysiologique du tremblement à l'harmaline chez le chat. Rev. canad. Biol. *31:* 181–191 (1972).

12 LAROCHELLE, L.; BEDARD, P.; BOUCHER, R., and POIRIER, L. J.: The rubro-olivo-cerebello-rubral loop and postural tremor in the monkey. J. neurol. Sci. *11:* 53–64 (1970).

13 OHYE, C.; BOUCHARD, R.; LAROCHELLE, L.; BEDARD, P.; BOUCHER, R.; RAPHY, B., and POIRIER, L. J.: Effect of dorsal rhizotomy on postural tremor in the monkey. Exp. Brain Res. *10:* 140–150 (1970).

14 POIRIER, L. J.; SOURKES, T. L.; BOUVIER, G.; BOUCHER, R., and CARABIN, S.: Striatal amines, experimental tremor and the effect of harmaline in the monkey. Brain *89:* 37–52 (1966).

15 VILLABLANCA, J. and RIOBO, F.: Electroencephalographic and behavioral effects of harmaline in intact cats and in cats with chronic mesencephalic transection. Psychopharmacologia *17:* 302–313 (1970).

Author's address: Dr. Y. LAMARRE, Département de Physiologie, Faculté de Médecine, Université de Montréal, CP 6128, *Montréal 101, Québec* (Canada)

Medical Primatology 1972. Proc. 3rd Conf. exp. Med. Surg. Primates, Lyon 1972, part II, pp. 282–288 (Karger, Basel 1972)

Epilepsy in the Photosensitive Baboon, *Papio papio*, and Drugs Acting on Cerebral GABA and Serotonin Mechanisms

B. S. MELDRUM and E. BALZAMO

Medical Research Council Neuropsychiatry Unit, Carshalton, Surrey, and Laboratoire de Neurophysiologie appliquée, Institut de Neurophysiologie et Psychophysiologie, CNRS, Marseille

In 1966 KILLAM *et al.* first described photically-induced epilepsy in the Senegalese baboon, *Papio papio*. This naturally-occurring syndrome has subsequently been extensively studied as a model of human epilepsy. Its clinical and neurophysiological features are reviewed by NAQUET and MELDRUM (1972).

This paper summarizes pharmacological experiments we have conducted in the last four years, designed to investigate the synaptic mechanisms underlying these epileptic responses.

An initial study employing atropine and eserine indicated that alteration of activity in cholinergic pathways, such as those forming part of the ascending reticular formation, did not significantly modify photically-induced myoclonic responses [MELDRUM *et al.*, 1970a].

There is good evidence that γ-aminobutyric acid (GABA) is an inhibitory transmitter in the cerebellum, hippocampus and neocortex. It also appears to be responsible for primary afferent depolarisation in the spinal cord, and thus for the phenomena known as 'presynaptic inhibition'. Many drugs which inhibit glutamic acid decarboxylase, thereby reducing the rate of GABA synthesis in synaptic endings, provoke seizures. Glutamic acid decarboxylase (GAD) requires pyridoxal phosphate as coenzyme and drugs which interfere with the synthesis or coenzymic function of pyridoxal phosphate are potent GAD inhibitors. Such 'pyridoxine antagonists' produce, after a moderate latent interval, a characteristic pattern of regularly-recurring seizures. In *P. papio* EEG records show that such seizures characteristically originate unilaterally in the posterior parietal or occipital cortex [MELDRUM

Table I. Epilepsy in *Papio papio.* Some convulsant agents and their relationship to GABA

	'Synaptic' effect	Dose enhancing photo-sensitivity mmol/kg	Dose producing seizures mmol/kg	Seizure latency min	Duration of seizures h
Isoniazid[1]	Pyp: GAD	0.5–0.7	0.7–1.1	7–40	1–2
Thiosemicarbazide[1]	Pyp: GAD	0.04–0.08	0.04–0.11	60–240	1–3
4-Deoxypyridoxine[2]	Pyp: GAD	0.2–0.5	0.53–0.87	15–45	1–4
Allylglycine[3]	?: GAD	0.9–3.1	4.0	60–120	4–20
3-Mercaptopropionic acid[3]	Comp: GAD	0.09–0.28	0.28–0.75	4–6	0.1–0.7
Bicuculline[2]	GABA receptor	–	0.0006–0.0018	0.1	<0.1–5.0
Picrotoxin[4]	GABA receptor	0.008–0.0013	0.0017–0.0025	2–10	0.1–2.5
Pentylenetetrazol[4]	?	0.03–0.09	0.09	1–2	<0.1–0.4

1 MELDRUM *et al.*, 1970b.
2 MELDRUM and HORTON, 1971.
3 HORTON and MELDRUM, in press.
4 MELDRUM, unpublished.

et al., 1970b; MELDRUM and HORTON, 1971]. Data relating to these and other drugs are summarized in table I. For both isoniazid and 4-deoxypyridoxine there is a dose range that is insufficient to induce convulsions directly but which greatly enhances photically-induced myoclonic responses, so that a weakly photosensitive animal, normally responding only with eyelid myoclonus, will readily show sustained generalized myoclonus persisting beyond the end of photic stimulation. The EEG discharges during such responses conform to the topographic pattern seen in highly photosensitive animals, but near the convulsant dose-level occipital after-discharges may occur.

Allylglycine inhibits cerebral GAD activity 30–60 min after administration in mice and produces seizures in baboons after 1–2 h [HORTON and MELDRUM, in press]. As after pyridoxine antagonists, these seizures recur regularly over a period of several hours; electrographically they often originate asymmetrically in the posterior neocortex. Also, there is a wide subconvulsant dose range that produces enhancement of photically-induced responses; such enhancement can last 8 h or more. However, addition of pyridoxal phosphate to cerebral homogenates does not relieve the GAD inhibition produced by allylglycine. Kinetic studies show that allylglycine is not a competitive inhibitor of GAD. On the other hand, 3-mercaptopropionic acid has been shown to inhibit GAD by substrate competition, and produce seizures after a latency of only a few minutes. These seizures originate synchronously in the frontorolandic cortex. Between 5 and 10 min after the i.v. injection of 3-mercaptopropionic acid photosensitivity is transiently enhanced [HORTON and MELDRUM, in press].

There are two convulsant alkaloids which, in spite of some initial experimental difficulties, have now been shown (a) to block the post-synaptic inhibitory action of GABA on neurones in the neocortex, and (b) to reduce presynaptic inhibition in the spinal cord. Bicuculline and picrotoxin are effective systemically at very low dose levels (see table I). Seizures begin within 4–10 sec of the i.v. injection of bicuculline, are generalized from the onset and are characterized by extremely sustained myoclonic activity. With subconvulsant doses it is difficult to demonstrate any enhancement of photosensitivity [MELDRUM and HORTON, 1971]. In contrast, after subconvulsant doses of picrotoxin (0.5–0.8 mg/kg) brief symmetrical bursts of frontorolandic spikes and waves are seen at intervals and there is enhanced photosensitivity lasting 1–3 h. The onset of seizure activity after convulsant doses is much slower than after bicuculline; the first generalized seizure may occur as late as 9 or 10 min after i.v. injection of picrotoxin. This is followed by further brief seizures at intervals, leading either to recovery or to more sustained seizure activity. The epileptic phenomena seen after subconvulsant or convulsant doses of picrotoxin are virtually indistinguishable from those seen after comparable doses of 3-mercaptopropionic acid.

Metrazol is included in table I because of its well-known capacity to enhance photically-induced epileptic responses in man. It is less potent than the other two drugs and its mode of action at the synaptic level is not known.

The principal method of inactivation of extracellular GABA is probably uptake into neurons or glia. However, drugs inhibiting GABA-transaminase (the enzyme initiating the further metabolism of GABA) increase cerebral GABA content and can modify photically-induced epileptic responses. Data relating to two such drugs are given in table II. Optimal doses of amino-oxyacetic acid or di-N-propyl acetate completely abolish myoclonic responses even in highly photosensitive animals. A slight excess of amino-oxyacetic acid leads to the appearance of 'spontaneous' seizures, but these are probably not related to the inhibition of GABA-transaminase.

Table II. GABA-transaminase inhibitors and epilepsy in *Papio papio*

	Dose diminishing photosensitivity mmol/kg	Duration of effect h	Latency of effect min	Dose producing seizures mmol/kg
Amino-oxyacetic acid[1]	0.05–0.14	6–24	30–40	0.19
di-N-propyl acetate[2]	0.15–0.9	1–3	5–15	–

1 MELDRUM *et al.*, 1970b.
2 PATRY and NAQUET, 1971.

GABA is presumed to be the transmitter substance in several inhibitory systems which are not merely anatomically and physiologically distinct but which differ in their biochemical and pharmacological properties. Our results suggest, for example, that the intracortical inhibitory system is particularly vulnerable to pyridoxine antagonists, whereas bicuculline and picrotoxin are effective on subcortical inhibitory mechanisms.

The other neurohumoral systems that are susceptible to pharmacological manipulation are monoaminergic. Experiments with drugs acting on catecholamine synthesis and storage, such as L-Dopa, *a*-methyl-*p*-tyrosine and reserpine have been largely negative [BALZAMO and NAQUET, 1970; MELDRUM and BALZAMO, in press; MELDRUM *et al.*, in press]. However, numerous drugs believed to act on serotoninergic mechanisms powerfully modify photically-induced epileptic responses. Lysergic acid diethylamide and some other chemically-unrelated compounds such as psilocybin and dimethyltryptamine, that are hallucinogenic and block the action of 5-hydroxytryptophan (5-HT) on smooth muscle, completely abolish both myoclonic responses and photically-induced spikes and waves [WALTER *et al.*, 1971; MELDRUM and NAQUET, 1971]. This is in spite of the fact that effective doses of psilocybin or dimethyltryptamine produce a marked increase in the occurrence of spontaneous spikes and waves in the resting EEG. The blockade of the photically-induced epileptic responses may be due to a reduction of transmission in the visual afferent pathway. Both LSD 25 and psilocybin reduce the amplitude of evoked responses to single flashes recorded at the level of the lateral geniculate [WALTER *et al.*, 1971; MELDRUM and NAQUET, 1971]. Table III shows that some non-hallucinogenic drugs related chemically to lysergic acid can also block myoclonic responses to photic stimulation.

Table III. Hallucinogenic and other drugs which abolish myoclonic responses to photic stimulation in *Papio papio*

	Dose μmol base/kg	Duration of effect h
LSD 25[1]	0.09–0.23	0.5–1.5
Psilocybin[2]	3.5–14	1–2
Dimethyltryptamine[2]	11–21	0.5
Methysergide[2]	4–10	1–2
Methylergometrine[3]	0.9–4	2–3
Ergotamine[3]	0.4–0.8	2–3

1 WALTER *et al.*, 1971.
2 MELDRUM and NAQUET, 1971.
3 VUILLON-CACCIUTTOLO and BALZAMO, 1972.

Methysergide, like psilocybin and dimethyltryptamine, enhances the incidence of spontaneous spikes and waves.

The probable relationship of these effects to activity at serotoninergic synapses has been illuminated by studies with appropriate precursors and enzyme inhibitors. The physiological precursor of 5-HT, L-tryptophan, is without effect on photically-induced responses. However 5-HT, at 15–30 mg/kg, produces effects similar to those of psilocybin or dimethyl-tryptamine. There is an increase in delta activity and in spontaneous spikes and waves, and photically-induced myoclonus is abolished for up to 3 h. Possibly L-tryptophan does not show these effects because 5-HT content is increased at storage sites where it is physiologically inactive. The physiological effectiveness of 5-HT can be increased by giving a monoamine oxidase inhibitor. Tranylcypromine produces a block of photically-induced myoclonic responses that is slow in onset but lasts up to 24 h. The EEG does not show the increase in delta activity seen after 5-HT, but if a 'sub-threshold' dose of tranylcypromine (10 mg/kg) is followed by a moderate dose of L-tryptophan (100 mg/kg) delta activity appears and myoclonic responses to photic stimulation are blocked [MELDRUM *et al.*, in press].

The rate limiting enzyme in the synthesis of 5-HT in the brain is tryptophan hydroxylase; this enzyme can be inhibited by *p*-chlorophenylalanine. Intraperitoneal injections of this drug produce a biphasic effect. Mildly photosensitive baboons become more responsive to photic stimulation after

Table IV. Drugs modifying the metabolism of serotonin and photic epilepsy in *Papio papio*

	Dose mg/kg	Effect on myoclonic responses	Latency h	Duration h
L-Tryptophan[1]	100–600	0		
L-5-HT[2]	10–30	–	0.5	2–3
p-Chlorophenylalanine[2]	200–600	+	36–48	24
		–	96	120
Tranylcypromine[1]	15–30	–	1–4	8–24

1 MELDRUM *et al.*, in press.
2 WADA *et al.*, in press.

about 48 h. Subsequently, between four and eight days after drug administration all animals lose their responsiveness [WADA *et al.*, in press].

In interpreting these drug effects it is necessary to consider separately the increase in slow wave activity and spontaneous discharges and the decrease in photically-induced myoclonic responses. The former is presumably related to activity in the serotoninergic neurons originating in the nuclei of the median raphe and terminating diffusely in the neocortex. The latter appears to be related to changes in afferent transmission in the visual pathway, perhaps to changes at the lateral geniculate produced by the 5-HT containing terminals which have been demonstrated by histofluorescence.

In vivo pharmacological studies provide only indirect evidence, but consideration of the results described above in conjunction with recent biochemical and micro-iontophoretic studies, provides good grounds for believing that synapses employing GABA and 5-HT are critically involved in epileptic responses to photic stimulation.

References

BALZAMO, E. et NAQUET, R.: Action de la réserpine chez le *Papio papio* photosensible. Modifications comportementales et électroencéphalographiques. Physiol. Behav. *5:* 561–569 (1970).

HORTON, R. W. and MELDRUM, B. S.: Seizures induced by allylglycine, 3-mercaptopropionic acid and 4-deoxypyridoxine in mice and photosensitive baboons, and different modes of inhibition of cerebral glutamic acid decarboxylase. Brit. J. Pharmacol. (in press).

KILLAM, K. F.; KILLAM, E. K. et NAQUET, R.: Mise en évidence chez certains singes d'un sydrome photomyoclonique. C. R. Acad. Sci. *262:* 1010–1012 (1966).

MELDRUM, B. S. et BALZAMO, E.: Etude des effects de l'alpha méthyl-*p*-tyrosine chez le *Papio papio*. C. R. Soc. Biol. (in press).

MELDRUM, B. S. and HORTON, R. W.: Convulsive effects of 4-deoxypyridoxine and of bicuculline in photosensitive baboons *(Papio papio)* and in rhesus monkeys *(Macaca mulatta)*. Brain Res. *35:* 419–436 (1971).

MELDRUM, B. S. and NAQUET, R.: Effects of psilocybin, dimethyltryptamine, mescaline and various lysergic acid derivatives on the EEG and on photocally induced epilepsy in the baboon *(Papio papio)*. Electroenceph. clin. Neurophysiol. *31:* 563–572 (1971).

MELDRUM, B. S.; NAQUET, R. and BALZAMO, E.: Effects of atropine and eserine on the electroencephalogram, on behaviour and on light-induced epilepsy in the adolescent baboon *(Papio papio)*. Electroenceph. clin. Neurophysiol. *28:* 449–458 (1970a).

MELDRUM, B. S.; BALZAMO, E.; GADEA, M., and NAQUET, R.: Photic and drug-induced epilepsy in the baboon *(Papio papio)*. The effects of isoniazid, thiosemicarbazide, pyridoxine and amino-oxyacetic acid. Electroenceph. clin. Neurophysiol. *29:* 333–347 (1970b).

MELDRUM, B. S.; BALZAMO, E.; WADA, J. A., and VUILLON-CACCIUTTOLO, G.: Effects of L-tryptophan, L-3,4-dihydroxyphenylalanine and tranylcypromine on the electroencephalogram and on photically-induced epilepsy in the baboon, *Papio papio*. Physiol. Behav. (in press).

NAQUET, R. and MELDRUM, B. S.: Pnotogenic seizures in baboons; PURPURA Experimental models of epilepsy Rouen Press, N.Y. (1972).

PATRY, G. et NAQUET, R.: Action de l'acide dipropylacétique chez le *Papio papio* photosensible. Canad. J. Physiol. Pharmacol. *49:* 568–572 (1971).

VUILLON-CACCIUTTOLO, G. et BALZAMO, E.: Action de quatre dérivés de l'ergot sur la photosensibilité et l'EEG du *Papio papio*. J. Pharmacol., Paris *3:* 31–45 (1972).

WADA, J. A.; BALZAMO, E.; MELDRUM, B. S., and NAQUET, R.: Effects of L-5-hydroxytryptophan and D,L-parachlorophenylalanine on the photomyoclonic syndrome in the Senegalese baboon *Papio papio*. Electroenceph. clin. Neurophysiol. (in press).

WALTER, S.; BALZAMO, E.; VUILLON-CACCIUTTOLO, G. et NAQUET, R.: Effects comportementaux et électrographiques du diéthylamide de l'acide D-lysergique (LSD 25) sur le *Papio papio* photosensible. Electroenceph, clin. Neurophysiol. *30:* 294–305 (1971).

Author's address: Dr. B. S. MELDRUM, MRC Neuropsychiatry Unit, Woodmansterne Road, *Carshalton, Surrey* (England)

Medical Primatology 1972. Proc. 3rd Conf. exp. Med. Surg. Primates, Lyon 1972, part II, pp. 289–296 (Karger, Basel 1972)

Behavioral, Neurophysiologic and Histologic Aspects of Experimental Amblyopia in Rhesus Monkeys

G. K. von Noorden

Wilmer Institute of Ophthalmology, Johns Hopkins University School of Medicine and Hospital, Baltimore, Md.

Amblyopia is defined as reduced visual acuity in one eye for which no obvious cause can be detected on physical examination of the eye. It occurs in patients with and without strabismus. Of special clinical and preventive interest are two of several existing forms of amblyopia: *strabismic amblyopia* and *amblyopia ex anopsia* (amblyopia of disuse). The first develops as an adaptive mechanism against visual discomfort caused by double vision and image confusion: the visual cortex is unable to integrate into a single mental image the impulses arising from disparate retinal images from the two eyes in stabismus. The second form, *amblyopia ex anopsia*, occurs in eyes that were deprived of normal stimulation early in life by corneal opacities, cataracts or other obstacles to normal visual stimulation [von Noorden, 1967, von Noorden and Maumenee, 1968].

Amblyopia affects approximately two percent of the population in the United States; it can be prevented or successfully treated in early childhood but not in the adult patient. Thus, it presents, at least potentially, a major health hazard for patients whose good eye becomes injured or diseased.

Clinical, psychophysical and electrophysiological aspects of amblyopia have been studied widely. Such studies have revealed a complex disturbance of sensory and motor function [Burian, 1969] without indicating, however, its basic seat and mechanism. Extension of our knowledge and, ultimately, the design of more effective therapy were impeded by lack of a non-human model for a more detailed exploration of the visual system. Thus, the need arose to produce an animal model for amblyopia in a species whose visual system is morphologically and functionally comparable to that of humans. For the past seven years we have been studying experimental amblyopia in the rhesus monkey *(Macaca mulatta)* and will briefly summarize our behavioral, neurophysiological and anatomical findings in this report.

Unilateral lid closure was performed in monkeys of various ages for periods ranging from two weeks to two years in order to define the age of susceptibility to interference with normal form vision stimulation. The animals were behaviorally trained, using the non-deprived eye, to respond to Landolt rings of different sizes, until visual acuity could be measured. The previously closed eye was then opened, the sound eye was sutured, and testing of the deprived eye was begun [VON NOORDEN *et al.*, 1970].

The results reveal that severe amblyopia ranging from bare light perception to the recognition of large forms could be produced if lid closure occurred during the first six weeks of life. A milder form of amblyopia (6/40) was present when the lid was closed at $7^1/_2$ weeks of age. The visual system was found insensitive to unilateral form vision deprivation in animals 12 weeks or older. It is of interest that in humans the upper limit of susceptibility to unilateral visual deprivation extends from birth to as late as the end of the fourth year of life.

Deprivation amblyopia is irreversible; the deprived eye of amblyopic animals was repeatedly rested after the non-deprived eye had remained closed for periods ranging from one to two years and visual recovery did not occur. This behaviour is strikingly similar to that observed in human patients. A cataract present at birth or shortly thereafter and not removed until the child is older will cause severe amblyopia in spite of postoperatively clear ocular media. On the other hand, a cataract developing in an older child or adult can safely be left in place for many years; visual acuity will return to normal after successful surgery.

We also studied effect of intermittent lid closure during the period of maximal susceptibility to visual deprivation. Our preliminary results indicate that unilateral lid closure of four weeks between the fourth and eighth week of life, and of only two weeks between the fifth and seventh week of life, caused irreversible and severe amblyopia. On the other hand, when one

---

Fig. 1. Cross-section through *a* right and *b* left lateral geniculate body of monkey whose right eye was closed after birth. Note lighter staining of all layers that receive input from the deprived eye (r) if compared with those connected with non-deprived eye (l).

Fig. 2. Higher magnification (×130) of layers 6, 5 and 4 from *a* right lateral geniculate body of the same animal. Note shrinkage of cells in layer 5 (connected with deprived eye) when compared with layers 6 and 4 (receiving input from non-deprived eye). *b* Left lateral geniculate body. Shrinkage of cells in layers 6 and 4 (from deprived eye), normal appearance of layer 5 (from non-deprived eye).

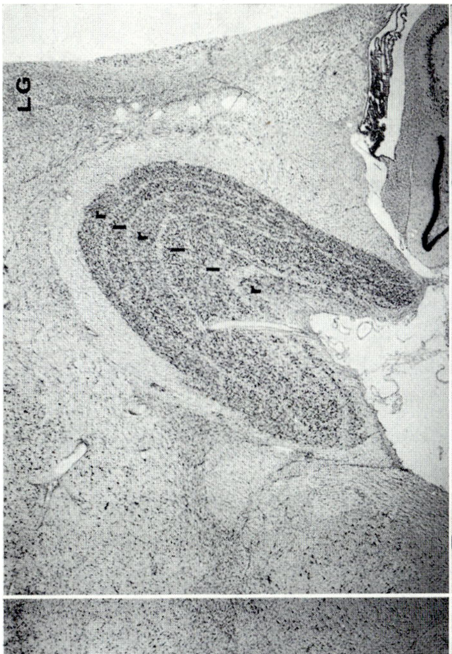

1a 1b

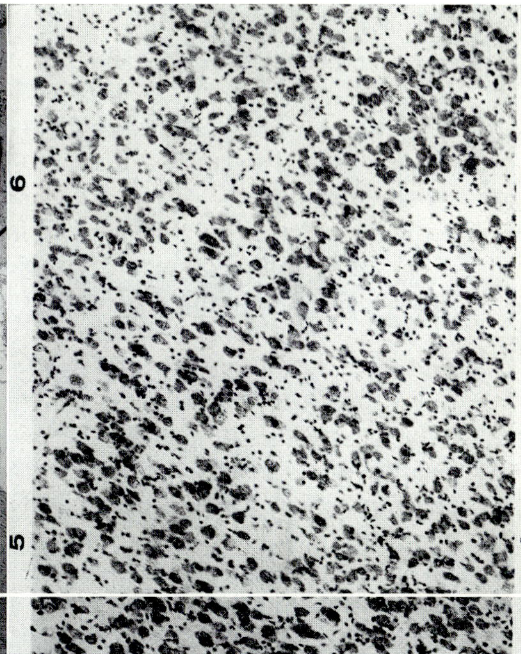

2a 2b

eye was closed for a month in an eight-week-old animal, complete visual recovery occurred after reopening.

The clinical relevance of this experiment is obvious: even temporary unilateral lid closure in infants may be extremely hazardous for visual development and should be avoided. Several patients with otherwise unexplained amblyopia have come to my attention who had had one eye closed for various reasons and for a period of only two to three weeks during their first year of life. HUBEL and WIESEL [1970] could show in kittens that unilateral eye closure during the period of high susceptibility for as little as two days may cause severe neurophysiological changes in the visual cortex.

Another aspect of our experiments concerned attempts to produce strabismic amblyopia in monkeys [VON NOORDEN and DOWLING, 1970]. We were able to produce strabismic amblyopia in four animals who were made esotropic during the first two weeks of life. This was accomplished by extirpating the lateral rectus and advancing and shortening the medial rectus muscle. A fifth animal, that was made esotropic at the age of 17 months, and two exotropic monkeys that exhibited an alternating fixation pattern, did not develop amblyopia. Thus, in these experiments it appears that the time of onset of esotropia and the fixation pattern (unilateral vs. alternating) are significant factors in determining whether amblyopia develops. Again, this behavior is quite similar to that observed in the human strabismic patient, where a direct relation exists between the occurrence of amblyopia and the age of onset of esotropia on the one hand, and the type of fixation pattern on the other.

With these experiments a useful animal model with behaviorally proven amblyopia became available for further laboratory studies. These studies have been concerned with neurophysiological and neuroanatomical factors.

My collaborators, BAKER, GREGG and MALPELI determined the nature of receptive fields of cortical neurons in areas 17, 18 and 19 in awake monkeys with behaviorally proven deprivation amblyopia and strabismic amblyopia. These findings were contrasted with those obtained in previous experiments performed in visually normal monkeys. The data show that in visually normal monkeys the majority of cortical neurons receive input from both eyes.

Fig. 3. a Right (RG) and *b* left (LG) lateral geniculate bodies from normal monkey. No difference in appearance of layers connected with the right (r) and left (l) eye. Compare with figure 1a and b.

Fig. 4. Higher magnification *a* of layers 6 and 5 of the right and *b* of layers 6, 5 and 4 from the left lateral geniculate body from a normal monkey. Compare with figure 2a and b.

3b

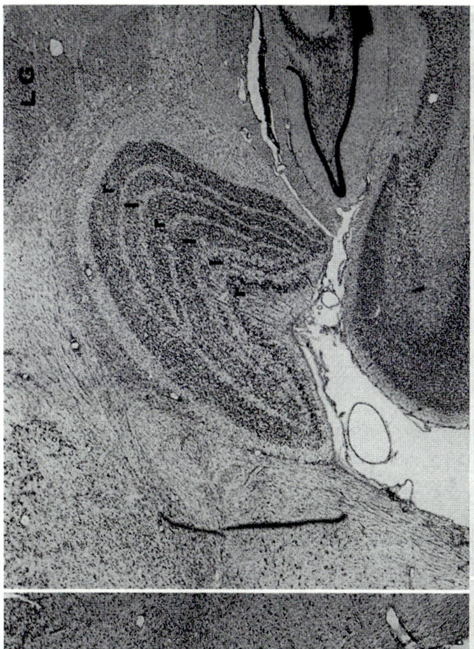

3a

4b

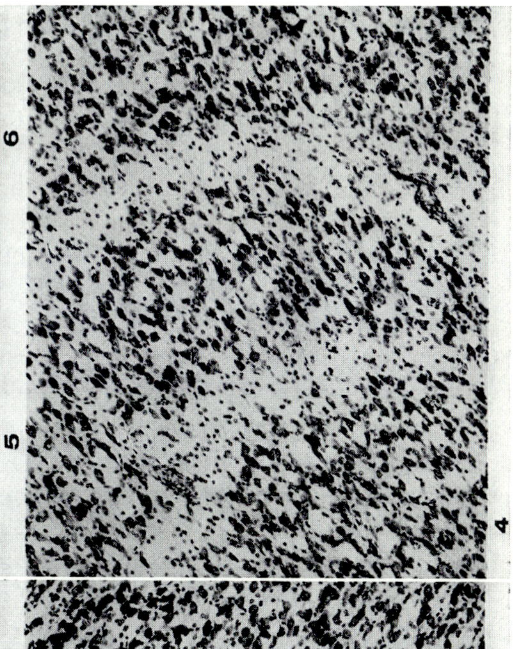

4a

However, a unilaterally visually-deprived animal received input primarily from the non-deprived eye. In esotropic animals with amblyopia a paucity of 'binocular' cells was noted; most cortical neurons in both areas 17 and 18–19 were driven by one eye or the other. Similar findings were obtained in an exotropic animal with normal vision in each eye [BAKER *et al.*, to be published]. It would go beyond the scope of this paper to further analyze the significance of these findings other than to state at this point that during visual immaturity a lack of congruity in input from the two eyes causes a profound disruption in the connections subserving binocular interaction. WIESEL and HUBEL [1963a and b, 1965 a and b] had previously obtained almost identical findings in kittens after unilateral lid closure, or with artifically produced exotropia.

In view of the severity of these behaviorally and neurophysiologically proven defects of cortical function, one would expect to find their morphological equivalent by histological studies of the tissues involved. It is reasonable to assume that synapses, as the primary sites of neuronal activity in the visual cortex, would be most likely to show morphological alterations in amblyopia. Indeed, such changes have been reported by others [GLOBUS and SCHEIBEL, 1967; VALVERDE, 1967, CRAGG, 1969; FIFKOVA, 1970a and b]. We have recently completed a preliminary electron microscopic study of synaptic density and morphology in the visual cortex of a monkey with unilateral visual deprivation amblyopia and have compared our observations with data obtained from comparable tissue of a normal monkey [VON NOORDEN and DOWLING, to be published]. We found that density and morphology of synapses were identical in comparable sections from cortical areas 17 and 18 of both specimens. Thus, we have so far been unable to identify the morphologic substrate for the severe functional cortical anomalies demonstrated by neurophysiological methods in animals deprived of form vision during visual immaturity, or to confirm anomalies of the cortical morphology described in such animals by other authors. During the past year we have extended this study to the inner plexiform layer of the retina in a deprived and non-deprived eye from the same animal. Comparing the ratio of bipolar cell ribbon synapses to amacrine cell conventional synapses we were unable to detect a significant difference [VON NOORDEN, to be published]. Obviously, this study will have to be extended to other aspects of retinal microstructure before any final conclusions can be drawn.

The only positive findings were obtained from the lateral geniculate bodies, which show a striking difference between the layers connected with the deprived and non-deprived eye (fig. 1 and 2). In an animal who had the right eye closed from birth to the end of his first year of life, layers 5, 3 and 2 of the right lateral geniculate body and layers 6, 4 and 1 of the left lateral geniculate body show a decrease in cell density and a shrinkage of the cell size. In normal monkeys no histologic differences are seen upon comparing the layers originating from the two eyes (fig. 3 and 4). Similar histological characteristics were described by WIESEL and HUBEL [1963a and b] in the lateral geniculate bodies of cats. Since we could demonstrate that such anomalies also occur in the lateral geniculate body of monkeys, it is reasonable to assume that stimulus deprivation amblyopia in humans causes histological alterations identical to those just described.

It is obvious that more time will pass before sufficient information is available to fully understand the basic physiological and morphological aspects of amblyopia. On the other hand, we can be reasonably optimistic that some of the riddles posed by this complex condition will eventually be solved, now that a more useful animal model is available. One hopes that these endeavors will ultimately serve to improve preventive, diagnostic, and therapeutic approaches to amblyopia and strabismus in our human patients.

References

BAKER, F. H.; GREGG, P.; MALPELI, J. G., and VON NOORDEN, G. K.: Effects of monocular deprivation and strabismus on response of visual cortical neurons. Presented at Association for Research in Vision and Ophthalmology, Spring Meeting 1972 (to be published).

BURIAN, H. M.: Pathophysiologic basis of amblyopia and of its treatment. Amer. J. Ophthal. *61:* 1–12 (1969).

CRAGG, B. G.: The effects of vision and dark-rearing on the size and density of synapses in the lateral geniculate nucleus measured by electron microscopy. Brain Res. *13:* 53–67 (1969).

FIFKOVA, E.: The effect of unilateral deprivation on visual centers in rats. J. comp. Neurol. *140:* 431–438 (1970).

FIFKOVA, E.: Changes of axosomatic synapses in the visual cortex of monocularly deprived rats. J. Neurobiol. *2:* 61–71 (1970).

GLOBUS, A. and SCHEIBEL, A. B.: The effect of visual deprivation on cortical neurons. A Golgi study. Exp. Neurol. *19:* 331–345 (1967).

HUBEL, D. H. and WIESEL, T. N.: The period of susceptibility to the physiological effects of unilateral eye closure in kittens. J. Physiol. *206:* 419–436 (1970).

NOORDEN, G. K. VON: Classification of amblyopia. Amer. J. Ophthal. *63:* 238–244 (1967).

NOORDEN, G. K. VON: Behavioral studies of form vision deprivation amblyopia in rhesus monkeys. Further observations. Presented at Association for Research in Vision and Ophthalmology. Spring Meeting 1972 (to be published).

NOORDEN, G. K. VON and DOWLING, J. E.: Experimental amblyopia in monkeys. II. Behavioral studies in strabismus amblyopia. Arch. Ophthal. *84:* 215–220 (1970).

NOORDEN, G. K. VON and DOWLING, J. E.: Electron microscopic findings in the visual cortex of a form vision deprived monkey. Presented at Association for Research in Vision and Ophthalmology, Spring Meeting 1971 (to be published).

NOORDEN, G. K. VON and MAUMENEE, A. E.: Clinical observations on stimulus deprivation amblyopia. Amer. J. Ophthal. *65:* 220–224 (1968).

NOORDEN, G. K. VON; DOWLING, J. E., and FERGUSON, D. C.: Experimental amblyopia in monkeys. I. Behavioral studies of stimulus deprivation amblyopia. Arch. Ophthal. *84:* 206–214 (1970).

VALVERDE, F.: Apical dendritic spines of the visual cortex and light deprivation in the mouse. Exp. Brain Res. *3:* 337–352 (1967).

WIESEL, T. N. and HUBEL, D. H.: Effects of visual deprivation on morphology and physiology of cells in the cat's lateral geniculate body. J. Neurophysiol. *26:* 978–993 (1963).

WIESEL, T. N. and HUBEL, D. H.: Single-cell responses in striate cortex of kittens deprived of vision in one eye. J. Neurophysiol. *26:* 1003–1017 (1963b).

WIESEL, T. N. and HUBEL, D. H.: Comparison of the effects of unilateral and bilateral eye closure on cortical unit responses in kittens. J. Neurophysiol. *28:* 1029–1040 (1965a).

WIESEL, T. N. and HUBEL, D. H.: Binocular interaction in striate cortex of kittens reared with artificial squint. J. Neurophysiol. *28:* 1041–1059 (1965b).

Author's address: Dr. G. K. VON NOORDEN, Johns Hopkins University, School of Medicine, *Baltimore, MD 21205* (USA)

Medical Primatology 1972. Proc. 3rd Conf. exp. Med. Surg. Primates, Lyon 1972, part II, pp. 297–304 (Karger, Basel 1972)

Lack of Precision of Thalamic Stereotaxy Based on Cranial Landmarks in Some Species of Macaca

G. PERCHERON[1], N. LACOURLY and D. ALBE-FESSARD

Université de Paris, Faculté des Sciences, Paris

The usual stereotaxy performed for neurophysiologic, psychologic or anatomical research uses the classical Horsley-Clarke principles [HORSLEY and CLARKE, 1908], where the triangular space is oriented by special cranial points, and in which two points are of great importance, the auricular point and the orbital one. Because for some easily available species no atlases exist and because, even when there is an atlas, important errors of placement appear (a point emphasized by OLSZEWSKI [1957] himself), we wondered if some types of corrections can be applied to the existing atlases ameliorating their use in the species described and permitting their use when other species are employed.

Materials

Two groups of data have been used as shown in figure 1.

(1) *The first group* uses data coming from eight atlases (published or not published) to which we added histological sections from five brains prepared especially for this purpose. For all this group, two studies have been performed: (a) *The first study* deals with the measurements of the positions of cranial and diencephalic points in Horsley-Clarke co-ordinates. (b) *The second study* deals with measurements made using intradiencephalic co-ordinates based essentially on the position of different points in relation to anterior and posterior commissures (CA, CP), or to the line joining them.

(2) *The second group* of data was taken from X-ray ventriculographies made on 27 living monkeys belonging to four species. In this group, studies similar to those made for the first group (third and fourth studies) have been performed.

We will deal here essentially with the results obtained from the data of the first and third studies made in the two groups (cranial co-ordinates).

1 Chargé de Recherches at INSERM.

Genus	First group, from atlases and histological sections		Second group, from ventriculography	
	first study, from cranial co-ordinates	second study, from commissural co-ordinates	third study, from cranial co-ordinates	fourth study, from commissural co-ordinates
Macaca				
mulatta	15	12	7	7
arctoides (speciosa)	2	2	10	10
fascicularis (irus, cynomolgus)	4	3	6	8
nemestrina	2	2	4	4
fuscata	1	1	0	0
Total	24	20	27	29

Fig, 1. Table showing species and number of animals used and the type of study performed. In the first study, the following existing stereotaxic atlases were used: *M. mulatta* [CLARKE and HENDERSON, 1920; MUSSEN, 1923; ATLAS and INGRAM, 1937; OLSZEWSKI, 1952]; *M. fascicularis* [SHANTA *et al.*, 1968]; *M. nemestrina* [CHATELIER, unpubl.; WINTERS *et al.*, 1969]; *M. fuscata* [KUSAMA and MABUCHI, 1970].

Methods and Results

For each animal, 14 measurements were made, essentially in anteroposterior or vertical dimensions. Body weight was also considered. A matrix comprising 24 or 27 lines (numbers of animals) and 15 columns (numbers of measures) was thereby formed.

(1) A preliminary analysis, based on mean values, standard deviations and ranges, permitted an evaluation of the dispersion of values.

If, for example, we examine the variations in the anteroposterior position of the CA, we obtain, for *Macaca mulatta*, values varying between ant. 13 and ant. 22.7. Thus the range for *M. mulatta* is 9.7 mm.

In the second group (third study), the position of this same point varies from 15.8 to 26.7 mm, the range for all the species being 10.9 mm. If one spe-

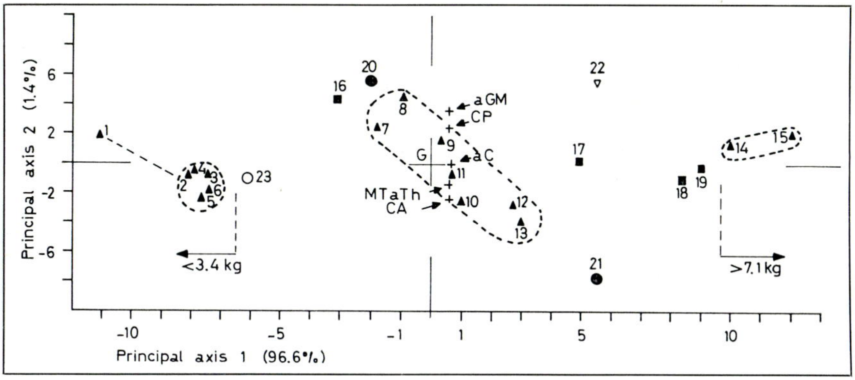

Fig. 2. Representation of animals in the principal plane obtained by the principal component analysis of the matrix of values obtained in the first study. For the signification of the different letters, see text. □ *M. mulatta; Δ M. fascicularis;* ● *M. arctoides;* ○ *M. nemestrina;* ▲ *M fuscata.*

cies only is considered, the range is slightly smaller; however, they are of 6.8 mm for a population of four *M. nemestrina.*

In the vertical dimension, variations are smaller, but still remain important. For example, CA co-ordinates vary in the first study from 3.4 to 8.2 mm, and in the third, from 10.4 to 16.9, the ranges being 4.8 and 6.5 mm, respectively. If we consider that the mean length of a macaca thalamus is 18 mm, this first analysis clearly shows that, without corrections, the Horsley-Clarke technique leads to large errors.

(2) A more sophisticated approach was used: the principal component analysis [see ANDERSON, 1966]. This technique is based on a mathematical transformation of the co-ordinates to determine, from a space of n dimensions, the plane in which the points representing each animal are projected with the minimum loss of information (plane from where the sum of the squared distances is minimal). This plane is defined by two principal axes crossing at the centre of gravity of all the points. Each of these axes represents a maximal percentage of the characteristics.

This technique does not give precise values of variations; its advantage is a vizualisation of the respective positions of the different animals, permitting one to determine which animals are the closest and whether an interspecies difference can be observed.

The results of the principal component analysis performed for the first study are presented in figure 2. To simplify, we have reduced in this case the

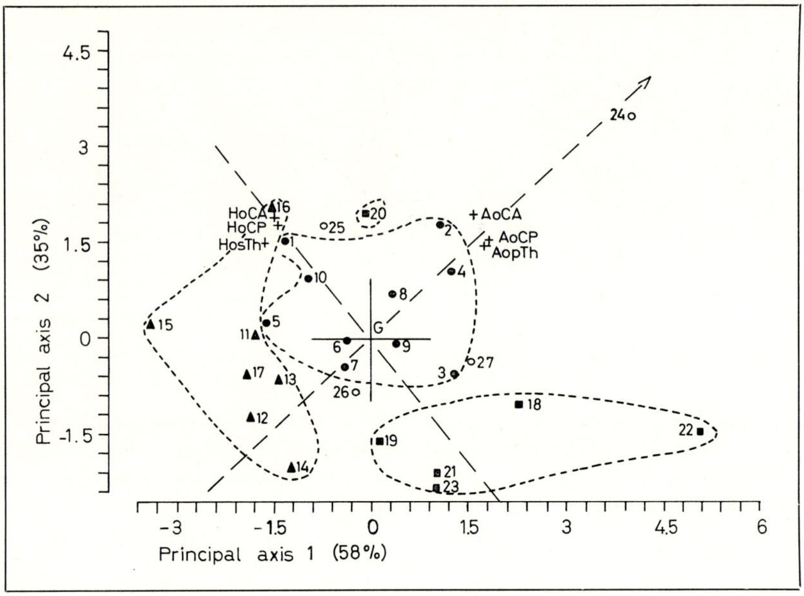

Fig. 3. Same representation as in figure 2, but dealing with values obtained in the third study (see text). ● *M. speciosa;* ▲ *M. mulatta;* Δ *M. irus;* ○ *M. nemestrina.*

number of characteristics measured for each monkey to the anterior position of six diencephalic points: anterior pole of nucleus geniculatus medialis (GM), anterior pole of CP, anterior pole of centrum medianum (aC), anterior part of fasciculus mamillo-thalamicus (MT), anterior pole of thalamus (aTh), posterior pole of CA. The weight was not introduced into this matrix.

The distribution of the points representing the animals in the principal planes shows:

(a) That the variations are explained almost entirely by a variation along the principal axis 1.

(b) The *M. mulatta* used in this study are grouped in three clusters.

(c) The macaca of other species are distributed in between these three groups.

When we looked for the factor which best explained this type of distribution for the *M. mulatta*, it appeared that the animals 1–6 had the smallest body weight (<3.4 kg), animals 14 and 15 had the largest body weight (>7 kg), and animals 7–13 fell in the middle. Similar observations were made for the other species.

Thus, on the basis of these observations, we came to the conclusion that a given atlas could perhaps be used for the different species, if a correction, based on the weight, was made with a formula yet to be determined. We have to note also that the majority of the animals used in this study were immature [less than 4.5 kg; see Napier and Napier, 1967]. Thus the weight factor represents, very likely, a development factor.

To verify these first results, we have done the same sort of analysis on data coming from X-ray ventriculographies (third study, fig. 1). The results are presented in figure 3. In this analysis, the characteristics used included, as in the preceding one, position in anteriority of some diencephalic points (AoCA, AoCP and posterior pole of the thalamus AopTh), but also vertical positions of three points (HoCA, HoCP and superior limit of the thalamus, HosTH).

In this second principal component analysis, (a) a clear grouping appears between animals of the same species and (b) the vector unity on each original axis of the six-dimensionl space has been projected on the plane (the line joining the gravity center to these points gives an idea of the position of these axes in space and the direction of variation along these axes). It can be noticed with this representation that all the characteristics based on anterior dimensions are closely correlated and that the vertical dimensions also are correlated, but the anterior and vertical dimensions are independent from each other (their axes being perpendicular).

If we consider the grouping of individuals of the same species, according to these two axes, we can see that the main *M. fascicularis* group is different from the two others due to the small values of their vertical dimensions. On the contrary, the *M. mulatta* group presents smaller anterior dimensions than the two others.

At this point, it was of interest to verify if the variations in anterior and vertical position of diencephalic points were correlated with other variables.

A correlation matrix (fig. 4) was performed using the preceding values and additional ones (cranial measures and weight). This matrix shows clearly that significant correlations exist between weight and anterior dimension values and that, on the contrary, the correlations with vertical dimensions are not significant.

The same conclusion can be reached in considering figure 5, where coordinates of the points represent the correlation of the different characters with the principal components axes 1 and 2. This representation shows that all the anterior variations are correlated, that they are the most important factor, and that they are correlated with the weight (W). The vertical dimension,

	W	AoFr	AoOc	AoCla	AoClp	HoCla	HoClp	HoFr	HoHc	AoCA	AoCP	AopTh	HoCA	HoCP	HosTh
W															
AoFr	0.56														
AoOc	0.04	0.37													
AoCla	0.75	0.64	-0.19												
AoClp	0.77	0.61	-0.24	0.94											
HoCla	-0.17	-0.23	0.22	-0.37	-0.37										
HoClp	-0.24	0.07	0.30	-0.13	-0.17	0.70									
HoFr	0.48	0.18	0.36	0.34	0.33	0.24	0.15								
HoHc	0.16	0.62	0.65	0.03	0.02	0.21	0.29	0.26							
AoCA	0.70	0.72	-0.23	0.95	0.94	-0.41	-0.14	-0.22	0.10						
AoCP	0.63	0.55	0.42	0.92	0.93	-0.48	-0.23	0.11	-0.10	0.95					
AopTh	0.56	0.42	-0.55	0.78	0.80	-0.36	0.21	0.07	-0.12	0.85	0.89				
HoCA	0.07	0.23	0.60	-0.20	-0.19	0.72	0.73	0.39	0.69	-0.21	-0.37	-0.37			
HoCP	0.11	0.35	0.51	-0.13	-0.15	0.60	0.74	0.23	0.77	-0.07	-0.25	-0.21	0.90		
HosTh	0.03	0.50	0.65	-0.13	-0.16	0.43	0.58	0.14	0.88	-0.07	-0.27	-0.26	0.78	0.90	

Fig. 4. Correlation matrix of the measures made in the third study (ventriculography group).

which is less correlated with the principal axis 1, is not correlated with the anterior dimension, nor with the weight.

Knowing this, it is easier now to have an interpretation of the distributions as a function of species in figure 3. The fact that the *M. fascicularis* group is isolated along an axis which does not depend on weight seems to indicate that the vertical position of their diencephalic points is a specific characteristic. On the contrary, all the anterior dimension variations are linked with weight, and the grouping as a function of species that appears on this figure is due to the fact that the *M. mulatta* group consisted of very young animals of low weight.

The complexity of the relation between weight and the anterior dimension axis is shown in figure 6 where the rank of monkeys on the anterior dimension axis is compared with the weight curve.

(3) From the two preceding types of analysis, it is clear that anterior dimensions are correlated with weight. It was thus of interest to examine if we could find a correction formula based on weight. For this purpose, we calculated for different thalamic points, what kind of precision can be obtained with this type of correction at $p = 0.05$. For example, in one case the preci-

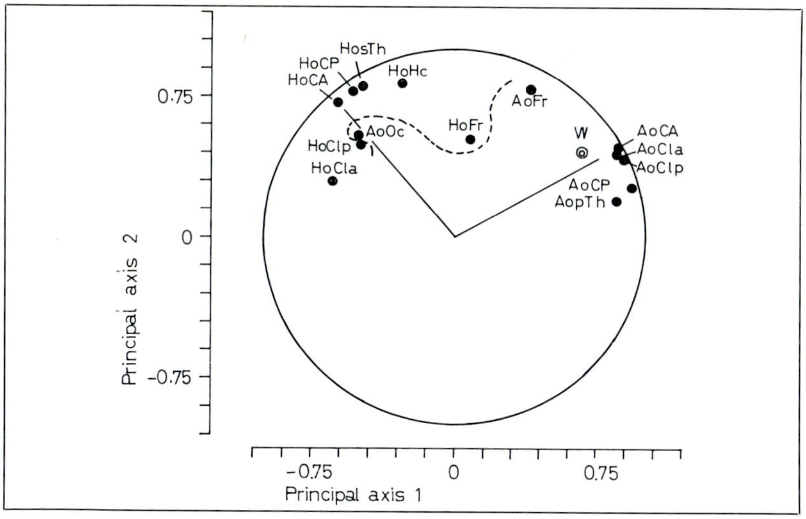

Fig. 5. Representation of characters W and cranial dimensions by their correlation with principal components. Third study, ventriculography group.

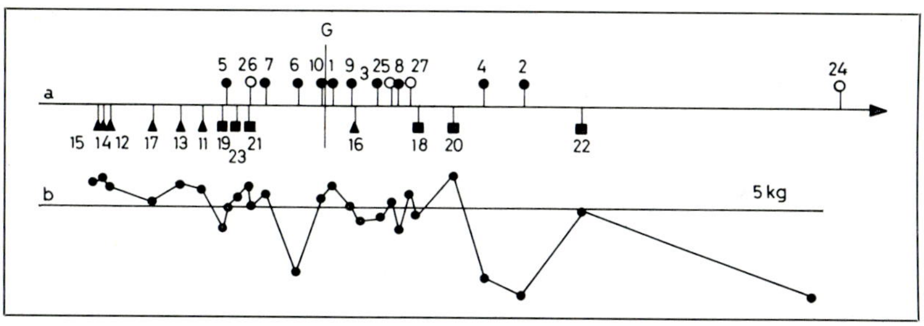

Fig. 6. a Projection on the axis of anteroposterior variations of the different monkeys positions. Third study, ventriculography group. *b* Curve of body weights of the same monkeys. ▲ *M. mulatta;* Δ *M. irus;* ● *M. speciosa;* ○ *M. nemestrina.*

sion of the determination was of ±4.1 mm in the first study and of ±4.2 mm in the second. These results thus show that, even with a weight correction, Horsley-Clarke co-ordinates are impossible to use for precise stereotaxy. Correlations with cranial dimensions or sex were found to be not significant.

(4) We have looked for other special cranial points, which, taken as a reference, could give better results. Analyses similar to the preceding ones, but using the position of the anterior and posterior clinoid processes, lead to a

better statistical precision of the position of the commissures (± 1.9); however, the variations are still important, particularly for vertical positions (±3.2).

(5) A final analysis was thus performed, using a commissural system of references. The results, while not presented here, show again interspecies differences, *M. fascicularis* group being separated from the others. However, they demonstrate that for *M. speciosa*, perhaps for *M. nemestrina*, and even *M. fuscata*, we could use the same atlas as for *M. mulatta*, under the condition that anteriorities are measured in reference to the commissural points. Thus, at the moment and for these latter four species, the OLSZEWSKI [1952] atlas can be used with a correction based on the position of the posterior commissural points determined by X-ray technique.

Due to the lack of space, we have not in this paper emphasized that X-ray ventricular determinations must be done with special care (as they are done in man), in order to prevent distortions (optically sighted teleradiography).

References

ANDERSON, T. W.: Introduction to multivariate statistical analysis, 6th ed. (Wiley, New York 1966).

ATLAS, D. and INGRAM, W. R.: Topography of the brain stem of the rhesus monkey with special reference of the diencephalon. J. comp. Neurol. *66:* 263–283 (1937).

CLARKE, R. H. and HENDERSON, E. E.: Investigation of the central nervous system. II. Atlas of photographs of the frontal sections of the cranium and the brain of the rhesus monkey (Macacus Rhesus). Johns Hopkins Hosp. Rep. 163–165 (1920).

HORSLEY, V. and CLARKE, R. H.: The structure and functions of the cerebellum examined by a new method. Brain 54–124 (1908).

KUSAMA, T. and MABUCHI, T.: Stereotaxic atlas of the brain of *Macaca fuscata* (University of Tokyo Press, Tokyo 1970).

MUSSEN, A. T.: A cytoarchitectural atlas of the brain stem of the macacus rhesus. J. Psychol. Neurol. *29:* 451–518 (1923).

NAPIER, J. R. and NAPIER, P. H.: A handbook of living primates (Academic Press, London 1967).

OLSZEWSKI, J.: The thalamus of *Macaca mulatta.* An atlas for use with the stereotaxic instruments (Karger, Basel 1952).

SHANTA, T. R.; MANOCHA S. L., and BOURNE, G. H.: A stereotaxic atlas of the Java monkey brain *(Macaca irus)* (Karger, Basel 1968).

WINTERS, W. D.; KADO, R. T., and ADEY, W. R.: A stereotaxic brain atlas for *Macaca nemestrina* (University of California Press, Berkeley 1969).

Authors' address: Dr. G. PERCHERON, Dr. N. LACOURLY and Dr. ALBE-FESSARD, 4, rue Gordon-Bennett, *F-75016 Paris* (France)

Medical Primatology 1972. Proc. 3rd Conf. exp. Med. Surg. Primates, Lyon 1972, part II, pp. 305–314 (Karger, Basel 1972)

Comparative Study of Proprioceptive Reflexes in Man and Baboon *(Papio papio)*

J. P. ROLL, M. BONNET and M. HUGON

Université de Provence, Département de Psychophysiologie, Centre de Saint-Jérôme, Marseille

This work deals with problems of motor behaviour mechanisms in primates. Skeletal muscle activity is under supraspinal controls, but it is also controlled by segmental influences originating from proprio and exteroceptive sensors. These different inputs result in phasic or tonic activity for movement or posture. Reflex activity in triceps surae [HOFFMANN, 1922; MAGLADERY *et al.*, 1951 and 1952; PAILLARD, 1955; and others (see DELWAIDE, 1970)] is a physiological index of the net integration of supraspinal and segmental control, at rest or during muscular activity. But the elements involved in this reflex activity provoked in normal man are rather obscure and the means of analysis somewhat limited. ROLL *et al.* [1971] have demonstrated similar reflexes in normal unanaesthetized baboon. In this paper we describe differences between man's and baboon's reflexes, comment upon them from a physiological point of view, and present *Papio papio* as a model for reflexes of man.

Methods

Ten normal healthy humans, 20 normal baboons and four chronic decorticated baboons were used for this study. The technique in baboon has been adapted from techniques for reflexology of man [PAILLARD, 1955] and described by ROLL *et al.* (1971). A restraining chair was specially designed in order to elicit and record the reflexes in a definite and constant way. The baboon sat, legs half-flexed (knee angle: 120°), soleus slightly stretched (ankle angle: 90°), the whole limb restrained by means of a plaster cast. The ambiant light was dim, and noise kept at low level. We used mechanical and electrical stimulations in order to elicit tendon reflexes (T-reflexes) or Hoffmann's reflexes (H-reflexes). T-reflexes were provoked by an electromagnetic device, giving single shocks or trains (up to 150 strokes/sec) to the Achilles tendon. The date and frequency of shocks were recorded, but not the amplitude of the muscular stretching responsible for reflexes. H-reflexes arose from electrical pulses delivered to the tibialis pcsterior nerve by epicutaneous electrodes. Bipolar recording through gross intramuscular electrodes gave a picture of electromyographic (EMG) responses. Measurement of the peak amplitude of the EMG response was considered as a measurement of the reflexes, and used for computation. Peripheral behavioural activity, which is easily controlled in man, occurs at impredictible intervals in baboon. Therefore, it was necessary to select *a posteriori* the responses which occurred during resting phases or, conversely, during activity.

In a few baboons precentral[1] ablation was performed by cortical suction under general anesthesia; in this procedure area 4, 6 and 8 were concerned, in part or totally. Excisions were bilateral. Testing was performed a week and more after surgery.

Results

Hoffmann's Reflex: Main Characteristics

In baboon, as in man, liminal stimulation elicits a H-reflex response restricted to the soleus muscle (fig. 1). Higher stimulation results in direct motor response (M) preceding in time the H-reflex. Usually, H and M responses are similar, in baboon and man, and clearly triphasic. Latencies at the peak of the medial negative phase have an average of 35 msec in man and 16 msec in baboon, owing to the difference of height and proportions in both species; the duration is respectively 13 and 6 msec, related to the length of muscular fibers. The essential point is the remarkable stability of these time characteristics in a given situation (fig. 1). This stability is an argument allowing to consider H-reflex as a monosynaptic one [PAILLARD, 1955], as well synchronized as the direct motor response. It is worth noting that the amplitude of reflexes is variable, and usually more so in baboon than in man, who is able to minimize his own behavioural instability.

H-Reflex Recruitment (fig. 2)

As previously pointed out, nerve stimulation results in M- and H-responses. M-response increases with the intensity of stimulus. A S-shaped curve depicts the related recruitment of alpha fibers in the soleus nerve, up to

1 We greatly thank Dr. M. MIHIC for his valuable assistance in this part of the experiment.

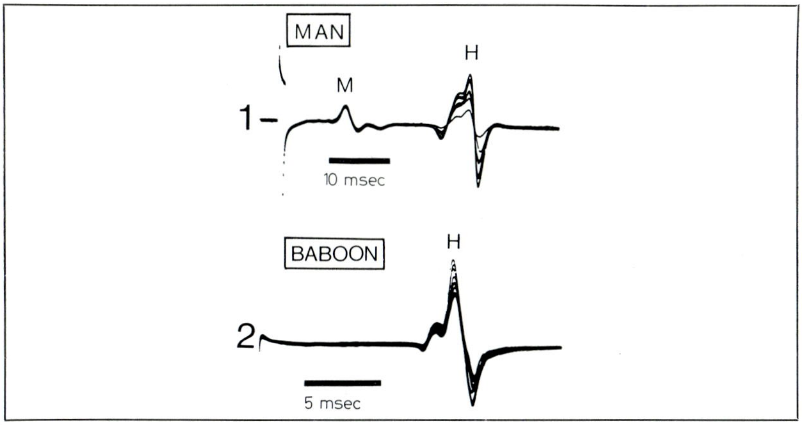

Fig. 1. H-reflex/M-response. *1* Direct motor (M) and Hoffmann reflex response (H) recorded from man's soleus muscle. *2* Hoffmann reflex response from baboon's soleus muscle 1 and 2 are the superposition of ten successive responses.

a total recruitment expressed by a maximal M-response (M-max.). The H-reflex always develops at lower intensity than M and increases with the stimulus. But the H-reflex soon reaches a constant maximal value (H-max.), whereas the stimulation usually remains subliminal for M-response. Then M-response develops and a decrease of H-reflex develops which parallels the M-response increase. As M-max. builds up, H-reflex becomes weak or null. Figure 2 shows M- and H-curves in man, normal baboon and decorticated baboon.

The (H-max./M-max.) × 100 ratio (we call it 'recruitment ratio') is about 50 in man, 10 in normal baboon, 25–50 in decorticated baboon. H-threshold is low in decorticated baboon.

H-reflex Recovery

Transient modifications of the H-response pathway excitability can be demonstrated by the ability of a second stimulus to induce a second H-response. Averaged curves (fig. 3) from man [Paillard, 1955; Delwaide, 1970] and baboon are compared and appear similar.

According to Paillard [1955], five phases can be identified in man:

Phase I: A phase of total inexcitability of the system; its duration does not exceed 2 msec.

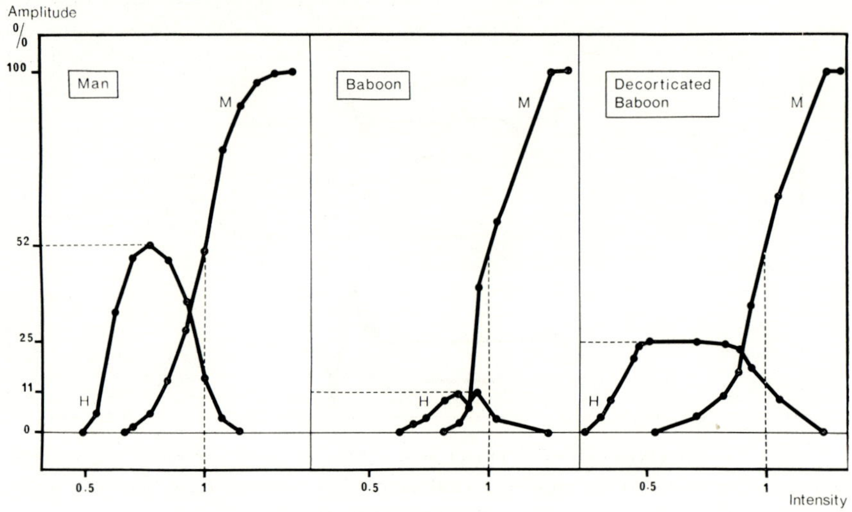

Fig. 2. Recruitment curves. Ordinate: Amplitude of H-reflex (H) and motor response (M) as a percent of normalized M-max. response. Abscissa: Electrical stimulus intensity in arbitrary units. Value of a unit represents the intensity producing a response 50% of M-max. response.

Phase II: This phase is 1–10 msec long and shows a short-lasting period of partial restoration of the excitability.

Phase III: Displays a total or subtotal inhibition of the test response, which lasts about 80–100 msec.

Phase IV: This is a progressive return towards a normal or supernormal state of excitability (120–300 msec).

Phase V: Long phase of important subnormal excitability lasting several hundred milliseconds.

Phase III in baboon is slightly shorter (lasting about 60 msec). Phase IV is marked by a clear supernormality of reflectivity. Phase V (late inhibition phase) is weak and short, compared to man. These characteristics remind us of the recovery curves in hyperreflexia, described by DELWAIDE [1970].

Periodic Tendon Stimulation (fig. 4 and 5)

In baboon repetitive mechanical stimulation produced by successive mechanical impulses from a magnetic coil results in discrete synchronous

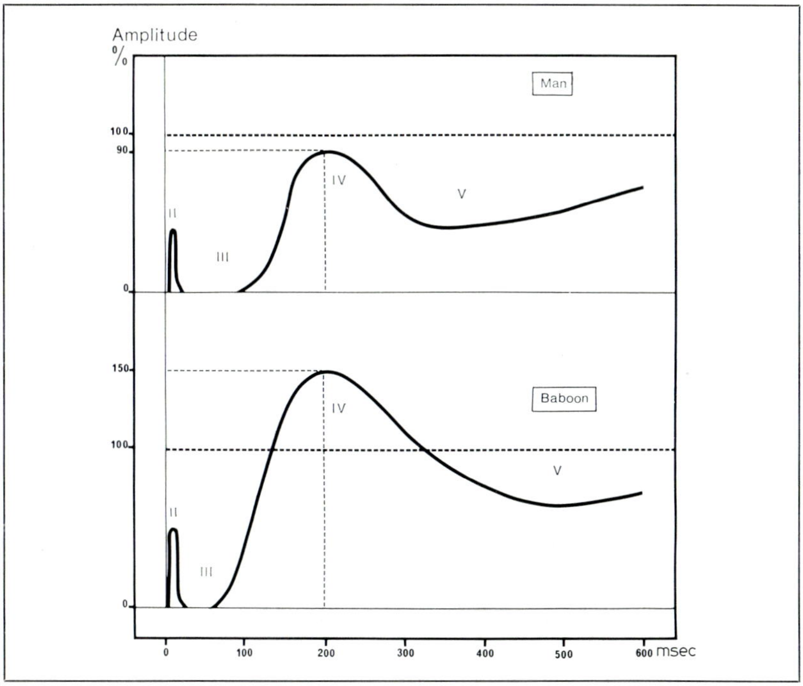

Fig. 3. Restoration cycle. Effect of double stimulation on the H-reflex of the soleus muscle in man (upper) and baboon (lower). Conditioning and test stimuli are electrical shocks which, by themselves, give identical reflex responses of about 50% maximum amplitude. Ordinate: Amplitude of H-reflex evoked by test stimulus as a percent of conditioning response. Abscissa: Interstimulus delay between conditioning and test stimuli.

responses, phase-locked to the mechanical impulses, with the T-reflex latency. At frequencies lower than 20 strokes/sec (interstimulus period over 50 msec), each stimulus elicits a rather large T-reflex. At a higher frequency, the first T-response is high. The following ones are null or small, fluctuating and clearly separated. Later, some recovery occurs and the responses fluctuate between zero and maximal T-reflex amplitude. In no case is there any augmenting response, neither in baboon nor in man. Usually a small tonic activity persists, depressed at first by the expected silent period which follows the last T-response. Over 75 strokes/sec in man and 100 in baboon, the successive T-reflexes interfere (fig. 5d), the appearance being that of a sustained tonic asynchronous response. H-reflexes, elicited at unpredictible

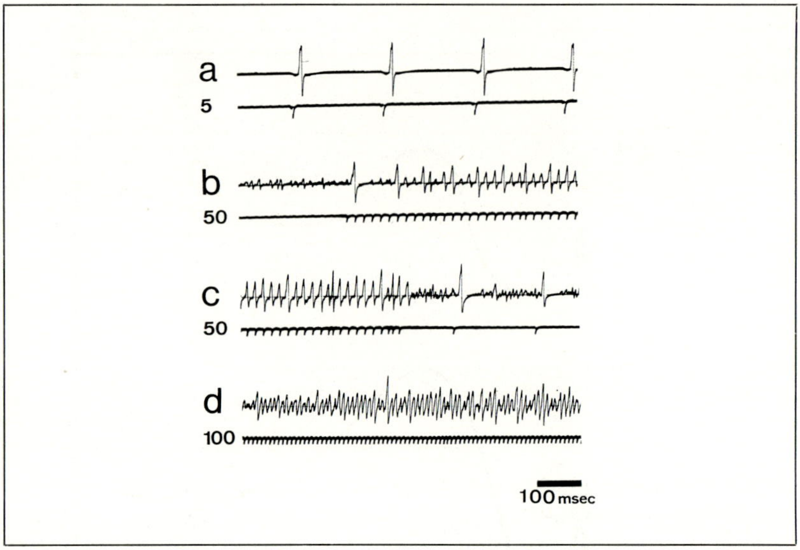

Fig. 4. Tendon reflex in baboon. Upper trace of each pair: EMG recording in soleus muscle during tendon mechanical stimulation. Lower trace of each pair: mechanical stimulus.

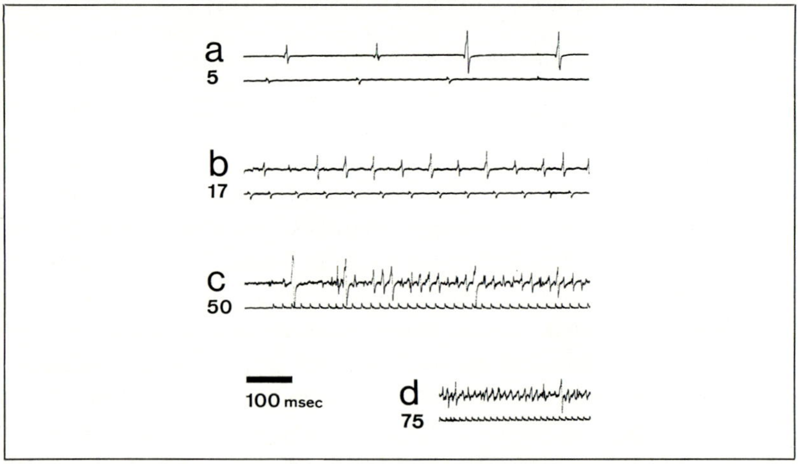

Fig. 5. Tendon reflex in man. EMG recording from man soleus muscle. Same presentation as in figure 4.

times, and not phase-locked to the mechanical stimulus, display a strong depression in man and baboon. A post-vibratory facilitation can be seen on H-reflex, even though the periodic mechanical stimulation has terminated.

Discussion

Origin, Development and Significance of the M- and H-responses

S-shaped curves depict the progressive recruitment in alpha fiber pool of the soleus nerve demonstrated through the EMG and related recruitment of motor units. Similar S-shaped curve can be built by cumulation from the very histogram of motor fibers in soleus and gastrocnemius nerve in cat. This similarity suggests that the M-curves express the cumulative recruitment of alpha fibers as the stimulation increases. If a logarithmic scale is employed in order to express the electrical intensity, M-curves in humans and monkeys display the very same shape, whatever the absolute values of stimulation are. M-curves are valuable interindividual and interspecies reference when we need a significant definition of the monosynaptic excitability, which can be expressed relatively to the 50% M-max. In man, H-reflex is believed to be monosynaptic and produced by the Ia afferent volley. In baboon, H-response has a fixed latency, a fixed duration, as in man, and we admit the H-reflex as a well-synchronized monosynaptic response, without any further comment. But a few physiological question arise from consideration of the H-curve.

The declining branch is usually believed to be due to collisions between the antidromic alpha activities and the orthodromic reflex alpha activities in the soleus nerve fibers: Collisions result in decreasing of the H-response, decreasing is complete with M-max. due to a general blocking on the motor fibers. The climbing branch of the H-curve is believed to be due to an increasing recruitment in the Ia pool of soleus nerve fibers, provoking a related spatial summation on motoneurons. Ib fibers can be also excited by the electrical stimulus and a disynaptic inhibition can develop, which will be responsible for the H-saturation. As a rule, the T-reflex in baboon is higher than the H-reflex, demonstrating a good motoneuronal excitability. Weakness of the H-reflex can be interpreted as being due to the Ib afferent volley inhibition, through 'Ib interneurons' which are responsible for this disynaptic inhibition. There are at least four other determinants for the non-development of the H-reflex: recurrent activity through Renshaw cells, presynaptic autoinhibition of afferent volley, blocking effect of the antidromic alpha activity, and, finally, internuncial activity, diffuse or specific for intermotor nucleus co-ordination.

Supraspinal controls are suggested by the increase of the recruitment ratio in man, during central activation of the soleus, or in baboon after decortication. These controls may consist in facilitation or disinhibition of spinal reflexes. Supraspinal centers do control the different determinants mentioned above.

The general purpose of the experiment in the baboon is to determine the respective importance of these different factors. But, as a whole, direct or indirect corticospinal control has to be considered in the light of developmental and functional considerations.

It is known that in the infant, the H-reflex can be evoked in several muscles. This finding could be explained by the presence of a greater control of the motor nuclei by the local proprioceptive reflexes. During development the H-reflex disappears from all but the soleus muscle as a result of motor maturation. In other words, cortical systems take over the commands of the motor functions decreasing the efficacy of the peripheral automatic mechanisms which are dominant in the infant. H-reflex is weak or null in fore-limb muscles. This is another illustration of some 'cortical domination', which may exist in the baboon soleus (four-handed animal). Mechanisms for such a control have to be explored through further experiments.

Recovery Cycle of the Spinal Monosynaptic Reflex

PAILLARD [1955], TABORIKOVA and SAX [1969], and DELWAIDE [1970] have studied the restoration cycle of the monosynaptic reflex in man. Their results will not be discussed here. However, one may notice in man and in baboon the similarity of curves of the direction and time course of the effects induced from a preceding reflex activation. This similarity seems to be somewhat paradoxal if one takes into account that the H-reflex in baboon has half the latency value of that in man. These findings seem to indicate that the recovery cycle is not tightly linked to conduction time, but rather to physical characteristics of contraction or physiological characteristics of central activities. Two excitability cycle characteristics seem to be identical in baboon and in spastic man: first the existence of a large phase IV; second, an attenuation of the late inhibitory phase. Among all the factors supposed to be responsible for phase IV, the reactivation of the spindles during the relaxing process of the soleus muscle is often taken into account. In spastic human subjects – and in normal baboon – high sensitivity of spindles can be considered as a factor of 'rebound', which thus would not be a true 'post-inhibitory' rebound. BOUAZIZ *et al.* [1971] demonstrated in baboon that excitation of the distally severed gastrocnemius nerve elicits heterosynaptic influence which results in a five-phase recovery reflex cycle, similar to the homosynaptic one in many aspects. Old and new results both suggest that central and peripheral factors associate their activity to produce the shape of the curve.

Finally we may point out the rebound of phase IV and peripheral relaxing process are both presumably linked to the clonic soleus activity in monkey and in spastic man. This is to be related to the high dynamic sensitivity of spindles in spinal animal, related to a gamma dynamic spinal activity [ALNAES *et al.*, 1965].

Periodic Tendon Stimulation

Studies of the tonic vibration reflex (TVR) in man [HAGBARTH and EKLUND, 1966; DE GAIL *et al.*, 1966; HOMMA *et al.*, 1971] have developed during the last years. The vibratory stimulus was shown to be adequate for the testing of muscle spindle dynamic sensitivity [LANCE *et al.*, 1968]. In man, TVR is rather weak and develops through an asynchronous muscular activity which depends on muscle stretching and central control. A sinusoïdal periodic stimulation is commonly used to produce such a stimulation. In our study we use repetitive mechanical impulses characterized by a constant rise time at any frequency in order to elicit T-reflexes. These reflexes develop at any frequency up to 75–100 strokes/sec. At higher frequencies, there is 'interference' of the EMG responses. It should be pointed out that the 'silent period' which follows the monosynaptic (H or T) reflex is about 60–80 msec. In man, the recovery curve of T-reflex (after a preceding T-reflex) displays a phase of inhibition which lasts 300 msec [PAILLARD, 1955]. Those two kinds of inhibition do not prevent the T-reflex to develop, but only depress it, and possibly also depress H-reflex when evoked at random time during the periodic mechanical stimulation. LANCE *et al.* [1968] and DELWAIDE [1970] propose a different theory based on presynaptic inhibition as a cause of reflex inhibition during the TVR in man.

Sinusoïdal mechanically-induced responses are slowly increasing up to the tonic vibration response. Impulse stimulus immediatly do provoke high responses. These differences seem to be due to difference in dynamics of the two stimuli which are not equivalent, at low frequencies at least.

Conclusion and Summary

In the triceps surae, the recruitment of the Hoffmann monosynaptic reflex is comparable in man and baboon. The restoration cycles are also comparable. The neurophysiological mechanisms involved in vibration reflex are not markedly different in the two species. However, the observed differences allow us to postulate functional hypotheses which involve more complex mechanisms than the elementary ones. Surgical and pharmacodynamical studies in baboon should permit further exploration which would be significant for man as well.

References

ALNAES, E.; JANSEN, J. K., and RUDJORD, T.: Fusimotor activity in the spinal cat. Acta physiol. scand. *63:* 197–212 (1965).

BOUAZIZ, Z.; BOUAZIZ, M. et HUGON, M.: Influence de la stimulation des nerfs gastrocnémiens sur l'activité tonique du soléaire et le réflexe de Hoffmann chez l'homme. J. Physiol. *63:* 16–17 (1971).

DELWAIDE, P. J.: Etude expérimentale de l'hyperréflexie tendineuse en clinique neurologique (Arscia, Bruxelles, 1970).

GAIL, P. DE; LANCE, J. W., and NEILSON, P. D.: Differential effects on tonic and phasic reflex mechanisms produced by vibration of muscles in man. J. Neurol. Neurosurg. Psychiat. *29:* 1–11 (1966).

HAGBARTH, K. E. and EKLUND, G.: Motor effects of vibratory stimuli in man; in GRANIT, ALMQUIST and WIKSELL Muscular afferents and motor control. Nobel Symp. I, Stockholm 1966.

HOFFMANN, P.: Untersuchungen über die Eigenreflexe (Sehnenreflexe) menschlicher Muskeln (Springer, Berlin 1922).

HOMMA, S.; KANDA, K., and WATANABE, S.: Tonic vibration reflex in human and monkey subjects. Jap. J. Physiol. *21:* 419–430 (1971).

LANCE, J. W.; NEILSON, P. D., and TASSINARI, C. A.: Suppression of the H-reflex by peripheral vibration. Proc. Austr. Ass. Neurol. *5:* 45–49 (1968).

MAGLADERY, J. W.; TEASDALL, R. D.; PARK, A. M., and PORTER, W. E.: Electrophysiological studies of nerve and reflex activity in normal man. V. Excitation and inhibition of two neurone reflexes by afferent impulses in the same nerve trunk. Bull. Johns Hopk. Hosp. *88:* 520–537 (1951).

MAGLADERY, J. W.; TEASDALL, R. D.; PARK, A. M., and LANGUTH, H. W.: Electrophysiological studies on reflex activity in patients with lesions of the nervous system. I. A comparison of spinal motoneurone excitability following afferent nerve volleys in normal persons and patients with upper motor neurone lesions. Bull. Johns Hopk. Hosp. *91:* 219–244 (1952).

PAILLARD, J.: Réflexes et régulations d'origine proprioceptive chez l'homme. Etude neurophysiologique et psychophysiologique (Arnette, Paris 1955).

ROLL, J. P.; BONNET, M., and HUGON, M.: The baboon as a model for human spinal reflexes studies; in New developments in EMG. 4th Int. Congr. of EMG, Bruxelles 1971 (in press).

TABORIKOVA, H. and SAX, D. S.: Conditioning of H-reflex by a preceding subthreshold H-reflex stimulus. Brain *92:* 203–212 (1969).

Author's address: Dr. J. P. ROLL, Laboratoire de Neurophysiologie comparée, Université de Provence, *F–13013 Marseille* (France)

Medical Primatology 1972. Proc. 3rd Conf. exp. Med. Surg. Primates, Lyon 1972, part II, pp. 315–327 (Karger, Basel 1972)

Eye Movements During Learning of Color and Form Discrimination Problems by Monkeys[1]

A. M. SCHRIER and J. VAUGHAN

Brown University, Providence, R.I.

Experimental psychologists have made extensive use of visual discrimination tasks in studying animal learning [SCHRIER *et al.*, 1965; SUTHERLAND and MACKINTOSH, 1971]. Obviously, in order to learn such tasks, the animals must consistently look at the discriminative stimuli (the stimuli correlated with reward or reinforcement), but relatively little is known about how they go about this, partly because of the difficulty of measuring their eye movements. In recent years, however, several techniques have been developed for studying eye movements in monkeys [BAGSHAW *et al.*, 1970; FUCHS, 1967]. though little systematic work has been reported as yet on eye movements in connection with visual learning tasks. OSCAR-BERMAN *et al.* [1971] studied eye movements of rhesus monkeys during learning of two pattern discrimination problems. Although the method of recording eye movements was fairly crude, they reported some systematic changes as learning progressed. An early tendency to spend more time looking at the stimulus display on one side rather than the other tended to diminish with practice. Also, the animals tended to respond where they were looking, and, increasingly, tended to spend more time looking at the positive stimulus than at the negative. SCHRIER *et al.* [1970, 1971] developed a computer-controlled system for recording eye movements of monkeys, which is based on the corneal reflection. They reported preliminary findings for a rhesus monkey performing an easy and a difficult brightness discrimination task. This work was repeated and extended by SCHRIER and WING [1972], using stump-tailed monkeys. Temporal distribution of fixations was one parameter of eye

1 Supported by a NSF research grant (GB-18953X) to the first author. Part of the research served for the second author's doctoral dissertation [VAUGHAN, 1970].

movements examined in these studies. It was found, for example, that the animals were using a scanning pattern on incorrect trials that would have been appropriate if the negative stimulus had been the positive. The results were similar to those of OSCAR-BERMAN *et al.* [1971] in that the animals tended to respond predominantly to the discriminative stimulus that they last fixated and after practice this was primarily the positive stimulus. The duration of individual fixations was also measured in these studies. Depending on the animal, the average fixation duration ranged from 147 to 186 msec and, generally speaking, was stable across tasks and relatively insensitive to many of the variables examined. One notable exception [SCHRIER and WING, 1972] was a marked, though temporary, increase in fixation duration following reversal of a difficult brightness problem.

The apparatus and techniques used in the work on brightness discrimination were used in the present experiment, which was also concerned with eye movements of macaques. In the present case, the animals were given a series of discrimination problems involving color or form as the relevant dimension and two types of shifts. In one type, called 'reversal shifts', the previously positive stimulus became negative and the previously negative stimulus became positive, and, in the other type, called 'nonreversal shifts', the relevant dimension was changed.

Method

1. Subjects

Two adult female macaques, one a rhesus *(Macaca mulatta)* and one a stump-tailed monkey *(M. arctoides)*, served as the subjects. Both animals had had extensive training in brightness discrimination studies which involved the use of the present eye-movement recording apparatus [SCHRIER *et al.*, 1970, 1971; WING and SCHRIER, 1972]. The stump-tail was tested several months after the rhesus completed training. Apparatus modifications during the intervening period and the pressure of time necessitated some differences in apparatus and procedure as noted below.

2. Apparatus

The major aspects of the apparatus have been described in detail in prior publications [SCHRIER *et al.*, 1970, 1971]. There were three primary components: (a) a restraining chair, (b) a helmet, held in place by a mouth brace, on which were mounted the stimulus presentation and other devices for obtaining the corneal reflection, and (c) a computer-controlled system for presenting stimuli, recording eye orientation, and monitoring choice responses. As before, during the testing, the animal was seated in the chair facing a panel containing three response 'keys' which the animal could press. These are referred to hereafter as the center and the side response keys.

Three stimulus presentation devices were supported by the helmet about 20 cm in front of the left eye of the animal. The devices were arranged as an inverted equilateral triangle with 5 cm sides (about 14° of visual angle). The 'center light', as in the earlier work, was a small pilot lamp mounted in a housing which had a 0.6-mm aperture. Color and form stimuli for two-choice discrimination problems were presented by means of two Micro-Miniature In-Line display units (Industrial Electronic Engineers, Van Nuys, Calif., Model 340) which formed the base of the triangle. The stimuli that were projected differed for the two monkeys, though in both cases they were four color-form compounds. For the rhesus, each unit dispayed either a white horizontal bar or a white cross centered on a red or green field (diameter approximately 7 mm or 2°). For the stump-tail each In-Line unit displayed either a red circle, a green circle, a red cross, or a green cross (the diameter of each circle was approximately 7 mm or 1° 56′ and each bar of the cross was 3 mm thick and 8 mm long or 48′ by 2° 18′).

Eye movements were recorded, as before, by monitoring the corneal reflection of a light source mounted on the helmet and focused on the animal's right eye. The corneal reflection was transmitted from a lens, also mounted on the helmet, to a special TV camera by means of a coherent fiber optics cable. The on-line computer sampled the location of the corneal reflection, as represented in the output signals of the TV camera, 60 times a second. For recording the location of the reflection, the computer-representation of the TV field was divided into a 16-by-16 recording grid. Each cell in the region of the grid corresponding to the locations of the experimental stimuli represented horizontal and vertical distances of approximately 3° of visual angle.

3. Procedure

The animal was placed in the restraining chair and the helmet put on its head Sessions were run daily, five to seven days a week. Each session was continued until a total of about 400 trials had been completed, or until the animal's restlessness prohibited reliable data recording.

The side that was positive on each trial was randomly determined with the restriction that each be positive about equally often in each session. A variable intertrial interval was selected at random from among four values, 4, 6, 8 and 10 sec. Each response during the intertrial interval initiated a new intertrial interval of 10 sec.

Each trial consisted of two phases, a 'calibration' phase and a 'discrimination' phase. The calibration phase was designed to permit the equipment to be maintained in spatial calibration from trial to trial and from session to session. Each trial began with a 400-Hz tone for a maximum of 4 sec. When the duration of any fixation exceeded a minimum value during the tone, the center light was illuminated for a maximum of 320 msec, while the tone remained on. The minimum duration of the fixation that was required was randomly chosen on each trial from the values 1.00, 1.33, 1.67 and 2.00 sec, so that the animal could not use the passage of time as a cue, rather than the illumination of the center light. If the animal pressed the center key within 1.6 sec after the onset of the center light, the discrimination phase of the trial was initiated, and the tone was terminated. The trial was cancelled if the animal responded to the center push-button during the tone, but before the onset of the center light, or if it did not respond to the center key in time.

The purpose of the variable fixation duration requirement, the short duration of the center light, and the short time allowed for the response to the center key was to increase the frequency of relatively long fixations on the locus of the center light before its onset. And, indeed, the 1- to 2-sec fixations that were obtained are up to ten times longer than the average duration otherwise observed in this situation. These fixations were used to maintain calibration during a session, as explained in an earlier paper [SCHRIER *et al.*, 1971]. Training the animals to make long fixations during this phase does not appear to influence duration of fixation during the subsequent discrimination phase [SCHRIER and WING, 1972].

During the discrimination phase of the trial, the discriminative stimuli were presented and the animal allowed to make a choice response. The discriminative stimuli were presented as soon as a saccadic eye movement occurred. The discriminative stimuli remained illuminated for a maximum of 4 sec. If the animal pressed the key on the same side as the positive stimulus, it was reinfored; a response to the other side, or failure to respond within the time limit, initiated the intertrial interval.

Because the data analyses involved only the period from the onset of the discriminative stimuli to the occurrence of a reinforced or nonreinforced choice response for the remainder of this paper, a 'trial' will refer only to this period. Any discrimination phase which did not end in a choice response was excluded from the analyses.

4. Learning Tasks

Brightness problems. The animals were first run for two days (about 1,000 trials) on a simple brightness problem in which the positive stimulus was illuminated with a homogeneous yellow color while the other stimulus was not illuminated. The animals received this problem again for one session at the end of the experiment.

Redundant relevant cue problem (RRC). Following the brightness problem, the dimensional preference of each animal was determined. They were presented with a problem consisting of redundant form and color cues (for the rhesus, the bar on a red background was positive and the cross on a green background was negative and, for the stump-tail, the green circle was positive and the red cross negative). After reaching a criterion of 90 correct trials of 100, which required 100–150 trials, a series of reinforced probe trials were intermingled with the regular RRC trials. The probe trials consisted of the opposition of the color and form cues; that is, for example, for the rhesus one stimulus was the cross on a red background and the other, the bar on a green background.

For the rhesus monkey, following criterion performance during each of two consecutive test sessions, one probe trial was given every five trials for a total of 20 probe trials each session. The stump-tailed monkey was given probe trials for only one session, one every ten trials for a total of ten probe trials. The results are presented now because they determined the next part of the procedure. Both animals clearly preferred the color dimension. Thirty of 38 responses by the rhesus and nine of ten responses by the stump-tail on probe trials followed the positive color component of the RRC problem, rather than positive form component. A two-tailed binomial test indicated the preference of both the rhesus ($p<0.001$) and the stump-tail ($p=0.044$) was significant.

Shift problems. Following the RRC problem, the animals were given a series of 16 problems, on each trial of which the two color and the two form stimuli were presented as color-form compounds. Half the problems were 'color problems', that is color was relevant (one of the colors was positive and the other negative) and form irrelevant (each of the forms was on the same side as the positive color on only a random half of the trials). The other half of the problems were 'form problems', during which form was relevant and color irrelevant. Eight of the problems involved reversal shifts, that is, the relevant stimuli (color or form) were the same as on the previous problem, but their relation to reinforcement was reversed. For example, if form 1 was positive and form 2 was negative on a particular problem, then, following the reversal shift, form 2 was positive and form 1 was negative. The remaining eight problems involved nonreversal shifts, that is the relevant stimuli were shifted from one dimension (color or form) to the other. On nonreversal shifts, the positive stimulus was the same as it had been when the dimension was last relevant. Table I summarizes the conditions for each of the 16 shift problems. One of two

Table I. Summary of stimulus conditions during the RRC problem and the shift problems

Problem	Relevant stimuli	Type of shift	Combination A		Combination B	
			C_1F_1	C_2F_2	C_2F_1	C_1F_2
RRC	C, F		+	–	not presented	
1	F	N	+	–	+	–
2	F	R	–	+	–	+
3	C	N	+	–	–	+
4	C	R	–	+	+	–
5	F	N	–	+	–	+
6	F	R	+	–	+	–
7	C	N	–	+	+	–
8	C	R	+	–	–	+
9	F	N	+	–	+	–
10	F	R	–	+	–	+
11	C	N	+	–	–	+
12	C	R	–	+	+	–
13	F	N	–	+	–	+
14	F	R	+	–	+	–
15	C	N	–	+	+	–
16	C	R	+	–	–	+

C=color; F=form; N=nonreversal shift; R= reversal shift; + =positive stimulus compound; – =negative stimulus compound.

combinations of the four stimulus compounds (combination A: color 1- was form 1- vs. color 2-form 2; and combination B: color 1-form 2 vs. color 2-form 1) used on half the trials of a problem and the other combination on the remaining half. It seems reasonable to classify the first problem of the shift series as a nonreversal shift because the animals were responding primarily on the basis of color differences during the RRC problem and, hence, had to shift to responding on the basis of form differences to solve the first shift problem.

Each problem was run until the animal attained a criterion of 90 correct responses in 100 trials (the 'criterion trials').

5. Data Analysis

For the rhesus monkey, the location of each discriminative stimulus was defined as a 3-cell by 3-cell area, which was a fixed number of cells above, and, depending on which stimulus, to the left or right of the recorded location of the long fixation of the center light, which, again, usually coincided with the calibration cell. (See table II, to be discussed later, for an illustration of the recording locations of the discriminative stimuli.) The discriminative stimulus areas for the stump-tailed monkey were each 3 × 4 cells, an additional cell having been added in the vertical direction because this animal showed somewhat greater variability of fixations in this direction. Each fixation was recorded as being a fixation of the left or right discriminative stimulus if its location fell within one of these two designated areas. Fixations elsewhere in the recording grid were ignored.

For each trial, the number of 'scans' of the discriminative stimuli was computed. A scan was defined as the fixation on one discriminative stimulus after a fixation on the other. For instance, fixation on only the left stimulus in a trial would be zero scans; fixations on the left, then the right stimulus would be one scan.

Results and Discussion

The discriminative performance of the two animals was quite similar, with the mean number of errors made before the criterion trials (errors to criterion) being 90 for the rhesus and 94 for the stump-tail. As might be expected from the results on the probe trials during the RRC problem, as well as innumerable previous findings with monkeys [MEYER *et al.*, 1965], the animals tended to learn color problems faster than the form problems (mean errors to criterion for both animals combined were 56 and 128 on color and form problems, respectively). Also, both animals learned nonreversal shifts faster than reversal shifts (mean errors to criterion for both animals combined were 78 and 106 on nonreversal and reversal shifts, respectively). This was true during both halves of the learning period. Overall performance was considerably better during the second half, suggesting

Table II. Spatial distribution of fixations by the rhesus monkey during reinforced trials of the shift problems. The percentage of fixations which were recorded in each cell of the central region of the recording grid is indicated. The defined locations of the left and right discriminative stimuli are in boldface. Recorded locations have been adjusted for variations in the location of the center-light fixation of the calibration phase of each trial

Vertical position (cell No.)	Horizontal position (cell No.)								
	1	2	3	4	5	6	7	8	9
1	0	1	1	0	0	0	0	0	0
2	0	**1**	**7**	**1**	0	**1**	**8**	**2**	0
3	0	**0**	**22**	**6**	0	**1**	**20**	**7**	0
4	0	**0**	**3**	**1**	0	**0**	**3**	**2**	0
5	0	0	0	0	0	0	0	0	0
6	0	0	0	0	0	0	0	0	0
7	0	0	0	0	1	0	0	0	0
8	0	0	0	0	6	0	0	0	0
9	0	0	0	0	0	0	0	0	0

shift-learning set formation (for problems 1–8, mean errors to criterion was 152, whereas, for problems 9–16 it was 32). A similar difference in speed of learning following the two kinds of shifts has been found in other studies, though there have been some in which no difference was observed [SUTHERLAND and MACKINTOSH, 1971, p. 307, design A]. Undoubtedly the results obtained will vary depending on the stimuli used and such aspects of training as the criterion of learning.

The spatial distribution of fixations during the shift problems is shown in table II for the rhesus monkey. The results were quite similar for the stump-tailed monkey, except, as noted earlier, for a somewhat greater variability in location of fixations in the vertical direction in the region of the discriminative stimuli. As table II suggests, almost all fixations that were recorded during the discrimination phase of trials were directed at the discriminative stimuli (specifically, 85% of the rhesus' fixations and 87% of the stump-tail's). Most of the remaining fixations were directed at the locus of the center light. Neither animal showed the marked position preference in observing the discriminative stimuli seen in an earlier study [SCHRIER *et al.*, 1970]. Approximately 52% of all of the rhesus' fixations on the discriminative stimuli were on the one on the right side and approximately 62% of all of the stumptail's were on the one on the right side. Perhaps the reason for the reduced observing response position preference is, as the findings of OSCAR-BERMAN *et al.* [1971] suggest, that both animals had had extensive experience in the test situation before the present experiment was begun.

The visual scanning behavior of the animals over the course of the experiment is summarized in figure 1. If the animals made only the minimal number of scans on each trial to fixate the positive stimulus once, the median number would be 0.5, since on half the trials the positive stimulus, and on the other half of the trials the negative stimulus, would be the first fixated. Such a model is clearly inadequate in the present situation since the median number of scans was consistently higher than this minimum for the criterion trials of each of the shift problems, and for most of the trials of the two brightness problems and of the RRC problems. Yet, it was more probable that the last fixation in any trial was on the positive stimulus, rather than the negative. For the series of shift problems as a whole, the proportion of reinforced criterion trials in which the positive stimulus was last fixated was 0.72. Among the individual problems, there was no consistent pattern of variation in this proportion, which presumably measures the degree to which the choice response was controlled by the positive rather than the negative stimulus of the individual problem.

The amount of scanning by both animals tended to increase over the series of problems starting with the initial brightness problem, although the rhesus monkey consistently scanned more than did the stump-tailed. The pattern of change in scanning over the series of problems was remarkably similar for the two animals, both showing an early peak in scanning activity during the third or fourth shift problem, followed by a decline and another increase in scanning beginning after the ninth or tenth shift problem. We have no immediate explanation for this particular pattern and it would be interesting to determine whether a similar pattern appears when the task does not involve a series of shifts of the type used here. The general upward trend in amount of scanning might be typical for the learning of a series of problems. WHITE and PLUM [1964], in an experiment with human children, reported an increase in amount of scanning as learning-set formation progressed. They also reported a decrease in amount of scanning as the children's performance continued to improve. Since there was still room for improvement in performance at the end of the present series of shift problems, it is quite possible that amount of scanning would have declined had the experiment been continued. WHITE and PLUM [1964] suggested that the changes

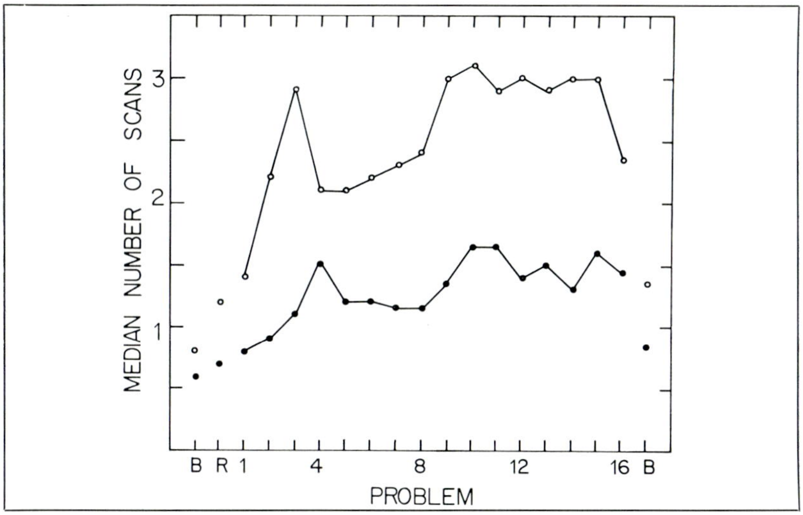

Fig. 1. Median number of scans of the discriminative stimuli for all reinforced trials of the two brightness and of the RRC problems, and for reinforced criterion trials of each shift problem. ● Stump-tail; ○ rhesus. B = Brightness problem; R = RRC problem.

in visual observing behavior during learning-set formation paralleled that seen during the learning of single discrimination problems, but their measure of eye movements was very crude and additional data are needed for both single and multiple problem learning.

There was no consistent relationship between number of scans during the criterion trials and the relevant dimension (median number of scans during color and form problems being 1.8 and 1.6, respectively) or between number of scans during the criterion trials and the type of shift (median number of scans on reversal and nonreversal shifts being 1.8 and 1.7, respectively). However, it cannot be assumed that the type of problem does not influence scanning because, as figure 1 suggests, there was more scanning during the shift problems and the RRC problem than during the brightness problem. That this was not merely a matter of practice is indicated by the marked drop in scanning activity when the brightness problem was reintroduced following the series of shift problems.

In our preliminary work [SCHRIER *et al.*, 1970], we found, not surprisingly, that the more time the animal spent scanning the stimuli, the longer it took to make its instrumental choice response, that is, the longer was its response latency. This was confirmed in the present experiment. In addition,

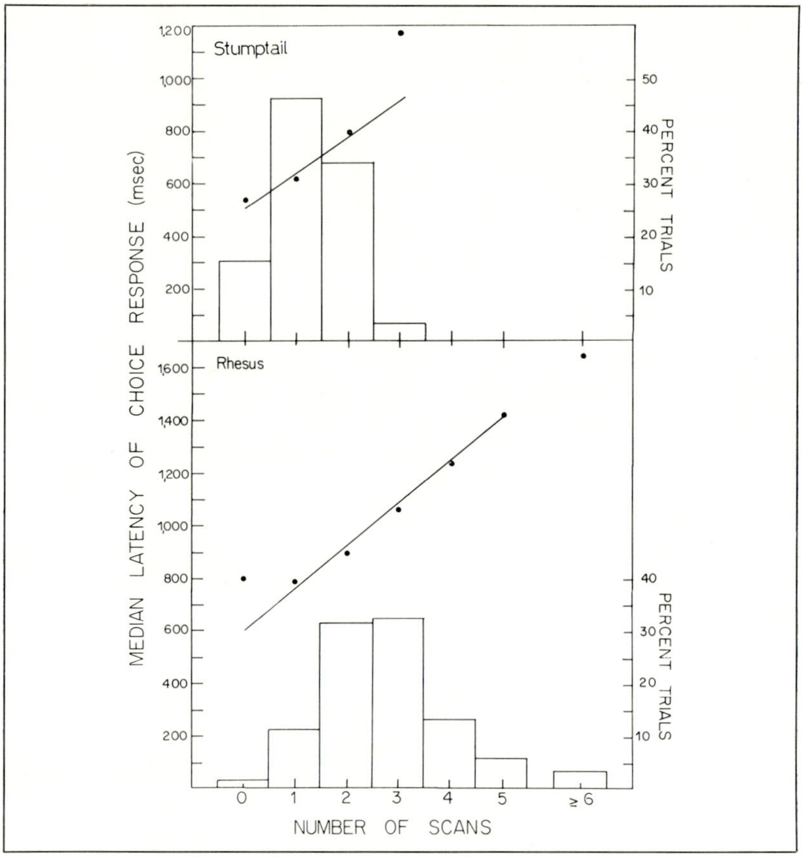

Fig. 2. Relationship between scanning activity and latency of choice response (circles and left-hand ordinate) during the criterion trials of the shift problems. The least-squares best-fitting straight line has been drawn through the data points. Only data points based on more than 75 trials were included in the least-squares calculations. Also shown (histogram and right-hand ordinate) is the percentage of trails in which each number of scans occurred.

there appeared to be a simple linear relationship between scanning and choice response latency, as can be seen in figure 2. In other words, each additional fixation resulted in a unit increase in latency of response[2]. Only two

2 The number of fixations is essentially equal to the number of scans plus 1 because the animals almost never look away from one of the discriminative stimuli and look at it again before first looking at the other one.

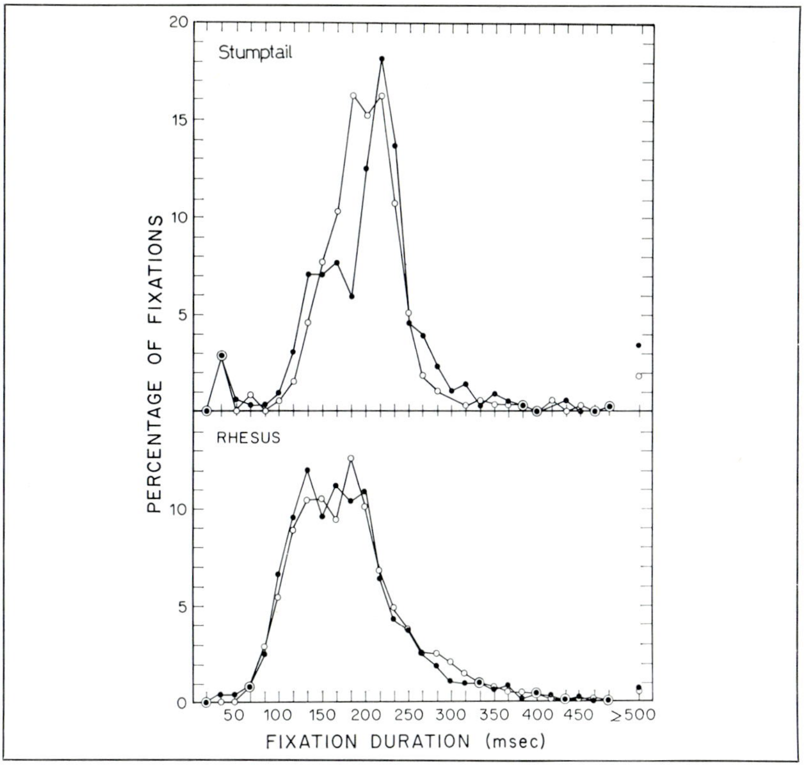

Fig. 3. Distribution of fixation durations in steps of 16.7 msec (1/60 sec) for reinforced criterion trails of color (○) and form (●) problems. Only fixations prior to the last complete fixation of a trail were included. The reason for this is related to aspects of the data analysis not reported here. The exclusion of the last complete fixation lowers the median duration, but has no appreciable effect on the relationship between the curves for color and form problems.

points deviated greatly from the least-squares line of best fit and these were based on a relatively small amount of data.

Another major parameter of eye movements that was analyzed was duration of fixation. Figure 3 shows the distribution of fixation durations separately for color and form problems. These particular distributions include only fixations preceding the last complete fixation on each trial, but there is little difference between the two distributions whether or not the last complete fixation is included. For the rhesus, the median duration of fixation during color problems was 177 msec and during form problems 171 msec (not counting the last complete fixation). For the stump-tail, the corresponding medians were 198 and 210 msec.

The general picture of eye movements that emerges from the present study is one of repetitive fixations alternating between the discriminative stimuli (scanning) followed by the choice response. Changes in the stimulus can influence scanning activity in the absence of coincident changes in the accuracy of the choice response (or in the density of reinforcement). For instance, for both animals, the rate of scanning was higher during the post criterion trials on the RRC problem than during the post criterion trials on the initial brightness problem. Yet, throughout these trials, performance levels were consistently high. Also, changes in the relation of the stimuli to reinforcement in the absence of any other changes in the stimuli or in accuracy of response can also affect the rate of scanning as, for example, occurred with respect to scanning activity during the criterion trials of the various shift problems (fig. 1). These observations are not consistent with secondary reinforcement hypotheses concerning the acquisition and maintenance of observing behavior, such as WYCKOFF'S [1952].

Though more extensive analyses are called for, our impression is, on the basis of our present and past work, that duration of fixation is a more stable aspect of eye orientation than is scanning activity. However, there is no doubt that [SCHRIER and WING, 1972] duration of fixation can change systematically under some conditions.

References

BAGSHAW, M. H.; MACKWORTH, N. H., and PRIBRAM, K. H.: The effect of inferotemporal cortex ablations on eye movements. Int. J. Neurosci. *1:* 153–158 (1970).

FUCHS, A. F.: Saccadic and smooth pursuit eye movements in the monkey. J. Physiol. *191:* 609–631 (1967).

MEYER, D. R.; TREICHLER, F. R., and MEYER, P. M.: Discrete-trial training techniques and stimulus variables; in SCHRIER, HARLOW, and STOLLNITZ Behavior of nonhuman primates, vol. 1., pp. 1–49 (Academic Press, New York 1965).

OSCAR-BERMAN, M.; HEYWOOD, S. P., and GROSS, C. G.: Eye orientation during visual discrimination learning by monkeys. Neuropsychologia *9:* 351–358 (1971).

SCHRIER, A. M.; HARLOW, H. F., and STOLLNITZ, F.: Behavior of nonhuman primates, vol. 1 (Academic Press, New York 1965).

SCHRIER, A. M.; POVAR, M. L., and VAUGHAN, J.: Measurement of eye orientation of monkeys during visual discrimination. Behav. res. Meth. Instr. *2:* 55–62 (1970).

SCHRIER, A. M.; POVAR, M. L., and VAUGHAN, J.: Primates in eye movement research; in GOLDSMITH and MOOR-JANKOWSKI Medical primatology 1970, pp. 847–858 (Karger, Basel 1971).

SCHRIER, A. M. and WING, T. G.: Eye movements of monkeys during brightness discrimination and discrimination reversal. In preparation, 1972.

SUTHERLAND, N. S. and MACKINTOSH, N. J.: Mechanisms of animal discrimination learning (Academic Press, New York 1971).

VAUGHAN, J.: A study of eye movements during learning of color and form visual discrimination problems by a rhesus monkey; unpubl. doct. diss. Brown Univer. (University Microfilms, Ann Arbon, Mich., 1970, No. 71–13947).

WHITE, S. H. and PLUM, G. E.: Eye movement photography during children's discrimination learning. J. exp. Child Psychol. *1:* 327–338 (1964).

WYCKOFF, L. B., jr.: The role of observing responses in discrimination learning. Psychol. Rev. *59:* 431–442 (1952).

Authors' addresses: Dr. ALLAN M. SCHRIER, Psychology Department, Brown University, *Providence, R. I. 02912;* Dr. J. VAUGHAN, Psychology Department, Hamilton College, *Clinton, N.Y. 13323* (USA)

Medical Primatology 1972. Proc. 3rd Conf. exp. Med. Surg. Primates, Lyon 1972, part II, pp. 328–331 (Karger, Basel 1972)

Effects of Chlordiazepoxide, Reserpine, and Imipramine on the Squirrel Monkey *(Saimiri sciureus)*

E. T. UYENO and H. NEWTON

Stanford Research Institute, Menlo Park, Calif.

In our previous studies we have evaluated the effects of several psychotropic compounds on the visual size discrimination performance of the squirrel monkey in the Wisconsin General Test Apparatus (WGTA) [UYENO, 1969]. In the present dose-response experiments, we used the same test procedure to evaluate reserpine, chlordiazepoxide hydrochloride and imipramine hydrochloride.

The WGTA, a special apparatus designed for discrimination training of rhesus monkeys [HARLOW, 1959], was modified for training the squirrel monkey. A pair of discs of different sizes was vertically placed on flat bases to cover two small round grooves, serving as food wells in a stimulus tray. The well under the correct disc had a piece of grape; the one under the incorrect disc was empty. The correct disc was placed either on the top of the right or left well in a random manner according to a GELLERMAN [1933] series. Three monkeys were trained to choose the smaller one of two discs, and the other four monkeys were trained to choose the larger one. A trial was initiated by placing a piece of grape under the correct disc, turning a switch to advance the tray, and simultaneously lifting the guillotine panel in front of the test cage to allow the animal to make a choice. When the switch was turned, it also activated a latency timer which was automatically stopped when the animal displaced one of the discs. The time which elapsed from the moment when the tray was advanced until the displacement of a disc, was considered to be the latency of the animal's response. The animals were given 30 trials a day on an easy discrimination problem and 30 trials on a difficult problem. In the easy discrimination problem the ratio of the area of the large disc to that of the smaller one was 1.96:1. In the difficult discrimination problem the ratio of the discs was 1.24:1. In each problem,

Table I. Drugs and doses evaluated

Drug	Dose					
	low		medium		high	
	mg/kg	μmole/kg	mg/kg	μmole/kg	mg/kg	μmole/kg
Reserpine	0.05	0.08	0.10	0.16	0.15	0.25
Imipramine HCl	2.5	7.88	5.0	15.77	10.0	31.52
Chlordiazepoxide HCl	10.0	29.74	20.0	59.48	30.0	89.22

the total number of correct responses of each monkey was expressed as a percent score. The performance on the easy problem stabilized at approximately 90% correct responses; whereas, that on the difficult problem stabilized at approximately 80%. The mean response latency on the easy problem stabilized at approximately 4.5 sec and that on the difficult problem at approximately 4.7 sec; the difference was not statistically significant.

After stability had been achieved, the animals were injected i.p. (1 ml/kg) with a coded control solution (0.9% saline) on Wednesdays to check the stability of the discrimination performance under conditions of control injection. On Thursdays they were injected i.p. (1 ml/kg) with a dose of a test compound to determine the effects of the compound on their size discrimination ability. The doses administered are shown in table I. Reserpine USP (Serpasil, supplied as parenteral solution by Ciba Pharmaceutical Co.) was injected i.m. 5 h before the test; whereas, chlordiazepoxide hydrochloride and imipramine hydrochloride dissolved in saline solution were administered i.p. 1½ and ½ h before the evaluation, respectively.

Performance on both the easy and difficult problems was not affected by the control compounds. However, the test compounds had marked effects at the high dose level, and some effects at the medium and low dose levels. Statistical analyses of the data were based on either of two criteria of impairment of performance during the drug trials: (1) a response latency of 10 sec or longer (i.e., more than twice the mean latency recorded on predrug or saline trials) or (2) a score less than 80% of the score obtained during the placebo trials. (In the drug trials, the percentage of animals that took 10 sec or longer to respond was greater than the percentage of animals whose scores were less than 80% of their scores on the saline trials.) On the basis of these criteria, the number of animals that showed a disruption in performance on the easy problem was obtained at each dose level of each

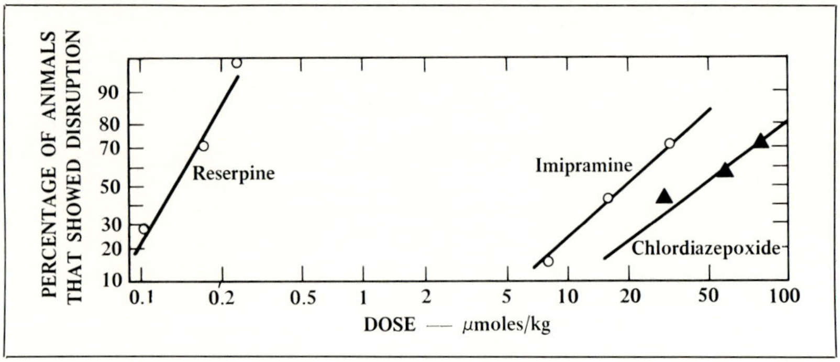

Fig. 1. Dose-response curves. Effects of the drugs on size discrimination performance.

Table II. ED_{50} and 95% confidence limits calculated from the dose-response data

Drug	ED_{50} μmole/kg	95% confidence limits μmole/kg
Reserpine	0.12	0.08– 0.17
Imipramine HCl	19.0	10.86–33.25
Chlordiazepoxide HCl	46.0	23.47–90.16

compound and expressed as a percentage of the number tested. Similarly, the percentages of animals that showed a decrease in performance on the difficult problem were obtained at each dose level of each compound. The percentages of animals that showed a decrease in performance on the easy problem were similar to the percentages of animals that showed a decrease in performance on the difficult problem. One set of percentages was plotted against dose (in μmoles/kg) on logarithmic probability paper (fig. 1). The dose-response curves show that disrupting effect of the compounds is dose-dependent.

The median effective dose (ED_{50}) was defined as the dose by which 50% of the animals were expected to show impairment in performance. The ED_{50} and the 95-percent confidence limits were computed according to the graphic method of LITCHFIELD and WILCOXON [1949] (table II). The ED_{50} values were used to calculate potency ratios to determine the relative potency of the compounds in the disruption of the learned performance. The analysis showed that reserpine was significantly more potent than imipramine, and imipramine was significantly more potent than chlordiazepoxide.

Previous research has shown that accuracy of discrimination performance was disrupted in a substantial percentage of squirrel monkeys treated with lysergic acid diethylamide (LSD-25). However, in the present experiment the disruption of the performance was mainly due to increase in latency, or failure to respond. The present results showing that the compounds greatly increased the latency, but hardly influenced the accuracy, are similar to the results obtained from the evaluation of phenothiazines and barbiturates [Uyeno, 1971].

References

Gellerman, L. W.: Chance orders of alternating stimuli in visual discrimination experiments. J. genet. Psychol. *42:* 206–208 (1933).

Harlow, H. F.: Learning set and error factor theory; in Koch Psychology. A study of science, vol. 2 (McGraw-Hill, New York 1959).

Litchfield, J. T., jr. and Wilcoxon, F.: A simplified method of evaluating dose-effect experiments. J. Pharmacol. exp. Ther. *96:* 99–113 (1949).

Uyeno, E. T.: Alteration of a learned response of the squirrel monkey by hallucinogens. Int. J. Neuropharmacol. *8:* 245–253 (1969).

Uyeno, E. T.: Behavioral effects of phenothiazines and barbiturates on the squirrel monkey. Proc. West. Pharmacol. Soc. *14:* 149–153 (1971).

Authors' address: Edward T. Uyeno and Harold Newton, Life Sciences Research, Stanford Research Institute, *Menlo Park, Calif.* (USA)

Medical Primatology 1972. Proc. 3rd Conf. exp. Med. Surg. Primates, Lyon 1972, part II, pp. 332–342 (Karger, Basel 1972)

Unitary Neuronal Activities of the Hypothalamus and Hydromineral Homeostasis in Non-Anaesthetized Monkeys *(Macaca mulatta)*[1]

J. D. VINCENT, E. ARNAULD[2], B. BIOULAC and J. M. A. FAURE

Laboratory of Physiopathology and Neurophysiology, Bordeaux

I. Introduction

The concentration and content of water in the body is precisely controlled despite influence of increasingly varying factors. This regulation is achieved through a dual neuroendocrine (ADH secretion) and behavioural (drinking) mechanism controlling respectively the renal output and the oral intake of free water.

These aspects are both controlled by hypothalamus, in response to change of the plasma osmolality which supplies a common stimulus for the two systems. Whether the hypothalamus contains osmosensitive neurons, whether their responses are specifically related to the osmotic or arousing nature of the osmotic stimulus, whether the osmoreceptors are anatomically separated from the neurosecretory cells and whether there is a coupling at the neuronal level between neuro-endocrine and behavioural mechanisms are some of the remaining problems investigated in the present study.

II. Materials and Methods

Twenty-one adult (4–4.5 kg) female rhesus monkeys were prepared using our techniques of chronic single unit recording, intracarotid (i.c.) injection and polygraphic recording of behavioural parameters of the conscious monkey [HAYWARD and VINCENT, 1970].

A chronic cylinder-hydraulic-microdrive arrangement mounted on the skull introduced the glass-coated plantinum micro-electrodes stereotaxically into the anterior hypotha-

1 Supported by grants from ATP No. 3 (INSERM), L.A. No. 39 (CNRS) and 'La Fondation pour la Recherche médicale française'.

2 Attachée de recherche à l'INSERM.

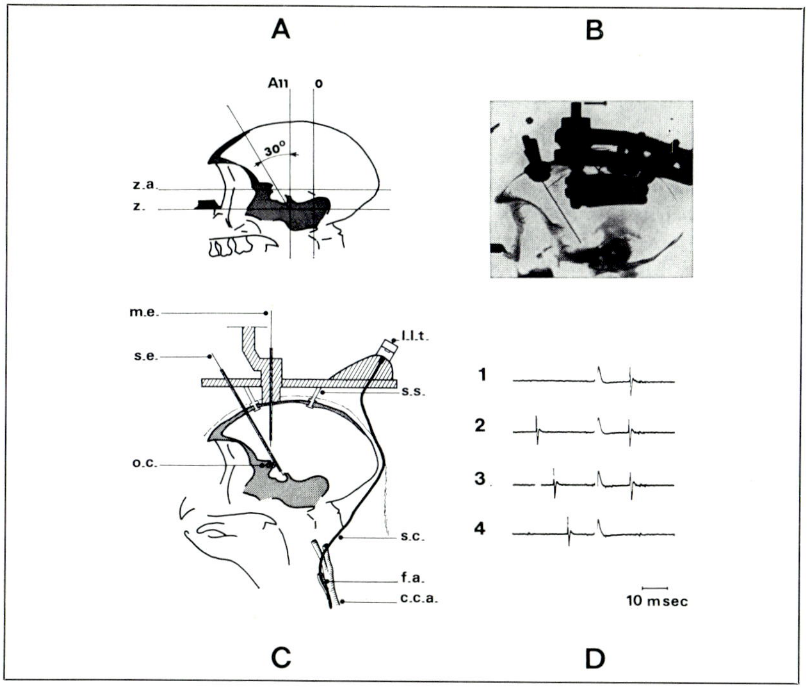

Fig. 1. A = Lateral diagrammatic view of an animal in the stereotaxic position showing the co-ordinates of the stimulating electrode. B = Lateral X-ray controlling the insertion of the stimulating electrode into the center of the posterior pituitary. C = Diagrammatic view of the complete head setting. D = extracellular recordings from an identified neurosecretory neuron in the NSO. 1–3 = Antidromic spike occurring at a fixed latency; in 2 and 3, spontaneous action potential occurring at a latency from the stimulation artefact superior to the antidromic latency. 4 = Cancellation of the antidromic spike by the orthodromic action potential occurring at a latency inferior to the antidromic latency. z = Zero plane; z.a. = zero atlas; m.e. = micro-electrode; s.e. = stimulating electrode; o.c. = optic chiasma; l.l.t. = luer lok tip with rubber diaphragm capping the artenal cannula; s.s. = stainless steel screew; s.c. = silicon rubber cannula; f.a. = facial artery; c.c.a. = common carotid artery.

lamus. A silicone rubber cannulla chronically implanted in the left common carotid allowed injections of saline solutions.

In five monkeys a concentric bipolar stainless steel electrode was introduced stereotaxically in the posterior pituitary under repeated teleradiographic control (fig. 1A–C).

The monkey had been previously conditioned to the primate restraining chair for several weeks and had been trained to drink from a tube fixed in front of the mouth. Unit recording was started five days after surgery. The animal was deprived of water eight hours prior to the beginning of each recording session. Water was delivered from a reservoir through a solenoid valve triggered by the observer.

When a stable neuron was detected, its sensitivity to i.c. injection of hypertonic sodium chloride solution was tested (1 ml during 5 sec, varying tonicity from 1.9 to 3.6%). The effects of different behavioural states (sleep, relaxation, arousal) on the spontaneous activity of the cell and its possible response to non-noxious stimuli (auditory, visual or tactile) were also examined.

Terminally the histology confirms the recording sites.

III. Results

A. The Osmosensitive Cells in the Hypothalamus

As a first step, our work has been exclusively concerned with the problem of osmoreceptors, in line with the Verney concept. VON EULER [1953] proposed that an 'osmoreceptors' cell should respond 'specifically' to an osmotic stimulus and not act additionally as an interneurone or a secretory cell. VERNEY [1947] had previously suggested that the osmoreceptors were contained in the neurosecretory cells of the supraoptic nucleus (NSO). On the other hand, BROOKS and his co-workers [1962] found their osmosensitive cells diffuse in the hypothalamus and responsive to both osmotic and sensory stimuli.

In our unanaesthetized animal, we had the opportunity to observe that the i.c. injection of hypertonic sodium chloride produced a complex behavioural response consisting of EEG arousal, lip-smacking and chewing. Since we had previously shown that 70% of hypothalamic neurons were sensitive to changes in sleep-waking behaviour, we were faced with the problem of determining whether a response of a cell is specifically related to the *osmotive* or to the *arousing* nature of hypertonic injection.

Of the 371 cells analyzed during hypertonic i.c. injection, 269 were osmosensitive. On the basis of their anatomical location, their pattern of discharge to i.c. osmotic stimuli and the response to arousing sensory stimuli, we divided the osmosensitive neurons into two major groups, the 'specific' and the 'non-specific' osmosensitive cells [HAYWARD and VINCENT, 1970].

The '*non-specific*' group (n = 132) corresponds to cells which responded to both osmotic and other arousing stimuli; for instance, opening the door of the cage, noise, etc. They may be depressed or activated by arousing stimuli or spontaneous arousal and they reacted in the same direction to osmotic stimulus. These cells lie diffusely scattered in the anterolateral hypothalamus.

137 osmosensitive cells were labelled '*specific*' because they responded to an hypertonic i.c., injection and did not respond to arousing stimuli. They were located in the NSO and of its immediate vicinity, the so-called perinuclear zone (PNZ). We found two subtypes of these 'specific' cells.

(1) *The first subtype* corresponds to cells (n = 56) with a *biphasic* response which consits of acceleration followed by inhibition. They did not respond to arousing stimuli or isotonic saline. The increasing of concentration of the osmotic stimulus produced a graded enhancement of both the acceleratory and the inhibitory phase. These neurons were located exclusively inside the NSO [VINCENT and HAYWARD, 1970]. We proposed to identify these cells as the neurosecretory cells and we interpreted the secondary phase of the response to saline as a recurrent inhibition (cf. part B).

(2) *The second subtype* of specific cells (n = 81) showed a monophasic response. They responded with a graded and constant acceleration in 51 cells (fig. 3 A1) and with an inhibition in 30 cells (fig. 3 B1). All these cells were located in the PNZ. On the basis of their location, the specificity, the constancy and the graduation of the response, we assumed that there were 'osmoreceptors' of excitatory and inhibitory type.

To confirm this hypothesis, we tried to identify the neurosecretory cells by their antidromic activation from the posterior pituitary.

B. Antidromic Identification of the Osmosensitive Cells

We recorded from 64 osmosentitive cells located in the NSO and the PNZ which exhibited a specific pattern of response to i.c. osmotic stimuli. These cells were also tested by stimulation of the posterior pituitary (fig. 1D).

(1) Twenty-seven specific osmosensitive cells (42%) responded to the osmotic stimulation with a *biphasic* pattern and were lying in the NSO. Twenty of them were conclusively identified as neurosecretory cells by their antidromic activation from the neurohypophysis on the basis of two criteria: stability of the latency of the antidromic spike and collision of the antidromic and orthodromic potentials (fig. 1D). We have previously suggested that the excitatory-inhibitory sequence produced in the NSO cell upon osmotic stimulation [VINCENT and HAYWARD, 1970] may result from a system of recurrent collaterals developing a post-synaptic inhibition. KELLY and

DREIFUSS [1970] and BARKER *et al.* [1971b] have recently demonstrated an inhibitory pathway to neurosecretory cells which appears to be mediated by neurosecreting axons. Evidence for this hypothesis is given by our observation in the primate of a long-lasting inhibition period (50–110 msec) following the antidromic invasion of the cell (fig. 2). The duration of this inhibition excludes the possibility of a direct influence of the antidromic action potential on the membrane permeability since the absolute refractory period of antidromic activation determined by repetitive stimulations never exceeded 5 msec. We were also able to reproduce by a train of repetitive antidromic stimulation (15 Hz during 5 sec) the following inhibition observed after the orthodromic activation of the cell by an osmotic stimulus. In both cases, the duration of the inhibition is proportional to the importance of the preceding activation.

NICOLL and BARKER [1971] have proposed that the neurosecretory product itself might also serve a synaptic function and mediate the recurrent inhibition. To confirm this hypothesis, we have tested the effect of an i.c. injection of ADH (1 mU in 1 ml of saline during 5 sec) on the firing rate of neurosecretory cells identified by their responses to osmotic and antidromic stimulations. Ten cells responded to the ADH injection with a rapid decrease and sometimes a total inhibition of their firing rate, and generally returned to the base line within 20–30 sec followed by a rebound effect (unpubl. results). This result supports the assumption of NICOLL and BARKER [1971] that ADH may serve as a transmitter in this recurrent inhibition; it is also an illustration of the so-called ultra-short feedback mechanism proposed for the parvocellular neurosecretory system by MOTTA *et al.* [1970].

(2) Thirty-seven cells (58%) responded specifically to osmotic stimulation with a *monophasic* excitation (n = 19) or inhibition (= 18) pattern. They were located in the PNZ. None of these cells were antidromically activated by the pituitary stimulation. We have suggested that they represent the Verney osmoreceptors [VINCENT and HAYWARD, 1970]. The lack of activation from the neurohypophysis could be a good argument for the separation of these receptors from the neurosecretory cells. Several findings support this suggestion that there is one or several synapses involved between reaction of the osmoreceptors and activation of the NSO cells [THORN, 1971]. Autonomic blocking agents block the release of ADH induced by the i.c. injection of hypertonic saline in rats showing that the neurosecretory cells are not directly activated by osmotic stimulation. The existence ot two types of response in our proposed osmoreceptors (inhibitory and excitatory) could be compared with the recent observation in the neurosecretory cells of two synaptic receptors

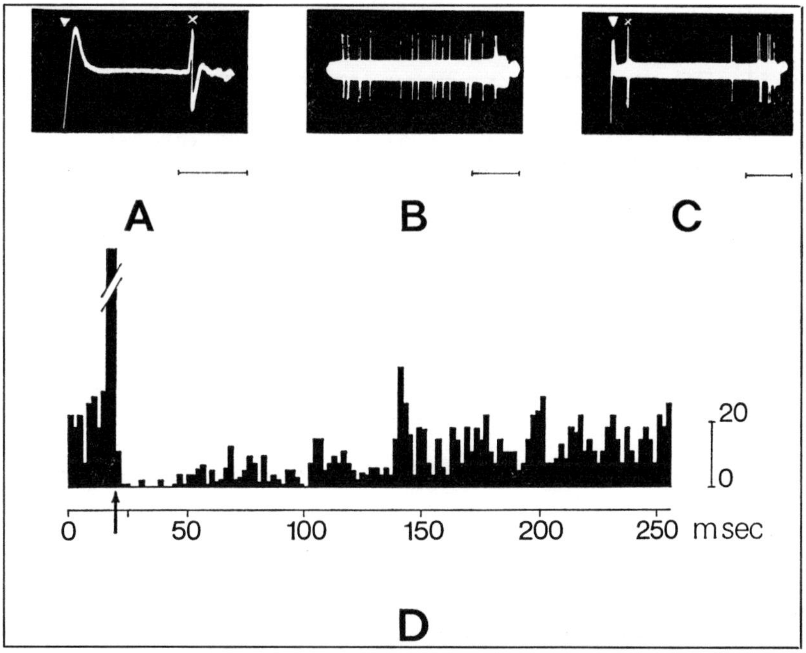

Fig. 2. Three examples from the same cell of 10 oscilloscope traces superimposed photographically. A = Superimposition of 10 antidromic potentials at a fixed latency, time calibration is 10 msec. B = Control superimposition of random discharges, the time calibration in 50 msec. C = Inhibitory pause in firing evoked by antidromic stimulation. D = Post-stimulus time histogram after stimulation of the posterior pituitary (500 stimuli). The stimulus artefacts are marked by arrows and the antidromic spikes by x.

(muscarinic and nicotinic) responding in opposite direction to iontophoretically applied Ach (BARKER *et al.*, 1971a].

C. Osmosensitive Cells during Drinking

The renal output and the oral intake of water are both controlled by the hypothalamus. A strong presumption exists of an active relationship between the two systems [NICOLAIDIS, 1969]. The present work was undertaken in order to determine in the unanaesthetized monkey the possible role in the control of drinking of the different types of osmosensitive hypothalamic cells [VINCENT *et al.*, in press]

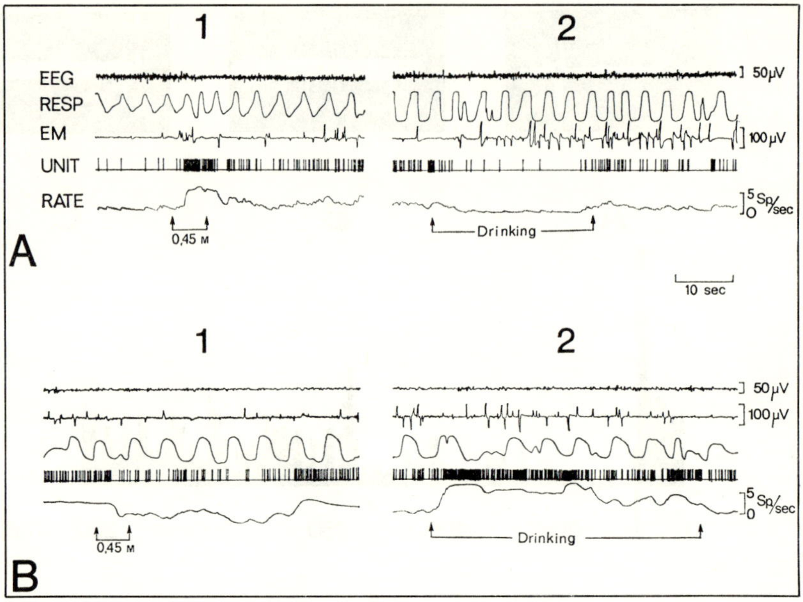

Fig. 3. Two 'specific monophasic' osmosensitive neurons in the PNZ. A = Monophasic excitatory pattern to a 5-sec i.c. injection of 1 ml 0.45 M hypertonic sodium chloride solution (1) and an opposite variation during drinking (2). B = Monophasic inhibitory pattern to the osmotic stimulus (1) with an acceleratory response to drinking (2). EEG = biparietal electrocorticogram; Resp. = nasal respiration; EM = eye movements; unit = pulse output from pulse height discriminator triggered by action potential of the cell; rate = analogous output proportional to the rate of unit discharge; Sp. = spike.

A total of 177 cells was analysed in the diencephalon during drinking. All these cells had been previously tested by i.c. injections and arousing stimuli. 100 cells were influenced by the osmotic stimulus. Our attention was directed toward the variation of the firing rate observed while the animal was drinking. No significant variation was observed in the non-osmosensitive cells; on the contrary, all the osmosensitive cells exhibited some changes which were different according to their type.

(1) In the '*biphasic*' *specific osmosensitive cells*, we observed a dramatic decrease of a few seconds after the beginning of drinking. The neurosecretory nature of these cells was confirmed by their antidromic activation. Their rapid inhibition during drinking cannot be explained by the long latency systemic changes following the digestive absorption of water and may be interpreted as a reflexive effect from peripheral receptors. This observa-

tion gives a neurophysiological support to the rapid increase of diuresis following drinking. The early renal response to drinking has been described in the rat as a pre-adaptive phenomenon anticipating the post-absorbtive effect of water [NICOLAIDIS, 1969].

(2) '*Monophasic' specific osmosensitive cells.* The cells which exhibited a monophasic accelerated response to hypertonic saline injection showed, during drinking, a dramatic decrease of firing (fig. 3A). The same inversion was observed in the inhibitory osmosensitive cells. These cells showed a specific inhibitory response to hypertonic saline injection, but were accelerated by drinking (fig. 3B).

We have suggested (part B) that the osmosensitive monophasic cells could be the osmoreceptors of the Verney type. The present data show that these cells respond in an opposite effect to drinking, suggesting they could represent an integrative control system with the summation of internal direct signals and peripheral feedback information. Receptors exist in the mouth, throat and stomach which may inform the osmoreceptor system of the volume of water drunk prior to the correction of the unbalance of the osmotic level in the plasma by the digestive absorption of water [LILJESTRAND and ZOTTERMAN, 1954].

Considering their anatomical location, the osmoreceptors in the PNZ are probably related to the supraoptic neurosecretory system and mediate antidiuretic response. On the other hand, no evidence exists that they are identical to those that mediate drinking, and it is strongly suggested that neurons lying outside NSO and PNZ play an important role in the osmotic control of drinking.

(3) *Non-specific osmosensitive cells.* These cells lie diffusely throughout the anterolateral hypothalamus and the preoptic area. They are sensitive to injections of hypertonic saline solution and react with identical response to mild arousing sensory stimuli. We have considered (part A) that these cells are only responding to the non-specific arousal component of the osmotic stimulus. All of these cells also react to drinking. Indeed drinking produced an increase in the level of arousal and the problem was again to identify the specific role that the drinking behaviour played in the change observed in single unit activity apart from the non-specific influence of sleep-wakefulness variations.

The majority of these cells react to drinking in the same manner, as to non-specific arousal, that is by acceleration (fig. 4A) or by inhibition (fig. 4B). But some cells (fig. 4C) react in an original manner to drinking; they are activated by the non-specific arousal and inhibited during drinking. It is

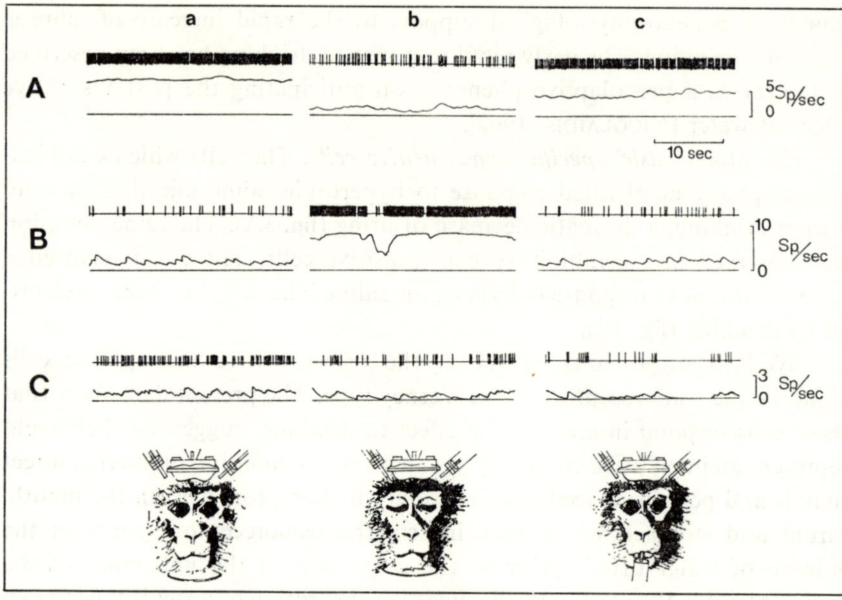

Fig. 4. Three 'non-specific osmosensitive' neurons during different behavioural states: arousal (a), drowsy state or relaxation (b), and drinking (c). A = Non-specific neuron which accelerates during arousal and drinking. B = Neuron which decelerates during arousal and drinking. C = Neuron located in the lateral hypothalamus which behaves with opposite variations during arousal and drinking. Sp. = spike.

interesting to observe that all of the cells of this latter type are located in the lateral and the dorsal hypothalamus in the vicinity of the globus pallidus. Investigations over many years have emphasized the role of the laterodorsal part of the hypothalamus as a critical focus and probably as the final integrating site of the central system controlling water intake [MORGANE, 1961; EPSTEIN and TEITELBAUM, 1964; MOGENSON and STEVENSON, 1966]. We suggest that these neurons are related to the arousal directed to drinking which supports the motivational aspect of this behavior. In the thirsty monkey, every arousing stimulus induces an activation of these cells; on the contrary, drinking induces an inhibition in spite of the state of arousal of the animal. It may be noted that only the drinking of water is able to produce such an inhibition independently of the motor behaviour; we observed no inhibition after a sham drinking induced by saline i.c. injection in the absence of available water. These neurons provide some support for an anticipatory mechanism of the satiation of drinking and can explain the cessation of a specific behaviour prior to the correction of the internal unbalance which has triggered the response.

Acknowledgements

We thank Dr. B. DUFY for his co-operation. Mr. R. MIGUELEZ prepared the figures; Mrs. BONHOMME-GAILLAUD, Mr. G. LABAYLE and Mr. F. RODRIGUEZ provided valuable technical assistance.

References

BARKER, J. L.; CRAYTON, J. W., and NICOLL, R. A.: Noradrenaline and acetylcholine responses of supraoptic neurosecretory cells. J. Physiol., Lond. *218:* 19–32 (1971a).

BARKER, J. L.; CRAYTON, J. W., and NICOLL, R. A.: Antidromic and orthodromic responses of paraventricular and supraoptic neurosecretory cells. Brain Res. *33:* 353–366 (1971b).

BROOKS, C. Mc C.; USHIYAMA, J., and LANGE, G.: Reactions of neurones in or near the supraoptic nuclei. Amer. J. Physiol. *202:* 487–490 (1962).

EPSTEIN, A. N. and TEITELBAUM, P.: Severe and persistant deficits in thirst produced by lateral hypothalamic damage; in Thirst in the regulation of body water (Pergamon Press, London 1964).

HAYWARD, J. N. and VINCENT, J. D.: Osmosensitive single neurones in the hypothalamus of unanesthetized monkeys. J. Physiol., Lond. *210:* 947–972 (1970).

KELLY, J. S. and DREIFUSS, J. J.: Antidromic inhibition of identified rat supraoptic neurones. Brain Res. *22:* 406–409 (1970).

LILJESTRAND, G. and ZOTTERMAN, Y.: The water taste in mammals. Acta physiol. *32:* 291–303 (1954).

MOGENSON, G. J. and STEVENSON, J. A. F.: Drinking and self-stimulation with electrical stimulation of the lateral hypothalamus. Physiol. Behav. *1:* 251–254 (1966).

MORGANE, P. J.: Evidence of an 'hunger motivational' system in lateral hypothalamus of the rat. Nature *191:* 672–674 (1961).

MOTTA, M.; PIVA, F., and MARTINI, L.: The hypothalamus as the center of endocrine feedback mechanisms; in The hypothalamus (Academic Press, New York 1970).

NICOLAIDIS, S.: Early systemic responses to orogastric stimulation in the regulation of food and water balance. Functional and electrophysiological data. Ann. N.Y. Acad. Sci. *157:* 1176–1203 (1969).

NICOLL, R. A. and BARKER, J. L.: The pharmacology of recurrent inhibition in the supraoptic neurosecretory system. Brain Res. *35:* 501–511 (1971).

THORN, N. A.: Osmoreceptors and the supraoptico-neurohypophysial system; in Proc. 25th Int. Congr. Int. Union of Physiological Sciences, vol. 8 (IUPS, Munich 1971).

VERNEY, E. B.: The antidiuretic hormone and the factors which determine its release. Philos. Trans. (B) *135:* 25–106 (1947).

VINCENT, J. D. and HAYWARD, J. N.: Activity of single cells in the osmoreceptor-supraoptic nuclear complex in the hypothalamus of the waking rhesus monkey. Brain Res. *23:* 105–108 (1970).

VINCENT, J. D.; ARNAULD, E., and BIOULAC, B.: Activity of osmosensitive single cells in the hypothalamus of the behaving monkey during drinking. Brain Res. (in press)

Author's address: Prof. J. D. VINCENT, Laboratoire de Neurophysiologie, Faculté de Médecine, Place de la Victoire, *F-33000 Bordeaux* (France)

Subject Index

Economy of space dictated that only a selective index be prepared. It is intended as an extension of the Table of Contents, to help the reader in search of information in his specific area. However, it is not a complete index *sensu stricto*. For example, 'transfusions in treatment of hepatic coma' is listed as 'cross-circulation', discussion of diets and of food contents is referred to by the single word 'nutritio', descriptions of climatic conditions is listed as 'climate', etc.